Law, Governance and Technology Series

Volume 31

The *Law, Governance and Technology Series* is intended to attract manuscripts arising from an interdisciplinary approach in law, artificial intelligence and information technologies. The idea is to bridge the gap between research in IT law and IT-applications for lawyers developing a unifying techno-legal perspective. The series will welcome proposals that have a fairly specific focus on problems or projects that will lead to innovative research charting the course for new interdisciplinary developments in law, legal theory, and law and society research as well as in computer technologies, artificial intelligence and cognitive sciences. In broad strokes, manuscripts for this series may be mainly located in the fields of the Internet law (data protection, intellectual property, Internet rights, etc.), Computational models of the legal contents and legal reasoning, Legal Information Retrieval, Electronic Data Discovery, Collaborative Tools (e.g. Online Dispute Resolution platforms), Metadata and XML Technologies (for Semantic Web Services), Technologies in Courtrooms and Judicial Offices (E-Court), Technologies for Governments and Administrations (E-Government), Legal Multimedia, and Legal Electronic Institutions (Multi-Agent Systems and Artificial Societies).

More information about this series at http://www.springer.com/series/8808

Mariarosaria Taddeo • Luciano Floridi
Editors

The Responsibilities of Online Service Providers

Springer

Editors
Mariarosaria Taddeo
Oxford Internet Institute
University of Oxford
Oxford, UK

Alan Turing Institute
London, UK

Luciano Floridi
Oxford Internet Institute
University of Oxford
Oxford, UK

Alan Turing Institute
London, UK

ISSN 2352-1902 ISSN 2352-1910 (electronic)
Law, Governance and Technology Series
ISBN 978-3-319-47851-7 ISBN 978-3-319-47852-4 (eBook)
DOI 10.1007/978-3-319-47852-4

Library of Congress Control Number: 2017931074

Printed on acid-free paper

This Springer imprint is published by Springer Nature
The registered company is Springer International Publishing AG
The registered company address is: Gewerbestrasse 11, 6330 Cham, Switzerland

Contents

About the Editors

Mariarosaria Taddeo works at the Oxford Internet Institute, University of Oxford and Faculty Fellow at the Alan Turing Institute. Her recent work focuses mainly on the ethical analysis of cyber security practices and information conflicts. Her area of expertise is Information and Computer Ethics, although she has worked on issues concerning Philosophy of Information, Epistemology, and Philosophy of AI. She published several papers focusing on online trust, cyber security and cyber warfare and guest-edited a number of special issues of peer-reviewed international journals: Ethics and Information Technology, Knowledge, Technology and Policy, Philosophy & Technology. She also edited (with L. Floridi) a volume on 'The Ethics of Information Warfare' (Springer 2014) and is currently writing a book on 'The Ethics of Cyber Conflicts' under contract for Routledge. Dr. Taddeo is the 2010 recipient of the Simon Award for Outstanding Research in Computing and Philosophy and of the 2013 World Technology Award for Ethics. She serves editor-in-chief of Minds & Machines, in the executive editorial board of Philosophy & Technology. Since 2016, Dr. Taddeo is Global Future Council Fellow for the Council on the Future of Cybersecurity of the World Economic Forum.

Luciano Floridi is professor of philosophy and ethics of information at the University of Oxford and director of research of the Oxford Internet Institute. His most recent book is *The Fourth Revolution: How the Infosphere Is Reshaping Human Reality* (2014 and 2016). He is a member of the EU's Ethics Advisory Group on ethical dimensions of data protection and of Google's advisory board on "the right to be forgotten" and chairman of the Ethics Advisory Board of the European Medical Information Framework. Among his recognitions, he has been elected Fernand Braudel senior fellow by the European University Institute, was awarded the Cátedras de Excelencia Prize by the University Carlos III of Madrid, and was the UNESCO chair in information and computer ethics and Gauss professor of the Academy of Sciences in Göttingen. He is a recipient of the MEA's J. Ong Award, the APA's Barwise Prize, the IACAP's Covey Award, and the INSEIT's Weizenbaum Award. He is a fellow of the AISB, the BCS, and the Académie Internationale de Philosophie des Sciences.

Chapter 1
New Civic Responsibilities for Online Service Providers

Mariarosaria Taddeo and Luciano Floridi

Online Service Providers (OSPs)—such as AOL, Facebook, Google, Microsoft, and Twitter—are increasingly expected to act as good citizens, by aligning their goals with the needs of societies, supporting the rights of their users (Madelin 2011; Taddeo and Floridi 2015), and performing their tasks according to "principles of efficiency, justice, fairness, and respect of current social and cultural values" (McQuail 1992, 47). These expectations raise questions as to what kind of responsibilities OSPs should bear, and which ethical principles should guide their actions.

Addressing these questions is a crucial step to understand and shape the role of OSPs in mature information societies (Floridi 2016). Without a clear understanding of their responsibilities, we risk ascribing to OSPs a role that is either too powerful or too little independent. The FBI vs. Apple case,[1] Google's and Yahoo!'s experiences in China,[2] or the involvement of OSPs within the NSA's PRISM program[3] offer good examples of the case in point. However, defining OSPs' responsibilities is challenging. Three aspects are particularly problematic: disentangling the implications of OSPs' gatekeeping role in information societies; defining fundamental principles to guide OSPs' conduct; and contextualising OSPs' role within the broader changes brought about by the information revolution.

The notion of 'gatekeepers' identifies those agents who have a central role in the management of resources and infrastructures that are crucial for societies (Lewin 1947). In our societies, OSPs are *information* gatekeepers (Calhoun 2002), as they control access to and flows of data and information (Shapiro 2000; Hinman 2005; Laidlaw 2008). As such, they exercise a regulatory function (Metoyer-Duran 1993), which entails moral responsibilities toward the public good. As Shapiro put it

> those who control the access to information have a responsibility to support the public interest. [...] and must assume an obligation as trustees of the greater good (Shapiro 2000, 225).

---

[1] https://en.wikipedia.org/wiki/FBI–Apple_encryption_dispute

[2] http://business.time.com/2014/01/08/are-google-yahoo-and-microsoft-living-up-to-their-promises-in-china/

[3] https://www.reformgovernmentsurveillance.com

M. Taddeo (✉) • L. Floridi
Oxford Internet Institute, University of Oxford, Oxford, UK

Alan Turing Institute, London, UK
e-mail: mariarosaria.taddeo@oii.ox.ac.uk

M. Taddeo, L. Floridi (eds.), *The Responsibilities of Online Service Providers*, Law, Governance and Technology Series 31, DOI 10.1007/978-3-319-47852-4_1

While there is a general consensus on OSPs' gatekeeping role and on their ability to influence the development of information societies, there is much less agreement on whether, as corporate agents, OSPs bear any responsibility toward the public good (Freeman 1999; Black 2001; Taddeo and Floridi 2015). As a result, the civic responsibilities of OSPs are often discharged *via* policies and practices unilaterally defined by OSPs themselves.

Things become more complicated once we consider the international and inevitably multicultural contexts in which OSPs operate, the transnational nature of their business, alongside the interdependency of the services that they offer in different regions of the world. In this context, the definition of the responsibilities of OSPs will be effective only insofar as it will rest on an ethical framework able to reconcile the different views and stakeholders' interests that they face.

Human rights have a central role in this debate, insofar as they identify fundamental universal principles, some of which expressly address Internet governance (Wettstein 2012b; Lucchi 2013). For example, a report[4] released by the UN in 2011stressed that

> [g]iven the Internet has become an indispensable tool for realizing a range of human rights, combating inequality, and accelerating development and human progress, ensuring universal access to the Internet should be a priority for all States.

In 2012, Internet freedom was declared a human right by the UN Human Rights Council, which called on states to promote and foster access to the Internet and to ensure that the rights to freedom of expression and information, as presented in Article 19 of the Universal Declaration of Human Rights, would be upheld online as well as offline.[5] However, both the Universal Declaration of Human Rights and the Resolution on the Promotion, Protection and Enjoyment of Human Rights on the Internet[12] mainly address state actors, making problematic the expectation that OSPs should be held responsible for respecting and fostering human rights (Karp 2009). This problem is not entirely new. The scope of human rights and the responsibilities that they pose to transnational corporations have already been analysed in the Declaration of Human Duties and Responsibilities (the so-called Valencia Declaration). The Declaration stresses the moral duties and legal responsibilities of all the members of the global community to observe and promote respect for human rights and fundamental freedoms. The global community encompasses state and non-state actors, individuals and groups of citizens, as well as the private and the public sector. Private companies are also expressly mentioned as responsible for promoting and securing human rights in the preamble of the UN Norms on the Responsibilities of Transnational Corporations and Other Business Enterprises.

Given OSPs' central role in shaping the informational environment and the societies depending on it, it is increasingly less acceptable to maintain that, as private companies, OSPs are only responsible to their employees and shareholders, and are

[4] http://www2.ohchr.org/english/bodies/hrcouncil/docs/17session/A.HRC.17.27_en.pdf

[5] Resolution on "The Promotion, Protection and Enjoyment of Human Rights on the Internet" (Human Rights Council of the United Nations 2012).

not bounded by human rights regulation (Chen 2009; Taddeo and Floridi 2015; Cath and Floridi 2016; Laidlaw Forthcoming). This is a point highlighted, for example, in the latest report of the Special Rapporteur on freedom of expression to the Human Rights Council, David Kaye, who stressed that

> Among the most important steps that private actors should take is the development and implementation of transparent human rights assessment procedures. They should develop and implement policies that take into account their potential impact on human rights.[6]

At the same time, however, it is problematic to ascribe to OSPs full responsibility for fostering and respecting human rights, and for deciding the circumstances in which these apply. For this prompts a privatization of the judging power and poses issues of transparency and accountability (Gerry and Berova 2014). Consider, for example, OSPs acting as both the "judge and the jury"[7] with respect to the decision of the European Court of Justice on the right to be forgotten (Rosen 2012; Floridi 2015). To this end, it is crucial to separate the responsibilities of OSPs from the duties and authority of the state.

Guidance on this matter has been provided by the Ruggie's framework.[8] The framework proposes a clear-cut distinction between the role of states and that of transnational corporations with respect to human rights (Wettstein 2012a). The distinction rests on three pillars: the duty of the state to *protect* against human rights abuses by third parties, including business; the corporate responsibility to *respect* human rights; and the responsibility of both states and corporates to provide victims with access to effective *remedy*, both judicial and non-judicial.

While the Ruggie's framework offers a valid tool to identify the responsibilities of transnational corporations, the proposed distinction between states' duties and corporates' responsibilities proves to be problematic when considering specifically the case of OSPs. Their crucial role in information societies, alongside their leading role in steering the information revolution and, hence, in shaping the informational environment make them political agents able to influence national politics and international relations (Broeders and Taylor forthcoming). As such, they differ quite radically from other transnational corporations and bear a wider set of responsibilities than other corporate agents (Scherer and Palazzo 2011). Broders and Taylor argue that, as political agents, OSPs should bear corporate *political* responsibilities:

> OSPs exercise power over their users and are a counter power to state power in all corners of the world. […] they are also political actors who merit serious diplomatic attention owing to their vital role in digital life, […] (Broeders and Taylor forthcoming).

The Ruggie's framework only partially addresses OSPs' political role and offers little insight to identify states' duties in cyberspace. Insofar as the framework rests on the Westphalian model of sovereign states, it struggles to address new forms of

---

[6] http://www.ohchr.org/EN/Issues/FreedomOpinion/Pages/Privatesectorinthedigitalage.aspx

[7] http://www.telegraph.co.uk/technology/google/10967211/Google-is-the-judge-and-jury-in-the-right-to-be-forgotten.html

[8] http://198.170.85.29/Ruggie-protect-respect-remedy-framework.pdf

political agents—like the EU and the UN—or NGOs (Nye 2004). At the same time, the model is challenged when considering sovereign states' authority in cyberspace, where it is problematic to draw national boundaries, and state's territoriality and sovereignty are difficult to define (Krasner 2001).

The limitations of the Ruggie's framework point to the third set of problems concerning the definition of OSPs' civic responsibilities, namely the understanding of the conceptual and moral changes (Floridi 2014a, b; Taddeo and Buchanan 2015) brought about by new phenomena such as, for example, Big Data analytics (Floridi 2012; Mittelstadt and Floridi 2015), individual and group privacy (Floridi 2014a, b), online trust (Taddeo 2010; Taddeo and Floridi 2011), cyber security and surveillance (Taddeo 2013; Taddeo 2014b), and cyber conflicts (Taddeo 2012; Taddeo 2014a; Floridi and Taddeo 2014). These changes concern the redefinition of crucial concepts, like those of political power and authority, as well as the distinction between real and virtual and the understanding of good and evil, and of the values on which present and future information societies rely.

The civic responsibilities of OSPs cannot be defined without considering these conceptual changes and without a foresight analysis of the future of information societies, toward which OSPs play a decisive role. In (Taddeo and Floridi 2015), we argued that the notion of information gatekeepers, the human rights framework, and the concepts of *respect* and *care for* the informational environment (Floridi 2013a) offer key milestones for an ethical framework able to identify and define both the civic responsibilities of OSPs and the right ethical infrastructure, i.e. the *infraethics* (Floridi 2013b), able to facilitate the discharging of such responsibilities. This requires identifying those

> expectations, attitudes, and practices that *can* facilitate and promote morally good decision and actions (Floridi 2013b, 738).

In the case of OSPs, the right infraethics encompasses trust, privacy, and transparency, as well as openness and pluralism.

It is clear that the definition of OSPs' responsibilities and of the infraethics supporting them will increasing shape the informational environment and future information societies (Floridi 2011; Cath and Floridi 2016). For this reason *ad hoc* approaches, tackling OSPs' responsibilities as they emerge, will be insufficient in the medium- and long-term, for they lack a meaningful reflection on current changes and any insight on future ones. Both can only be achieved by coordinating expertise and theorising to understand the values that will continue shaping our societies, the different stakeholders' views, alongside the role of OSPs and of other key agents in designing the informational environment.

This volume provides a step in this direction, by collecting eighteen contributions addressing the issue of OSPs' responsibilities from different angles. Each contribution is either invited or a paper presented during the workshop "*Understanding the Responsibilities of Online Service Providers in Information Societies*", held in 2015 at the Oxford Internet Institute, University of Oxford. The goal is to offer a multidisciplinary collection of essays spanning from ethics and corporate social responsibilities to policy and legal analyses of OSPs civic responsibilities. The volume is divided into three parts, focusing on 'Responsibilities & Liabilities', 'Business Ethics & Policies', and 'Users' Rights & International Regulations' respectively.

Chapter two—*The moral responsibilities of online service providers*—opens the first part of the book, with a reprinted version of a paper appeared in 2015 (Taddeo and Floridi 2015). It analyses the main contributions to the debate on the moral responsibilities of OSPs. By endorsing the method of the levels of abstraction (LoAs) (Floridi 2008), it first focuses on the moral responsibilities of OSPs *in* the web (LoA_{IN}). These concern the management of online information, which includes information filtering, Internet censorship, the circulation of harmful content, and the implementation and fostering of human rights. The chapter then delves into the moral responsibilities ascribed to OSPs *on* the web (LoA_{ON}) describing existing legal regulations of access to users' data. The analysis highlights two main results. First, OSPs' public role—especially their gatekeeping function, their corporate social responsibilities, and their role in implementing and fostering human rights—has acquired increasing relevance in the specialised literature. Second, there is a lack of an ethical framework that can (a) define OSPs' responsibilities, and (b) provide the fundamental sharable principles necessary to guide OSPs' conduct within the multicultural and international context in which they operate.

The analysis of OSPs responsibilities in the web continues in the third chapter—*The immunity of internet intermediaries reconsidered?*—which focuses on the role of OSPs in the digital world. The goals of this contribution are to identify the role of new power players in the digital world and to analyse the boundaries between immunity and liability of Internet intermediaries. Specifically, this contribution addresses questions such as: will OSPs become *quasi judges* controlling every Internet activity? Are they entitled to defend the rights of the users and, more in general, the legitimacy in cyberspace? Which principles and policies should be adopted in order to foster an ethos of compliance and social responsibility for these new gatekeepers?

Chapter four—*Is Google responsible for providing fair and unbiased results?*—focuses on the responsibility of search engines in general, and Google in particular, for providing unbiased search results. The chapter identifies areas of potential responsibilities of search engines by endorsing two approaches, one technical and the other societal. The technical approach considers the impact of decisions made at the design stage on users' search results. The societal approach informs discussion on the impact that biased search engines may have for information and knowledge acquisition in society.

Chapter five—*Speaking truth to/as victims, a jurisprudential analysis of data breach notification laws*—analyses the duty that OSPs have to inform customers when the company suspects security breaches and unauthorised access to customer data. This chapter focuses on data breach notification duties from a jurisprudential perspective and considers what duties can legitimately be attributed to victims of crime in a democratic society. It then applies this analysis to OSPs and their duty to inform either their customers or a state agency about data breaches.

Chapter six—*Did the Romans get it right? What Delfi, Google, eBay, and UPC TeleKabel Wien have in common*—concludes the first part of the volume by providing a description of the legal principles set out in recent landmark cases of both the EU Court of Justice in Luxembourg (CJEU) and the European Court of Human Rights (ECHR), including the Google France, Google Spain/Costeja, UPC TeleKabel Wien and Delfi cases. It examines the legal grounds and arguments put forward by the respective courts in order to judge OSPs' responsibilities.

The second part of the book collects analyses focusing on business ethics and corporate social responsibilities. Chapter seven—*Responsibilities of OSPs from a business ethics point of view*—opens this part by focusing on a central aspect of the business ethics debate, i.e. the balance between OSPs' responsibilities and their ability to be competitive in the market. This chapter argues, first, that the analysis of OSPs' business responsibilities should rest less on purely individual concepts of responsibility and more on the concepts of group agency; and, second, that there are ways of ascribing responsibility to companies compatible with competition, as the debate on corporate social responsibilities (CSR) shows.

The analysis on OSPs' CSR continues in chapter eight—*Myth or promise? The corporate social responsibilities of online service providers for human rights*. The chapter analyses a number of CSR frameworks shaping OSPs' conduct. In particular, it concentrates on the UN Guiding Principles and on the Global Network Initiative, one of the leading multi-stakeholder initiatives guiding CSR for technology companies. The goal is to address OSPs' accountability with respect to human rights. In doing so, the chapter focuses on key issues such the relevance of CSR frameworks for protecting human rights online.

Chapter nine-*Online Service Providers—a new and unique species of the firm?*—draws on cross-disciplinary literature from economic theory, international business theory, economic geography, and information technology, in order to analyse whether OSPs differ from other transnational corporations and whether this entails different CSRs. The chapter concludes that, albeit OSPs differ from other transnational corporations, such difference are not wide enough to consider OSPs a unique kind of firm. The analysis then focuses on the non-monetary relationship between OSPs and their end-users, and on the societal impact that this could have, especially in developing countries.

Chapter ten—*Online service providers as human rights arbiters*—continues the analysis of fair policies for OSPs and of their responsibilities with respect to human rights. The chapter focuses on case-studies of EU regulation such as the E-commerce directive, to assess whether and to what extent measures of blocking, filtering, and content removal interfere with the human rights standards related to freedom of expression and freedom of information. The chapter concludes that OSPs' self-defined guidelines are insufficient to counter the human rights challenges.

Chapter eleven—*Licensing of user-generated content: why less is more*—delves into the responsibilities of OSPs with respect to users-generated content that is subject to licensing clauses found in terms of service agreements. The chapter argues that contractual ambiguity existing in OSPs' terms of service agreements negatively affects users and OSPs alike.

The third part of the book is dedicated to 'Users' rights & international regulations' and begins with chapter twelve—*Online service providers' liability, copyright infringement, and freedom of expression. Could Europe learn from Canada?* This contribution analyses recent interpretations of the Dir. 2000/31, which have lifted the bar on providers' duties and highlights their impact on OSPs business and, most important, on freedom of expression. The chapter then focuses on the Canadian regulation, which is based on the so-called 'notice and action' principle, and suggests it offers a suitable framework for the regulation of OSPs' liabilities in Europe.

Chapter thirteen—*Non-financial disclosures in the tech sector: furthering the trend*—focuses on transparency and analyses the different ways through which corporate non-financial disclosure mechanisms can contribute to developing an ethical framework for OSPs. The chapter discusses three areas where transparency makes an impact on ethical standards for OSPs: corporate reporting on interaction with governments on privacy and free expression issues; disclosures of staff demographics; and sharing of information on digital security topics, including encryption and breach. Finally, it assesses the ways through which transparency mechanisms, and their consistent implementation, could help OSPs realize their new civic responsibilities.

Chapter fourteen—*Should we treat Big Data as a public good?*—discusses Big Data and the responsibilities of OSPs to ensure that the user-generated data continue to improve individual well-being, innovation, and sustainable development. To this end the chapter maintains that Big Data should be considered a public good and that platforms for public and private partnerships in managing Big Data should be in place.

Chapter fifteen—*Internet intermediaries as responsible actors? Why it is time to rethink the e-Commerce Directive as well*—questions the suitability of the e-Commerce Directive. The purpose of this chapter is twofold. It aims to show that there is a need to review Articles 12 to 15 of the e-Commerce Directive; and that the very rationale linking Articles 12 to 14 of the e-Commerce Directive is ill-suited to address the complexity and diversity of OSPs' activities.

Chapter sixteen—*Towards fostering compliance by design, drawing designers into the regulatory frame*—begins by considering the extent to which EU General Data Protection Regulations would redefine the governance of personal data in a series of key ways and focuses, in particular, on the 'by design and by data protection'. It argues that this notion shifts the responsibility away from the user and explicitly invokes the role of the designer within the regulatory frame. The chapter then describes *ideation cards* as a suitable method to foster cross-disciplinary collaborations. It maintains that, whilst such cards will not necessarily create experts in data protection, they have the potential to sensitise designers to existing regulation.

Three commentaries conclude the volume. The first one—*Does great power come with great responsibility? The need to talk about Corporate Political Responsibility*—analyses the role of OSPs in contemporary societies and argues that OSPs act as political agents, with a relevant role in both national politics and international relations. Such a role requires extending the scope of the CSR that OSPs bear in two ways:

> [...] more serious mechanisms for accountability and (b) a recognition of the political role of corporations (Broeders and Taylor forthcoming).

The second commentary—*The Economic Impact of Online Intermediaries* —focuses on the role of Internet intermediaries to drive economic, social, and political development and considers whether the consolidation of OSPs' economic power impacted conventional business trade models and changed firm-level competition. It concludes that, while OSPs have provided technologically superior market entrants, they have not yet disrupted supply-chains, with the exception of software, publishing, and professional services sectors, in which online intermediaries have provided tangible productivity gains.

The third commentary— *Online Service Providers and ethical disclosure in sales*—addresses the need to develop business norms for the commodification of user information by OSPs. This contribution maintains that in considering the responsibilities of OSPs, referring to business norms, rather than interpersonal moral norms, leads to draw ethical conclusions with clear normative force.

References

Black, J. (2001). Decentring regulation: Understanding the role of regulation and self regulation in a 'post-regulatory' world. *Current Legal Problems, 54*(1), 103–146.

Broeders, D., & Taylor, L. (Forthcoming). Does great power come with great responsibility? The need to talk about corporate political responsibility. In M. Taddeo & L. Floridi (Ed.), *Law, governance and technology series*. Berlin/Heidelberg/New York/London: Springer.

Calhoun, C. J. (Ed.). (2002). *Dictionary of the social sciences*. New York: Oxford University Press.

Cath, C., & Florid, L. (2016, May). The design of the internet's architecture by the Internet Engineering Task Force (IETF) and Human Rights. *Science and Engineering Ethics*.

Chen, S. (2009). Corporate responsibilities in internet-enabled social networks. *Journal of Business Ethics, 90*(4), 523–536.

Floridi, L. (2008). The method of levels of abstraction. *Minds and Machines, 18*(3), 303–329.

Floridi, L. (2011). A defence of constructionism: Philosophy as conceptual engineering. *Metaphilosophy, 42*(3), 282–304.

Floridi, L. (2012). Big data and their epistemological challenge. *Philosophy & Technology, 25*(4), 435–437.

Floridi, L. (2013a). *The ethics of information*. Oxford: Oxford University Press.

Floridi, L. (2013b). Distributed morality in an information society. *Science and Engineering Ethics, 19*(3), 727–743.

Floridi, L. (2014a). *The fourth revolution, how the infosphere is reshaping human reality*. Oxford: Oxford University Press.

Floridi, L. (2014b). Open data, data protection, and group privacy. *Philosophy & Technology, 27*(1), 1–3.

Floridi, L. (2015). Should you have the right to be forgotten on google? Nationally, Yes. globally, No. *New Perspectives Quarterly, 32*(2), 24–29.

Floridi, L. (2016). Mature information societies—A matter of expectations. *Philosophy & Technology, 29*(1), 1–4.

Floridi, L., & Taddeo, M. (Eds.). (2014). *The ethics of information warfare*. New York: Springer.

Freeman, J. (1999). *Private parties, public functions and the new administrative law, SSRN Scholarly Paper ID 165988*. Rochester: Social Science Research Network.

Gerry, F., & Berova, N. (2014). The rule of law online: Treating data like the sale of goods: Lessons for the internet from OECD and CISG and sacking google as the regulator. *Computer Law & Security Review, 30*(5), 465–481.

Hinman, L. (2005). Esse Est Indicato in Google: Ethical and political issues in search engines. *International Review of Information Ethics, 3*(6), 19–25.

Human Rights Council of the United Nations. (2012). U.N. Human Rights Council: First Resolution on Internet Free Speech. http://www.loc.gov/lawweb/servlet/lloc_news?disp3_l205403231_text

Karp, D. J. (2009). Transnational corporations in 'bad States': Human rights duties, legitimate authority and the rule of law in international political theory. *International Theory, 1*(01), 87.

Krasner, S. D. (2001). Rethinking the sovereign state model. *Review of International Studies, 27*(05).

Laidlaw, E. (Forthcoming). Myth or promise? The corporate social responsibilities of online service providers for human rights. In M. Taddeo & L. Floridi (Eds.), *The responsibilities of online service providers* Law, governance and technology series. Berlin/Heidelberg/New York/London: Springer.

Laidlaw, E. (2008). Private power, public interest: An examination of search engine accountability. *International Journal of Law and Information Technology, 17*(1), 113–145.

Lewin, K. (1947). Frontiers in group dynamics. Human Relations *1*(2), 143–153.

Lucchi, N. (2013). Internet content governance and human rights. *Vanderbilt Journal of Entertainment and Technology Law, 16*, 809.

Madelin, R. (2011). The evolving social responsibilities of internet corporate actors: Pointers past and present. *Philosophy & Technology, 24*(4), 455–461.

McQuail, D. (1992). *Media performance: Mass communication and the public interest*. London/Newbury Park: Sage Publications.

Metoyer-Duran, C. (1993). Information gatekeepers. *Annual Review of Information Science and Technology (ARIST), 28*, 111–150.

Mittelstadt, B. D., & Floridi, L. (2015). The ethics of big data: Current and foreseeable issues in biomedical contexts. *Science and Engineering Ethics*, May.

Nye, J. S. (2004). *Soft power: The means to success in world politics* (1st ed.). New York: Public Affairs.

Rosen, J. (2012). The right to be forgotten. *Stanford Law Review Online, 64*, 88.

Scherer, A. G., & Palazzo, G. (2011). The new political role of business in a globalized world: A review of a new perspective on CSR and its implications for the firm, governance, and democracy. *Journal of Management Studies, 48*(4), 899–931.

Shapiro, A. L. (2000). *The control revolution: How the internet is putting individuals in charge and changing the world we know* (2nd Printing ed.). New York: Public Affairs.

Taddeo, M. (2010). Modelling trust in artificial agents, A first step toward the analysis of E-trust. *Minds and Machines, 20*(2), 243–257.

Taddeo, M. (2012). Information warfare: A philosophical perspective. *Philosophy and Technology, 25*(1), 105–120.

Taddeo, M. (2013). Cyber security and individual rights, striking the right balance. *Philosophy & Technology, 26*(4), 353–356.

Taddeo, M. (2014a). Just information warfare. *Topoi*, April, 1–12.

Taddeo, M. (2014b). The struggle between liberties and authorities in the information age. *Science and Engineering Ethics*, September, 1–14.

Taddeo, M., & Buchanan, E. (2015). Information societies, ethical enquiries. *Philosophy & Technology, 28*(1), 5–10.

Taddeo, M., & Floridi, L. (2011). The case for E-trust. *Ethics and Information Technology, 13*(1), 1–3.

Taddeo, M., & Floridi, L. (2015). The debate on the moral responsibilities of online service providers. *Science and Engineering Ethics*.

Wettstein, F. (2012a). CSR and the debate on business and human rights: Bridging the great divide. *Business Ethics Quarterly, 22*(4), 739–770.

Wettstein, F. (2012b). Silence as complicity: Elements of a corporate duty to speak out against the violation of human rights. *Business Ethics Quarterly, 22*(01), 37–61.

Part I
Responsibilities and Liabilities

Chapter 2
The Moral Responsibilities of Online Service Providers

Mariarosaria Taddeo and Luciano Floridi

Abstract Online service providers (OSPs)— such as AOL, Facebook, Google, Microsoft, and Twitter—significantly shape the informational environment (infosphere) and influence users' experiences and interactions within it. There is a general agreement on the centrality of OSPs in information societies, but little consensus about what principles should shape their moral responsibilities and practices. In this article, we analyse the main contributions to the debate on the moral responsibilities of OSPs. By endorsing the method of the Levels of Abstract (LoAs), we first analyse the moral responsibilities of OSPs *in* the web (LoA_{IN}). These concern the management of online information, which includes information filtering, Internet censorship, the circulation of harmful content, and the implementation and fostering of human rights (including privacy). We then consider the moral responsibilities ascribed to OSPs *on* the web (LoA_{ON}) and focus on the existing legal regulation of access to users' data. The overall analysis provides an overview of the current state of the debate and highlights two main results. First, topics related to OSPs' public role—especially their gatekeeping function, their corporate social responsibilities, and their role in implementing and fostering human rights—have acquired increasing relevance in the specialised literature. Second, there is a lack of an ethical framework that can (a) define OSPs' responsibilities, and (b) provide the fundamental sharable principles necessary to guide OSPs' conduct within the multicultural and international context in which they operate. This article contributes to the ethical framework necessary to deal with (a) and (b) by endorsing a LoA enabling the definition of the responsibilities of OSPs with respect to the well-being of the infosphere and of the entities inhabiting it (LoA_{For}).

M. Taddeo (✉) • L. Floridi
Oxford Internet Institute, University of Oxford, Oxford, UK

Alan Turing Institute, London, UK
e-mail: mariarosaria.taddeo@oii.ox.ac.uk; luciano.floridi@oii.ox.ac.uk

M. Taddeo, L. Floridi (eds.), *The Responsibilities of Online Service Providers*,
Law, Governance and Technology Series 31, DOI 10.1007/978-3-319-47852-4_2

2.1 Introduction

Among the private companies involved in the discussion on Internet governance, online service providers (OSPs)—such as AOL, Facebook, Google, Microsoft, and Twitter—play a crucial role. Since the emerging of Web 2.0, OSPs have become major actors, which significantly shape the informational environment (infosphere) and influence users' experiences and interactions within it. OSPs went from offering connecting and information-sharing services to paying members to providing open, free infrastructure and applications that facilitate digital expression, interaction, and the communication of information. This evolution has put OSPs in a peculiar position. For they often stand between the protection of users' rights and government requests, as well as shareholders' expectations. It is not a coincidence that some of the major OSPs—AOL, Apple, Dropbox, Facebook, Google, LinkedIn, Microsoft, Twitter, and Yahoo—have joined forces and created the Reform Government Surveillance (RGS)[1] group to participate in the public debate on the regulation of Internet surveillance and the use of Information and Communication Technologies (ICTs) within governmental security strategies.

While there is a general agreement on the centrality of OSPs in information societies, there is still little consensus about what principles should shape OSPs' moral responsibilities and practices, over and above current legal requirements. These range from Google's generic motto "don't be evil" to much more specific guidelines concerning the protection of the public interest and the respect for basic democratic principles, e.g. openness, transparency, freedom of the Internet, security, and legal certainty, as identified in the 2011 G8 Deauville Declaration.[2] As a result, OSPs' efforts to act on societal issues are still problematic and often encounter shortcomings in design, implementation, and public recognition.

In this article we analyse the main moral responsibilities ascribed to OSPs during the past 15 years. In order to offer a systematic overview, we will look at OSPs' moral responsibilities using the method of the levels of abstraction (LoAs). This will enable us to distinguish OSPs' responsibilities on the basis of the different kinds of information that they control. Categories for Internet control have already been provided in the relevant literature. For example, (Eriksson and Giacomello 2009) distinguish three categories of Internet control: access to the Internet, functionality of the Internet, and activity on the Internet. The latter ranges from filtering and blocking content online, and surveillance, to shaping the political and social discourse. OSPs' actions belong to the 'activity on the Internet'. However, within this category, OSPs control and regulate different types of data and information and their responsibilities vary accordingly. The method of LoAs will help us to distinguish them.

Before proceeding, a brief introduction to the LoAs is required. Any given system, for example a car, can be observed by focusing on specific properties while disregarding others. The choice of these aspects, i.e. the observables, depends on the observer's purpose or goal. An engineer interested in maximising the aerodynamics of a car may focus upon the shape of its parts, their weight and the materials.

A customer interested in the aesthetics of the car may focus on its colour and on the overall look. The engineer and the customer observe the same car at different LoAs. Thus a LoA is a finite but non-empty set of observables accompanied by a statement of what feature of the system under consideration such a LoA stands for. A collection of LoAs constitutes an interface. An interface is used when analysing a system from various points of view, that is, at varying LoAs. It is important to stress that LoAs do not have to be hierarchical (though they can be): the engineer's and the user's LoAs are not one higher or lower than the other. And note that a single LoA does not reduce a car to merely the aerodynamics of its parts or to its overall look. Rather, a LoA is a tool that helps to make explicit the observation perspective and constrain it to only those elements that are relevant in a particular observation for the chosen purpose (Floridi 2008).[3]

In this article, we will focus on two LoAs. One will highlight the moral responsibilities of OSPs *in* the web (LoA_{IN}), while the other will focus on moral responsibilities *on* the web (LoA_{ON}). The former pertains to the regulation of the content available online. LoA_{IN} highlights issues concerning information filtering, freedom of speech, censorship, and privacy. At LoA_{ON}, the focus shifts to the access to the metadata concerning users' activities online. To illustrate the distinction, consider that, given the two LoAs, the debate on the role of OSPs in collaborating with the US government within the PRISM program concerns OSPs' responsibilities *on* the web; while the discussion on OSPs' compliance with the request of the Chinese government to censor some of the information available online is about the responsibilities of OSPs *in* the web.

The analysis of the literature reveals that, during the past 5 years, increasing attention has been devoted to OSPs' public role and impact on contemporary societies (Fig. 2.1). OSPs are often seen as *information gatekeepers* (Calhoun 2002) (more on this in Sect. 2.1), for they control the information available online by making it accessible to the users (Shapiro 2000; Hinman 2005; Laidlaw 2008). This position ascribes a public role to OSPs. This is an unprecedented role for OSPs, which unveils new opportunities along with new problems and responsibilities that are profound and often require OSPs to align their goals with the needs of contemporary information societies (Madelin 2011). As Shapiro put it

> in democratic societies, those who control the access to information have a responsibility to support the public interest. […] these gatekeepers must assume an obligation as trustees of the greater good (Shapiro 2000, 225).

Given the international and multicultural contexts in which OSPs operate, the specification of their moral responsibilities will be effective – i.e. it will be regarded as ethically sound, appropriate, and desirable and offering a suitable guidance to shape OSPs' conduct by the different stakeholders involved in this scenario – only insofar as it will rest on an ethical framework able to reconcile the different ethical views and stakeholders' interests that OSPs face while acting as information gatekeepers. The analysis we propose in this article has the goal of laying the groundwork for such a framework, the definition of which has been left to a second stage of our research. Let us begin by considering OSPs' responsibilities at LoA_{IN}.

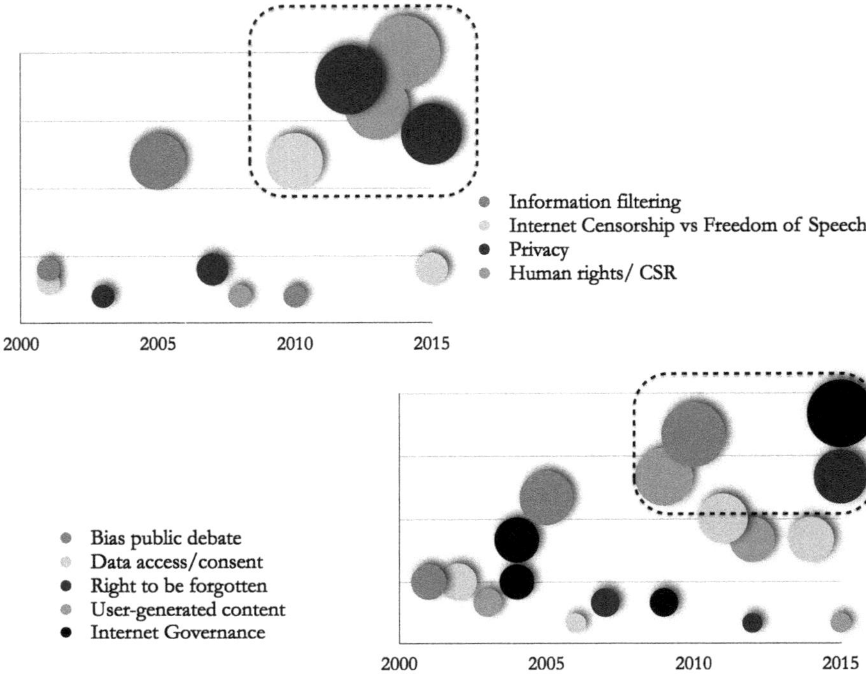

Fig. 2.1 The two graphs show some of the most relevant topics concerning the responsibilities of OSPs addressed in the literature in the past 15 years. The size of the circles is proportional to the number of research articles, books, and edited volumes that include either in the title or in the keywords one of the topics listed in the legends and which were published in the timespan indicated on the x-axis. While topics such as information filtering, user-generated content, and Internet governance have been central in the debate since 2000, other issues like OSPs' corporate social responsibilities and human rights, freedom of speech, and impact of OSPs on the public debate have attracted increasing attention in the past 5 years.[4]

2.2 LoA_{IN}: Moral Responsibilities of OSPs in the Web

The analysis of OSPs' moral responsibilities with respect to the management of the content made available online has been a central point of research in different fields, including information and computer ethics, corporate social responsibilities and business ethics, computer-mediated communication, law, and public policy. Three topics are particularly salient in this debate: the organisation and managing of access to information; censorship and freedom of speech; and users' privacy. These topics have overlapping aspects and implications, which make it difficult to conceive a clear-cut separation of each issue. However, they also identify three important sets of ethical problems worthy of dedicated analyses.[5] In the rest of this article we will focus on each set separately. This slightly artificial structuring has the advantage of providing a conceptual map that will then allow the reader to identify the

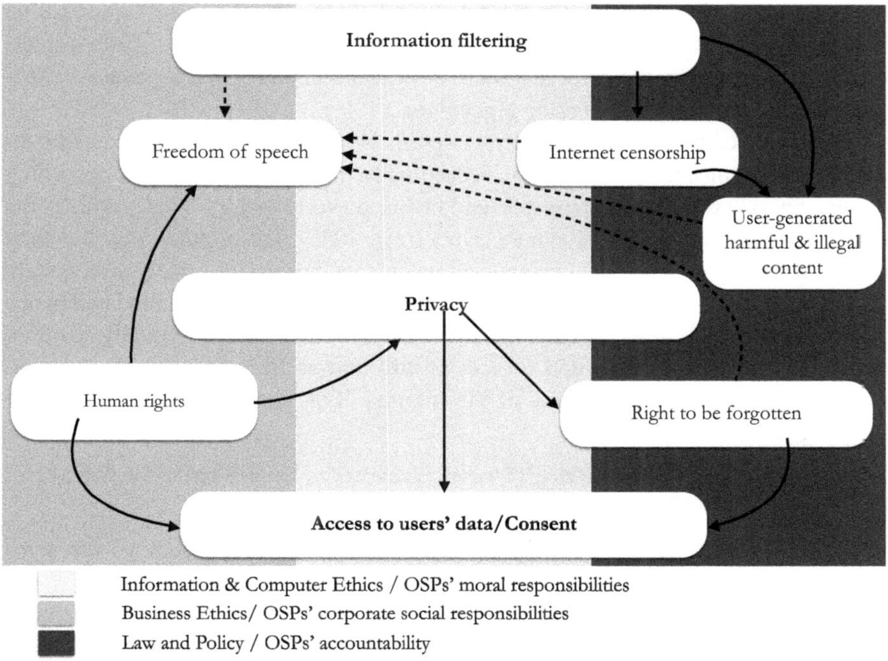

Fig. 2.2 This figure shows the key topics and the research areas in which the responsibilities of OSPs have been debated in the past 15 years. The *dotted arrows* indicate conflicting topics, while the *continuous arrows* link consistent topics. The direction of the *continuous arrows* signifies dependence relation between different topics, e.g. freedom of speech depends on the specification of human rights.

overlapping areas (Fig. 2.2) more easily. Let us begin by focusing on online information filtering.

2.2.1 *Managing Access to Information in the Web: Information Skewing*

The organisation and management of the access to information available online raises problems concerning the way in which search engines select and rank such information (Nagenborg 2005; Spink and Zimmer 2008; Tavani 2014). While the research on this topic initially focused exclusively on search engines, with the emergence of the Web 2.0 social networks and news aggregators also became objects of analysis, for these OSPs too can skew users' access to online information.

Introna and Nissenbaum's article (Introna and Nissenbaum 2006) is among the first publications on this topic. It analyses the role of search engines in defining the scope of access to online information and stresses the relation between such a scope and the development of a pluralistic democratic web. The article advocates diversity of the sources of information as a means to guarantee the fairness of information filtering processes and the democratic development of the Internet.[6] Both aspects can be jeopardised by the corporate, market-oriented interests of the private companies running indexing and ranking algorithms.

The article compares search engines to publishers and suggests that, like publishers, search engines filter information according to market conditions, i.e. according to consumers' tastes and preferences, and favour powerful actors. This promotes the so-called "rich gets richer" dynamic (Huberman 2003). For popular websites tend to be ranked higher hence acquiring even greater visibility. Conversely, this system makes less visible those websites that are already poorly linked or visited and hence ranked lower. This dynamic prompts a vicious circle, which eventually leads to expunging niche, less renowned sources of information from the web, thus endangering the plurality and diversity of the Internet. Two corrective mechanisms are then suggested: embedding the

> value of fairness as well as [a] suite of values represented by the ideology of the Web as a public good (Introna and Nissenbaum 2006, 182)

in the design of indexing and ranking algorithms, and transparency of the algorithms used by search engines.

A different position on transparency of search and ranking algorithms has been prosed in (Granka 2010)[7]. The article points out that disclosing the structure of these algorithms would facilitate ill-intentioned manipulations of search results, while not bringing any advantage to the average non-tech-savvy user. Granka's paper also disputes the idea that market regulation of the Internet threatens the diversity of the information sources. On the contrary, it maintains that, in a market-regulated environment, companies will devote their attention to the quality of the search results, which will have to meet the different needs and expectations of every user, thereby guaranteeing diversity of the sources and fairness of the ranking. In this respect, the article also objects to the analogy describing OSPs, search engines in particular, as publishers. Search engines

> parse through the massive quantities of available information [...], the mechanisms whereby content is selected for inclusion in a user's search result set is fundamentally different than in traditional media—search engines universally apply an algorithm, whereas traditional news media makes case-by-case decisions (Granka 2010, 365).

The problem remains, however, when a search engine has a virtual monopoly and hence no real competition within a whole market, as it is currently the case for Google in Europe.

OSPs' editorial role is also analysed in (Goldman 2006). The article describes search engine bias as a necessary consequence of OSPs' editorial work,

> to prevent anarchy and preserve credibility, search engines unavoidably must exercise some editorial control over their systems. In turn, this editorial control will create some bias (Goldman 2006, 119).

While the analysis recognises that such filtering may reinforce the existing power structure in the web and bias search results toward websites with economic power

(Elkin-Koren 2001), it also advocates that the correction of search bias will follow from the fine-tuning of the search results with users' preferences. No extra moral responsibilities should be ascribed to OSPs in this respect. A similar position has also been expressed in Lev-On and Manin's and Lev-On's articles (Lev-On and Manin 2007; Lev-On 2009). The articles suggest that, given the huge amount of data filtered by search engines, unintentional exposure to diverse and non-mainstream information cannot be excluded. The issue then arises as to whether incidental exposure to diverse information may suffice to maintain an open, pluralistic web.

The personalisation of search results—offering diversified results based on the preferences of each individual, rather than those of the majority—has also been proposed as a remedy to the concerns highlighted by Introna and Nissenbaum. For the tailoring of search results leads to an organic refinement of searching and ranking algorithms so as to accommodate users' preferences and, at the same time, it makes it possible to correct the distortion performed by OSPs while fostering diversity in the sources and information circulating in the web. This is, for example, the argument proposed by both Goldman's and Crawford's articles (Goldman 2006; Crawford 2005).

The personalization of search results is not uncontroversial. Far from being seen as a solution to the problems engendered by information filtering, it has been objected to as a threat to democratic discourse in contemporary societies. In this respect, issues have been raised by several scholars (Sunstein 2001; Anderson 2008; Spink and Zimmer 2008; Pariser 2012). Custom-tailoring of search results challenges the basic underpinning of a deliberative democracy insofar as it undermines the possibilities of sharing cultural background and experiences and reduces the chances of being exposed to sources, opinions, and information that may support or convey different world views. In particular, Sunstein's book (Sunstein 2001) criticises any approach relying on users' preferences and market dynamics to shape information access and communication:

> it is much too simple to say that any system of communication is desirable if and because it allows individuals to see and hear what they choose. Unanticipated, unchosen exposures, shared experiences are important too (Sunstein 2001, 131).

He argues that a custom-tailored access to information leads to a world fragmented into different versions of "the daily me" (Negroponte 1996),[8] in which each individual would be isolated in their *informational bubble* (Pariser 2012), from which conflicting views are excluded. A similar argument has also been proposed in Pariser's book (Pariser 2012). The book criticises the personalisation of access to online information, because it promotes personalised *informational ecosystems* and *echo-chambers* that undermine the emergence and fostering of democracy.

Over the years, the discussion concerning the responsibilities of OSPs has moved from defining the measures that OSPs should deploy to correct their market bias and ensure a pluralistic web, to understanding the impact that OSPs have on the Internet as well as on the flourishing of democratic values and on societies at large (Fig. 2.1). This shift is partly due to the ideal of a democratic web inspiring the design of the

Internet as a free, open network for the sharing of information (Toffler et al. 1995; Negroponte 1996; Diamond 2010). At the same time, the centrality of ICTs and in particular of the Internet in contemporary societies stresses the need to regulate access to online information so to protect and foster individual liberties and the democratic ideal. OSPs are major actors in this scenario, contributing to the shaping of both the informational environment and societies. For this reason, Sunstein's and Pariser's analyses ascribe to OSPs a civic responsibility to foster plurality and democracy.

Similar analyses leave unaddressed the identification of the principles that should guide OSPs when dealing with their civic responsibilities. Defining such principles proves to be a difficult task. OSPs are private companies to which academia, policy-makers, and society increasingly ascribe the role of *information gatekeepers*, generating the expectation that they will perform their tasks

> well and according to principles of efficiency, *justice*, *fairness*, and *respect* of current social and cultural values (McQuail 1992, 47) (emphasis added).

The notion of gatekeepers has been studied in business ethics, social sciences, and legal and communication studies since the 1940s. It characterizes those agents who have a central role in the management of resources and infrastructures that are crucial for societies. For example, in 1947, Lewin famously described mothers and wives as gatekeepers, for they were the ones deciding and managing the access and consumption of food for their families (Lewin 1947).

Metoyer-Duran (1993) offers a fruitful definition of gatekeepers according to which an agent is a gatekeeper if that agent

> (a) controls access to information, and acts as an inhibitor by limiting access to or restricting the scope of information; and (b) acts as an innovator, communication channel, link, intermediary, helper, adapter, opinion leader, broker, and facilitator.

Conditions (a) and (b) entail moral responsibilities, insofar as gatekeepers have a regulatory function. The private nature of gatekeepers, along with the responsibilities entailed by (a) and (b), is one of the cruxes generating the problems concerning their moral responsibilities (Freeman 1999; Black 2001).

Framing the discussion on the moral responsibilities of OSPs using the notion of gatekeepers unveils OSPs' public role, along with the accompanying friction that they may experience between corporate and public interests. However, this notion also risks biasing the discussion in an unfruitful way. Two major concerns arise in this respect.

The first concern emerges when considering the extant literature on corporate social responsibilities (CSR) (Crane et al. 2008), which focuses mainly on the duties towards societies that are inherent to the responsibilities of private companies having a gatekeeping function (Matten and Crane 2005; Palazzo and Scherer 2006; Scherer and Palazzo 2006; Albareda et al. 2007; Blowfield and Murray 2008; Okoye 2009; Helgesson and Mörth 2013). In this case, the analysis of the moral responsibilities is shaped by a deontological bias, addressing the moral duties that gatekeepers have *qua* controlling agents. This is not wrong *per se*. However, such a bias often leads to disregarding the rights of the *gated* (Barzilai-Nahon 2008), the receivers of the gatekeepers' actions, i.e. the *moral patients*.

The second concern arises from the attempt to overcome the first. In this case, users are usually identified as the ultimate moral patients. However, OSPs' gatekeeping function does not affect only users' online experiences, for OSPs' control over online information also makes them key agents shaping users' experience as well as the informational environment (Laidlaw 2010; Cerf 2011). The need then arises to define the moral responsibilities of OSPs with respect to both the users and the informational environment. Such a need becomes more pressing as one considers the extent of the control exercised by OSPs on the latter.[9] The regulation of user-generated content available online offers a good example of the case in point. The next section focuses on this topic.

2.2.2 *Internet Censorship and Harmful Content*

OSPs also manage access and circulation online of user-generated content. Part of this management implies preventing the dissemination of illegal content (e.g. child pornography), of hate speech, and of other material that may be deemed harmful to individuals and societies, e.g. pro-suicide, pro-anorexia or terrorism-promoting websites. Other forms of censorship may be prompted by governments to pursue political agendas beyond individual and social welfare.

Legally speaking, OSPs are generally not liable for the user-generated content that they host.[10] At the same time, OSPs have been encouraged to monitor and filter, to the extent that they can, the content circulating on the web (Hildebrandt 2013). Two main models have been endorsed to assess OSPs' liability with respect to third party content. The first one is the so-called "safe harbour" model.[11] In this case, the intermediary liability only applies to OSPs with respect to specific types of content, e.g. copyrighted material. In this model, OSPs are liable if they do not comply with the "notice and take down" procedure and hence do not act promptly to remove or disable access to illegal information when they obtain actual knowledge of such content. The second model guarantees broad immunity to OSPs by considering them as carriers of user-generated content for which they do not bear any liability, somewhat like a postal service. The question remains as to whether OSPs have any moral responsibilities to monitor and filter the web to prevent the dissemination of offensive and harmful material.[12]

Johnson has noted that, while it might be feasible to hold OSPs legally liable for the circulation of some contents, it would be much more difficult to argue that OSPs should be morally responsible for the behaviour of their users (Johnson 2009). This last point is quite uncontroversial, but it may also be misleading. The issue at stake is not whether OSPs should be held morally responsible for their users' actions. Rather, the problem is whether OSPs bear any moral responsibilities for circulating on their infrastructures third-party generated content that may prove harmful.[13] To some extent, similar responsibilities have already been ascribed to other media, like television and newspapers. Smoking advertisements have been banned in European countries because of their potential to induce harmful habits in their audience.[14] In this case, media are not held responsible for the actual smoking habits of the audience, nor are they held responsible for the tobacco industry's intention to promote smoking. But they are held responsible for the potentially harmful consequences of the information that they would disseminate.

Vedder's contribution (Vedder 2001) delves into this issue and suggests that OSPs should be held morally responsible for the dissemination of harmful content. The article distinguishes between *prospective* and *retrospective* moral responsibility and stresses that the two aspects go hand in hand. According to Vedder's analysis, OSPs are usually considered *prospectively* responsible insofar as they have the moral duty of avoiding possible future harm to their users. It is more problematic to ascribe *retrospective* responsibility to OSPs, for it presupposes guilt, and it has been maintained in the literature that such responsibilities cannot be attributed to communities or non-physical persons. However, Vedder's article argues that, since OSPs are considered prospectively morally responsible, they should also be held retrospectively responsible, and hence they bear full moral responsibility for the content that they circulate.

A similar position has also been supported in the analysis proposed by Tavani and Grodzinsky (Tavani and Grodzinsky 2002). The article analyses the case of Amy Boyer, a young woman who was first stalked and then killed by Liam Youens, a man who used the web to collect information about the victim that was relevant to his plan.[15] Following Vedder's argument, the paper puts the burden of the responsibility for the information circulating online about the victim on both OSPs and the users who shared such information with the killer.

In a commentary, Vinton Cerf (Cerf 2011) touched directly on the role of OSPs in preventing harmful uses of the web stating that

> it does seem to me that among the freedoms that are codified […] should be the right to expect freedom (or at least protection) from harm in the virtual world of the Internet. The opportunity and challenge that lies ahead is how Internet Actors will work together not only to do no harm, but to increase freedom from harm (Cerf 2011, 465).

Following Cerf's commentary, it may be desirable to ascribe moral responsibilities to OSPs with respect to the circulation of harmful material. However, this ascription raises further problems when considering the duties that these responsibilities may prompt, e.g. policing and filtering the content available online, and the possible breaches of individual rights, such as freedom of speech and information, and anonymity. This is a difficult balance to strike and to implement.[16] While OSPs should be held responsible for respecting this balance, and should be involved in the discussions aiming at striking a fair balance, it should not be their duty to define the balance and decide, for example, how much freedom of information can be sacrificed in the name of users' safety and security.

Reducing the harm on the Internet has put OSPs in a difficult position, standing between citizens' rights and expectations of a free, uncensored, access to information. OSPs are also caught in the friction between national and international powers.

Some national powers, for example, seek to limit their citizens' right to freedom of speech and anonymity, while the international community recognises these as fundamental human rights. The next section analyses this problem.

2.2.2.1 Internet Censorship and Freedom of Speech

In 2012, Internet freedom was declared a human right by the UN Human Rights Council, which called on states to promote and foster access to the Internet and to ensure that the rights to freedom of expression and information, as presented in Article 19 of the Universal Declaration of Human Rights, would be upheld online as well as offline.[17] Do OSPs have any responsibilities with respect to Internet freedom and with human rights in general? Some authors, like (Chen 2009), have argued that OSPs, and in particular social networks, bear both a legal and a moral responsibility to respect human rights, because of the centrality of their role on the web and of their knowledge of the actions undertaken by other agents, e.g. governmental actors, in the network. At the same time, both the Universal Declaration of Human Rights and the Resolution on the Promotion, Protection and Enjoyment of Human Rights on the Internet[12] mainly address states actors, making problematic the expectation that OSPs should be held responsible for respecting and fostering human rights (Karp 2009). This problem does not exclusively concern OSPs. It also involves several other private actors, especially those working in the international market (Anderson 2012), making this issue a central topic in the literature on business ethics. Consider, for example, the cases of human rights violations reported by Human Rights Watch concerning the energy industry, such as Royal Dutch/Shell's operations in Nigeria, British Petroleum in Colombia, and Total and Unocal's construction works in Burma and Thailand.[18]

Some authors, like Santoro and Brenkert, stress the need to consider the context in which companies act before assessing their moral responsibilities (Brenkert 2009; Santoro 1998). Santoro proposes a "fair share theory" to assess the moral responsibilities of multinational companies complying with the requests of an authoritarian state. According to this theory, the responsibilities for respecting and fostering human rights are ascribed differently depending on the capability of the company. Santoro poses two conditions for evaluating the capabilities of private companies and ascribing responsibility: (i) they have to be able to make a difference, i.e. change local government policies; and (ii) they have to be able to withstand the losses and damages that may follow from diverging from local governmental directions and laws. Both conditions highlighted in (Santoro 1998) are problematic. Condition (i) offers a justification to any private company that may engage in immoral, or unlawful, actions. For the inability to make the difference in governmental policies allows the company to claim no moral responsibility for any violation of the human rights in which it may partake while collaborating or complying with a local government's directives. Condition (ii) does not stand as a valid requirement *de facto*, at least when considering major OSPs. For instance, in 2010 Google withdrew from China and still managed to be one of the most competitive OSPs in the global market. More recently, Facebook's CEO commented on this point stating that

> Today we're blocked in several countries and our business is still doing fine. If we got blocked in a few more, it probably wouldn't hurt us much either.[19]

Other scholars support a different view and hold private actors morally responsible for the protection and fostering of human rights (Arnold 2010; Cragg 2010; Wettstein 2012). The preamble of the Universal Declaration of Human Rights is often mentioned to support this point. It states that

> every individual and every organ of society, keeping this Declaration constantly in mind, shall strive by teaching and education to promote respect for these rights and freedoms […].[20]

The responsibility of all members of societies to promote respect for human rights has been remarked and further elaborated in the Declaration of Human Duties and Responsibilities (the so-called Valencia Declaration),[21] which focuses on the moral duties and legal responsibilities of the members of the global community to observe and promote respect for human rights and fundamental freedoms. The global community encompasses state and non-state actors, individuals and groups of citizens, as well as the private and the public sector. Private companies are also expressly mentioned as

> responsible for promoting and securing the human rights set forth in the Universal Declaration of Human Rights

in the preamble of the UN Norms on the Responsibilities of Transnational Corporations and Other Business Enterprises.[22]

One of the cases about the moral responsibilities of OSPs and the respect of human rights (freedom of speech in particular) that has been most debated in the relevant literature concerns the complying of some OSPs, like Google, Microsoft, Yahoo!, and Facebook, with the requests made by the Chinese government on Internet censorship and surveillance.[23] OSPs have responded in different ways. Some, like Google (in 2010) and Yahoo! (in 2015), eventually decided not to comply with these requests and withdrew from the Chinese market. Others refer to the so-called consequentialist argument to justify their business in China. The argument was first provided by Google to support its initial compliance with the Chinese government requests. It holds that, while the Chinese people could not access some sources of information due to the local censorship, they could still use Google's services to access a whole lot more online information. In more sophisticated terms, it endorses the logic of a 'better than nothing' approach. More recently, Facebook and Microsoft have proposed the same argument. As Facebook's CEO states

> I believe we have a responsibility to the millions of people in these countries who rely on Facebook to stay in touch with their friends and family every day. If we ignored a lawful government order and then we were blocked, all of these people's voices would be muted, and whatever content the government believed was illegal would be blocked anyway.[24]

Those who maintain that private companies ought to comply with human rights, because these are preeminent to local governmental actions, criticise the consequentialist argument.

> Multinationals [...] should respect the international rights of those whom they affect, especially when those rights are of the most fundamental sort (Donaldson 1992, 68).

Such a position is also maintained in Dann and Haddow's article (Dann and Haddow 2007), whose article ascribes moral responsibility to company executives, who make the final decisions and shape a company's conduct.

A different account of the moral responsibilities of OSPs partaking in local governmental censorship has been provided in Brenkert's analysis (Brenkert 2009), where the notion of 'obedient complicity' is suggested,

> [t]his would occur when a business follows laws or regulations of a government to act in ways that support its activities that intentionally and significantly violate people's human rights (Brenkert 2009, 459).

The notion rests on the idea of permissible moral compromise. This is the compromise that agents make with themselves to forgo or even violate some of their moral principles to fulfil other, more important, values. OSPs operating in countries requiring Internet censorship face conflicting responsibilities towards different stakeholders, not just users, but also local employees and shareholders. For this reason, these OSPs may be justified in engaging in a moral compromise that may violate human rights, if this enables the achievement of more important objectives.

Brenkert's article proposes the so-called 'all things considered' approach to assess whether an OSP may be in a position to violate its moral principles or universal rights. The article considers the immediate context in which OSPs operate and the multiple responsibilities that this implies. For example, an OSP may be put in the position to compromise its moral values or to disregard human rights and comply with local laws lest its employees working in a given territory be held liable for the company's decision, or to avoid damaging the shareholders' interest. According to Brenkert's article, a moral compromise may be justified in these cases.

As any consequentialist approach, the 'all things considered' enables one to cover a wide range of responsibilities of private companies and assess them with regard to the company's maximum utility. This proves problematic, because the assessment of the moral responsibilities of a company depends on the scope of the context that is being considered. Recalling the LoA methodology, let us assume that one endorses a LoA to identify the company's interest. In doing so, one may focus solely on the local interests of the company, the risks that the company may take in refusing to respect local laws, and the benefits that may follow from complying with the requests of local authorities. This LoA may support the acceptance of moral compromise and justification of the possible breach of human rights. However, such a LoA proves to be too narrow to consider properly the interest of a company operating in the international market, such a company needs to consider more factors than its local interests. It would be a LoA adopted according to the wrong purpose. A less restricted LoA – adopted for a better purpose – could account for observables such as the company's global reputation, the impact that breaching human rights may have on the company's public image, as well as the company's local interest. It would thus unveil the relevance of respecting human rights even when this may conflict with the interest of the shareholders. It follows that while the 'all things considered' approach was intended to mitigate the burden of OSPs' moral responsibilities, it actually offers one more argument in favour of OSPs' duty to respect and foster human rights.

The debate on the responsibilities of OSPs with respect to human rights highlights the challenges that come from the multicultural and international context in which OSPs work. It also shows the global relevance and impact that OSPs have on information societies. While it is increasingly less acceptable to maintain that OSPs, as private companies, are only responsible to their employees and shareholders, it is also problematic to ascribe to OSPs full responsibility for the fostering and respecting of human rights. For this entails that OSPs can arbitrarily and independently decide the circumstances and the modes in which they need to respect such rights.

Two aspects have been under-estimated in this context. One concerns the role and responsibilities of actors like the UN or the European Union in regulating OSPs' conduct so to ensure that they effectively respect human rights in their activities, independently from the geographic regions in which such activities are conducted. This is quite a problematic topic, for it prompts questions concerning sovereignty, Internet governance, and the territoriality of jurisdiction. However, as remarked in this section, international compliance of private companies with human rights is not a new problem and some legal international procedures are already in place to tackle it.

The second aspect concerns the definition of an ethical framework that can address the problems at hand. The analyses considered in this section identify in human rights such a framework. However, this has been shown to be insufficient, for human rights restrict the focus to human moral patients. As stressed in Sect. 2.1, OSPs do not only affect human users, they also shape the informational environment. Overlooking OSPs' role within the environment that they build will impair any attempt to define their wider moral responsibilities towards the whole infosphere from an environment perspective. The risk is that the resulting analysis will be either too generic, i.e. OSPs should respect human rights in all circumstances, or too narrow, i.e. OSPs' responsibilities concern exclusively human agents and only in some circumstances.

The time has come to consider OSPs' responsibilities with respect to users' privacy at LoA_{IN}.

2.2.3 OSPs' Responsibilities and User Privacy

The voluntary sharing online of personal information raises several concerns with respect to the protection of users' privacy.[25] For one thing, the personal information that is voluntarily shared online often exposes online and offline personas (Taddeo 2014) beyond the original intention of the users, leading to unforeseen breaches of their privacy and to potentially harmful consequences. Cyber-stalking (Tavani and Grodzinsky 2002) and the use of social networks to check employees' and students' backgrounds (Qi and Edgar-Nevill 2011; Semitsu 2011) offer good examples of said harmful consequences.

Responses to these concerns address both users' habits and OSPs' attitudes towards privacy. Some refer to the so-called "privacy paradox" (Acquisti 2004; Barnes 2006; Norberg et al. 2007; Rosen 2015) to stress that individuals continue to disclose personal details online, albeit being aware of the risks that this habit poses to their privacy. Qi and Edgar-Nevill caution that

> as social networking search and investigation become more popular, the public needs to know the processes and the rules regulating these activities. Understanding the extent of data disclosure on the social network is the first step for all (Qi and Edgar-Nevill 2011, 74).

At the same time OSPs, and particularly social networks, are considered responsible for a *de facto* devaluation of privacy, for they nudge their users to share more personal information using both open statements, see for example Sun Microsystems' CEO, McNealy "you have zero privacy anyway. Get over it",[26] and architectural design or apps like Facebook's newsfeed and Beacon (Baym 2011; Lanier 2011). In the rest of this section, we will review the moral responsibilities ascribed to OSPs with respect to users' privacy at LoA_{IN}.

An interesting contribution to this debate has been offered in Spinello's article (Spinello 2011). The article rests on the understanding of privacy as an individual moral right, which OSPs have the moral responsibility to protect. Following the definition of privacy as "limited control of restricted access" to personal information proposed by Tavani and Moor (Tavani and Moor 2001), Spinello's paper advocates the need to give users the power to control and limit access to the information that they share. OSPs would comply with their moral responsibility to protect users' privacy by endorsing a proactive approach and measures that would ensure users the maximum level of control over their personal information.

The proactive approach mentioned in Spinello's analysis is recalled in Hull, Lipford, and Latulipe's paper (Hull et al. 2011), which also attributes to OSPs moral responsibility to protect users' privacy at LoA_{IN}. This article relies on Nissenbaum's analysis of privacy (Nissenbaum 2010) and criticizes OSPs' architectures, in particular social networks, for treating human relations as if they were all of the same kind. OSPs' platforms would be better suited to managing users' personal information if they respected context-sensitive privacy norms, considering a greater array of social settings, rather than just focusing on the distinction between public and private. It is worth noting that, to some extent, this criticism has been taken seriously. Some social networks, e.g. Google+ and Facebook, allow access to a user's personal information to be curbed depending on the kind of social relations that they enjoy with other users.

The analyses provided in the previous articles rest on an understanding of the protection of privacy as an individual choice, i.e. a user only needs to apply high privacy settings when sharing her information to protect her right, and OSPs need to offer and facilitate such a choice. However, when one takes into consideration the interpersonal nature of information sharing in contemporary hyper-connected societies, this approach proves to be too narrow to be effective and to cast light on the parties who bear moral responsibility for protecting privacy. Even if a user has a highly protective privacy setting, personal information could be accessed by unauthorised parties due to the setting of other users in his/her network (Schwartz 1999; Caudill and Murphy 2000). This raises new problems, insofar as the difficulty of monitoring other users and uncertainty about their behaviours poses the need for a more refined privacy management.[27]

Two approaches have been proposed to overcome such difficulties. One, the *communitarian approach*, shifts moral responsibility for controlling information and protecting privacy from individuals to the community (O'Hara 2010; Xu 2012). The other, the *proxy approach*, focuses on OSPs and other major private and public actors, which can enforce social controls through regulation and codes of conduct (Smith et al. 2011).

The communitarian approach frames privacy as a public good, the benefits of which concern the community and not just the individuals. Following the analysis provided by Etzioni (Etzioni 1999), O'Hara's article (O'Hara 2010) maintains that privacy, as an individual right, may undermine community welfare if it is not properly curbed and balanced against other social concerns. The choice of some individual to share personal information online may be harmful not just to that individual, but to the community at large. Hence, according to this view, the right to privacy implies the duty of the single person toward the community to share responsibly and to monitor the flow of personal information circulating in their network. Individuals, as part of a community, need to agree on strategies for collectively managing the shared information, e.g. establishing a friendship-based model for privacy protection (Besmer et al. 2009), or rules of thumb regarding sharing with other users, such as asking for approval before disclosing content from those involved (Lampinen et al. 2011).

Moral responsibility for the protection of individual privacy shifts back to OSPs when embracing the proxy control approach. In this case,

> people try by one means or another to get those who have access to resources or expertise or who wield influence and power to act at their behest to secure the outcomes they desire (Bandura 1999, 13).

OSPs bear responsibility because they are the depositaries of users' trust. Donaldson and Dunfee argue that there is an integrative social contract between users and OSPs (Donaldson and Dunfee 1999). Users provide their personal information to OSPs, which in turn offer some services to the users. One obligation, following this social contract, is that the OSPs accept the responsibility of managing consumers' personal information properly. This social contract rests on users' trust in the company's compliance with the contract. This trust is essential to overcome initial users' uncertainty and foster interactions online, and as such it is crucial for OSPs to preserve it (Weckert 2005; Taddeo 2010; Turilli et al. 2010). A significant difficulty is that the relation between users and OSPs seems to be modelled more on a version of a gift economy, which de-responsabilitises the gifter with respect to the giftee, rather than on any more or less metaphorical social contract (Floridi 2015a, b).

Trust in OSPs and the proxy approach have also shaped the application of the right to be forgotten in Europe, raising non-trivial ethical problems. The next section will delve into this issue.

2.2.3.1 The Right to Be Forgotten and OSPs' Responsibilities

The so-called right to be forgotten was announced in Europe in 2012:

> if an individual no longer wants his personal data to be processed or stored by a data controller, and if there is no legitimate reason for keeping it, the data should be removed from their system (Reding 2012).

It has its roots in the French right of oblivion, which allows an individual to object to the publication of information about his/her criminal conviction, once s/he has served the sentence and been rehabilitated. As Mayer-Schönberger argues, this right is also rooted in the European history of the XX century, when the collection and retaining of personal information often turned out to be a powerful tool in the hand of totalitarian regimes. In post-1989 Europe, the possibility to be forgotten is seen as an extra measure fostering democracy and plurality (Mayer-Schönberger 2011).

If at first sight this right may seem an uncontroversial means to empower citizens to protect their privacy by ensuring them the control over their personal data, a more attentive analysis unveils the friction between this right and the right to freedom of speech and information. Striking the correct balance between the two is not a simple matter. Things change, for example, depending on which side of the Atlantic one is.[28] According to the European approach, privacy trumps freedom of speech; whereas the American view is that freedom of speech is preeminent with respect to privacy (Rosen 2012). Hence, defining the responsibilities of OSPs with respect to the right to be forgotten turns out to be quite problematic, as it involves the balancing of different fundamental rights as well as considering the debate on the national versus international governance of the Internet (Floridi 2015a).

All these issues became evident with the ruling of the Court of Justice of the European Union (CJEU), which, in May 2014, decided that, given some circumstances, Google (or any other search engine operating in Europe) must remove from its search results the links to personal information if this is "*inaccurate, inadequate or no longer relevant*"(emphasis added).[29] The ruling opened a Pandora's box,[30] because anyone who thinks it inappropriate for some information concerning their personal life to be accessible online can now ask Google to delist it from its search results (note that the information would not be removed from the web, despite what it is indicated in Reding's quotation above).

According to European regulation, OSPs that are presented with a request to remove personal information "shall carry out the erasure without delay", unless the retention of the information is deemed essential for the right of freedom of expression. This ascribes to OSPs the responsibility to assess, on a case-by-case basis, the legitimacy of the sharing of the personal information online and to decide at which point the delisting of such information would be a case of undue censorship. Having to define the criteria for deciding which delisting requests to approve, Google sought the advice of a pool of international experts, who suggested a set of principles that should guide it in complying with the ruling of the CJEU.[31]

Very briefly the council advised Google to (a) apply the delisting decision across all its European websites (e.g. Google.de, Google.it, Google.es and so on), and (b) to notify publishers when a delinking procedure was initiated. Four more criteria were offered to guide Google in assessing the delisting requests: (i) evaluate the public role of the data subject, (ii) consider whether information to be delisted may impact private or public interests, (iii) consider the source of the information, and finally (iv) the timeframe of information was suggested as a criterion to assess its relevance. Both the suggestions and the criteria proposed by the advisory council unveil the responsibility to *judge* the information in question ascribed to Google and to several other OSPs (both Yahoo! and Microsoft have set forms available online to allow users delisting request) by the ruling of the CJEU.

The judging role of OSPs is controversial. For example, Rosen argues that, in this way, the power and responsibility of making public decisions shifts from judicial courts to private actors (Rosen 2012). The public role of OSPs in contemporary societies is undeniable, and so is the need to ensure that OSPs will act consistently with the public good (the reader may recall the discussion on the responsibilities of OSPs as information gatekeepers in Sect. 2.1.). Yet, the application of the right to be forgotten goes a step too far. It does not ascribe to OSPs the responsibility to *act* by respecting the criteria for protecting and fostering individual rights as well as societies' moral principles and welfare. It puts OSPs in the position to have to *decide* about those criteria and those principles and their implementation. Hence, OSPs become both "*the judge and the jury*".[32]

As remarked in (Gerry and Berova 2014), the ruling started a privatization of the judging power, which poses issues of transparency and accountability. OSPs, being private companies, do not have to comply with the same standards that apply to public institutions, nor are they expected to disclose any information about how they comply with the court's order. Yet absence of transparency and accountability risks paving the way to corruption, arbitrary decisions, and unfair applications of the ruling of the CJEU. Even more importantly, while the ruling may strengthen individuals' control over their personal information at LoA_{IN}, it does not do much to reinforce an individual's control at LoA_{ON}, that is, on the access that third party, e.g. corporate or government agents, may have to their data trails.

The reader may recall that, in Sect. 1, we mentioned the RGS group as an example of the reaction of OSPs to the difficult position in which OSPs find themselves. The RGS group signed a letter to the US President and Congress asking them to endorse five principles in revising US surveillance policies: limiting government authority in assessing users' data, oversight and accountability, transparency, respect for the free flow information, avoiding conflicts among governments.[33] The RGS group and its principles are OSPs' response to the debate prompted by the Snowden revelations on PRISM, the NSA massive surveillance programme. The PRISM scandal raised significant concerns internationally about surveillance in information societies. Most of the analyses developed on this topic frame the problem as the balance of surveillance and security with individual rights (Taddeo 2013). One crucial way in which such a balance is achieved is by regulating access to users' data. When considering this aspect, problems arise with respect to OSPs' role and responsibilities as information gatekeepers. The next section focuses on this topic.

2.3 LOA_{ON}: OSPs' Moral Responsibilities on the Web

In this section, we shall adopt the LoA_{ON} to consider OSPs' responsibilities in managing access to users' data.[34] In contemporary information societies, data are a crucial, resourceful asset, which can drive and support the economy, industry, scientific research, welfare as well as governance, surveillance, and security. Regulating access to data is not a trivial matter, as it involves balancing societal interests and progress with individual rights. Privacy plays a crucial role in this context, for users' data trails are quite revealing of their tastes, health, finance status, and social interactions. OSPs often stand between individuals' personal data and powerful agents aiming at gaining access to such data, e.g. governments as well as private companies, and OSPs themselves have a strong interest in collecting and mining users' data. The question then arises as to what principles should regulate access to users' personal data and information and what should OSPs' responsibilities be in accessing, controlling, and managing users' data.

Different positions have been held in the relevant literature in this respect. Some see in OSPs the depositary of users' trust. As such, OSPs have the responsibility to respect individual rights while managing their data (Donaldson and Dunfee 1999). Others strengthen this view by referring to a duty of *loyalty* of OSPs toward their users (Kerr 2002). The duty of loyalty demands that parties remain faithful to each other even when conflict arises between the interests of the peers. Kerr's article provides four criteria to identify those relationships in which the trusted party has a duty of loyalty to the trusting one: (i) if the former has some discretion or power and (ii) can unilaterally use this power to affect the trusting party, (iii) if the trusting party is vulnerable and/or at the mercy of the party holding the power, and (iv) if the trusting party is entitled to expect that the trusted party will act in her interest. Kerr's paper maintains that the relationship between OSPs and their users satisfies all four conditions. Condition (iv) is quite interesting, for it stresses a point that has also been highlighted in an opinion published by the Article 29 Working Party (Art. 29WP),

> [users] usually have an expectation about the purposes for which the data will be used. There is a value in honoring these expectations and preserving trust and legal certainty.[35]

Users' rights and *expectations* that users have about those rights play a central role in the regulation of data access developed during the past three decades as well as in the definition of OSPs' responsibilities at LoA_{ON}.

More recently, however, both academics and policy makers have criticised the effectiveness of focusing exclusively on users' rights when assessing data management (Acquisti and Grossklags 2005; Jolls and Sunstein 2005; World Economic Forum 2012; Cavoukian 2014; Cate et al. 2014; Kiss and Szőke 2015). The keystone of the criticism is the "notice and consent" model. The model rests on the assumption that users give consent to the treatment of their personal data after having read carefully the notice of each service to which they subscribe. However, this model ceases to be effective in contemporary societies. On the one hand, the more individuals use ICTs in their daily practises, the higher the number of privacy notices that they are expected to read, and the less attention users devote to the notice. On the other hand, the "notice and consent" model basically offers a Hobson's choice[36] and stands between the users and the services they want to access. Thus, in order to access a given service, users may unintentionally consent to types of data processing that in reality they do not want. Schermer, Custers, and Hof (Schermer et al. 2014) refer to the "crisis of consent" to indicate a phenomenon that has led the notice and consent model to be devoid of the role originally attributed to it.

An attempt to address this crisis of consent has been provided in 2013, when the Organisation for Economic Co-operation and Development (OECD) published an updated version of the *Guidelines on the Protection of Privacy and Transborder Flows of Personal Data* (OECD 2013). The guidelines were first implemented in 1980. [37] Since then, they have provided a common ground for national and international regulation of data access. The main problem addressed by the guidelines is the protection of user privacy. The main goal is to avoid users experiencing any physical or moral harm due to third-parties accessing their data. The update of the OECD guidelines has set a watershed in the definition of duties and responsibilities in managing users' data, given the 2013 version switched the focus from users' rights to the duties of data stewards (data controllers and data users, (Cate et al. 2014)), with an entire new section (Part three) devoted to guide data stewards in implementing the accountability principle. The principle states that "a data controller should be accountable for complying with measures which give effect to the principles stated [in section two]" (OECD 2013, 15), data stewards respect the principle if they meet the following three requirements: deploy privacy management procedures; can demonstrate that such procedures are appropriate; notifies the relevant authorities if and when a security breach affecting personal data occurs.

This shift fuelled the debate on the responsibilities of data stewards, e.g. OSPs, in contemporary information societies (Kiss and Szőke 2015).

In their report, Cate, Cullen, and Mayer-Schönberger (Cate et al. 2014) defend this shift, arguing that it offers a better framework to fine-tune privacy with the different uses of data in contemporary societies. The report also stresses that (i) the responsibilities pertain to the processing of data rather than to obtaining consent from users, and that (ii) concern should be focused more on the use of the collected data than on the collection itself. Point (ii) rests on the observation that the context in which data may be used in the future, as well as the value that they will have, is often unclear or unforeseeable at the moment of the collection. It is then the responsibility of data stewards to ensure that users' data will be processed in a way that respects individual rights—such as privacy, anonymity, and transparency—even when used in contexts and for purposes that were not foreseen at the moment of the collection.

A different approach has been proposed in Cavoukian's article (Cavoukian 2014). The article objects to the shift—from the rights of data subjects to the duties and responsibilities of data stewards—as being paternalistic and dangerous for the protection of privacy. The endorsement of privacy by design is suggested as an alternative method for managing data access so to respond to the needs of contemporary society without threatening users' privacy. According to Cavoukian's article, data stewards have the responsibility of implementing design measures that protect a user's privacy by default. In particular, the article stresses the value of de-identification methods, for they can protect individual privacy while exploiting the many benefits following from the use of personal data.

The OECD guidelines, Cate, Cullen, and Mayer-Schönberger's report, as well as Cavoukian's analysis offer guidance for policy solutions to address the crisis of the 'notice and consent' model. These guidelines find their limit in the absence of a conceptual framework that can account for the role that both data and data stewards, and OSPs in particular, play in contemporary societies. Developing such a framework along with ethical analyses to define the principles shaping the conduct of data stewards are preliminary and necessary steps towards a fair regulation of data access and management.

2.4 Conclusion

In this article we have discussed the current literature focusing on the moral responsibilities of OSPs. We have highlighted that the role of OSPs as information gatekeepers, the corporate social responsibility that this role entails, and the respect of human rights, are topics that have become increasingly relevant during the past 5 years and across the three research areas of information and computer ethics, business ethics, and law (Fig. 2.1).

The academic interest in these topics stems from the pressing need felt by society to regulate OSPs' conduct in order to ensure the respect of the public good and the fostering of societal welfare. Such a need is often addressed by endorsing an *ad hoc* approach and by delegating to OSPs normative decisions. A good example of the case in point is offered by Google, which is currently both the "judge and the jury" with respect to the application of the right to be forgotten in Europe.

Given the relative novelty and the very significant relevance of the role that OSPs play in contemporary societies, it does not come as a surprise that attempts to regulate OSPs rests on an *ad hoc* strategy to tackle problems as they emerge while, at the same time, the debate on the legal requirements for OSPs' conduct identifies long term solutions. However, the definition of such requirements proves to be difficult when considering OSPs' gatekeeping function, the multicultural, international context in which they operate, as well as the interdependency of the services that they offer in different regions of the world. The latter is a specific feature of OSPs, which requires careful consideration, ethical foresight, and long-term planning. All this makes the attempt to regulate OSPs' conduct by endorsing an *ad hoc* approach unsatisfactory if not unfeasible. These problems can be overcome once legal analyses rest on an ethical framework that can identify fundamental, shareable principles to shape OSPs' conduct.

As OSPs' gatekeeping role impacts both users' access to information and the dynamics of the informational environment, any ethical framework that defines such principles should account for the rights of both users and the environment. Recalling the two LoAs adopted in the previous analysis, this ethical framework should endorse a LoA_{FOR}, that is a LoA that can identify principles and OSPs' responsibilities *for* the informational environment, fostering its flourishing and the wellbeing of the entities inhabiting it.

An ethical framework endorsing such an environmental approach has been proposed in (Floridi 2013). An analysis of OSPs' responsibilities embracing Information Ethics has not yet been provided. However, some key aspects of this ethical framework—especially the concepts of 'care' and 'respect'; the flourishing of the environment as a function of its plurality and diversity; and ultimately the responsibility of human agents to care for the design and management of the informational environment so to ensure its wellbeing (Floridi and Taddeo 2014, Taddeo 2014) fit particularly well with the need to identify fundamental sharable ethical principles that may guide OSPs' conduct. Thus the previous analysis lays the ground for such a framework, the definition of which will be the focus of our future work.

Notes

1. https://www.reformgovernmentsurveillance.com
2. http://ec.europa.eu/archives/commission_2010-2014/president/news/speeches-statements/pdf/deauville-g8-declaration_en.pdf
3. The reader interested in the methodology of the LoA may find useful the following books (Heath et al. 1994; Diller 1994; Jacky 1997; Boca 2014). Philosophers interested in the concept of abstraction as used in this article may wish to see (Hoare 1972).
4. The graphs have been produced using atlas.ti a qualitative analysis software and a database built with the references returned by WEB OF SCIENCE™ when searching for any of the topics indicated in the graphs' legends.
5. Dissemination and access to copyrighted material has also been a topic of great interest in research concerned with OSPs. However, this problem falls outside the scope of this article, for it has more to do with liability and the application of laws protecting copyright online than with the moral duties of OSPs. The interested reader may find useful the analyses of copyright online provided in (Hanel 2006; Edwards 2011; Friedmann 2014).

6. Other relevant contributions on the diversity of the sources and information available on the web have been provided in the literature in information and communication studies, law, and public policy. The interested reader may find useful the following articles (Pandey et al. 2005; Pasquale 2006; Hargittai 2007; Van Couvering 2007; Diaz 2008; Hinman 2008; Lewandowski 2011).
7. The reader interested in the transparency and the copyright of code will find interesting the following articles (Reger 2004; Wolf et al. 2009).
8. Concerns for the implication that filtering of information may have for participative democracy and the nature of the web have also been expressed in (Lessig 1999).
9. The issue arises as to whether OSPs should be ascribed moral responsibilities with respect to societies at large or solely with respect to societies depending on ICTs. The answer depends on the way such responsibilities are defined. For example, if one considers the protection of privacy a duty to respect human rights, then one could argue that OSPs bear this responsibility independently from the level of distribution of their services in a given region. One could also argue that societies where Internet is not pervasive will sooner or later become information societies and hence that, even if OSPs do not massively affect these societies, they will in the foreseeable future. We would like to thank one of the anonymous reviewers for pointing out this aspect.
10. With the exception of countries like China and Thailand, where the strict liability model is endorsed and OSPs are liable for third-party content.
11. For a critical analysis of the 'safe harbour' model see (Pagallo 2011).
12. An interesting analysis of OSPs' legal responsibilities with respect to this has been provided in (Burk 2011).
13. A legal analysis of third-party liability under US tort law has been provided in (Ziniti 2008).
14. http://ec.europa.eu/health/tobacco/law/advertising/index_en.htm
15. http://articles.latimes.com/1999/dec/05/news/mn-40632
16. Internet censorship and freedom of speech have also been at the centre of a debate focusing on the balance between individual rights and state power. This topic does not fall within the scope of this article. The interested reader may find useful (Taddeo 2013, Taddeo 2014).
17. Resolution on "The Promotion, Protection and Enjoyment of Human Rights on the Internet" (Human Rights Council of the United Nations 2012).
18. http://www.hrw.org/reports/1999/enron/
19. https://m.facebook.com/zuck/posts/10101974380267911
20. http://www.un.org/Overview/rights.html
21. http://www.unesco.org/bpi/eng/unescopress/1999/99-92e.shtml
22. The document has been approved on August 13, 2003 by the United Nations Sub-Commission on the Promotion and Protection of Human Rights. http://business-humanrights.org/en/united-nations-sub-commission-norms-on-business-human-rights-explanatory-materials

23. Governmental censorship has spread throughout the globe with the Internet; the literature on OSPs' responsibilities in China casts an interesting light on a problem that concerns several other countries around the world (Aceto et al. 2015).
24. https://m.facebook.com/zuck/posts/10101974380267911
25. For a review of the most relevant contributions of the debate on information privacy the reader may refer to (Tavani and Moor 2001; Solove 2008; Smith et al. 2011).
26. Mark Zuckerberg, Facebook's CEO, declared in 2010 that privacy is not a social norm any more as "people have really gotten comfortable not only sharing more information and different kinds, but more openly and with more people". http://www.theguardian.com/technology/2010/jan/11/facebook-privacy
27. Research on privacy breaches occurring because of third-party access and of users' habits have been provided in (Brandimarte et al. 2010; Lampinen et al. 2011; Wang et al. 2011; Madden 2012; De Wolf et al. 2014).
28. An example of such a friction is discussed in section 3 with respect to the debate on the 'right to be forgotten'.
29. http://ec.europa.eu/justice/data-protection/files/factsheets/factsheet_data_protection_en.pdf
30. Julia Powles maintains an extensive bibliography online at http://www.cambridge-code.org/googlespain.html
31. Disclosure: one of the authors of this paper (L. F.) is a member of the Advisory Board.
32. http://www.telegraph.co.uk/technology/google/10967211/Google-is-the-judge-and-jury-in-the-right-to-be-forgotten.html
33. https://www.reformgovernmentsurveillance.com
34. Net neutrality also refers to responsibilities on the web. However, this problem concerns the backbone infrastructure of the web and hence it involves Internet Service Providers more than Online Service Providers. The interested reader may find useful the following articles: (Blumenthal 2001; Lessig 2007; Schahczenski 2008; Turilli et al. 2012).
35. Article 29 Data Protection Working Party, "Opinion 03/2013 on purpose limitation", p. 4.
36. This is a free choice in which only one option is offered, so it is really equivalent to a 'take it or leave it' choice.
37. A brief description of the history of the definition of international guidelines for the protection of privacy has been provided in (Gerry and Berova 2014; Kiss and Szőke 2015).

References

Aceto, G., Botta, A., Pescapè, A., Feamster, N., Awan, M. F., Ahmad, T., & Qaisar, S. (2015). Monitoring internet censorship with UBICA. In M. Steiner, P. Barlet-Ros, & O. Bonaventure (Ed.), *Traffic monitoring and analysis* (pp. 143–157). Lecture Notes in Computer Science 9053. Springer International Publishing. http://link.springer.com/chapter/10.1007/978-3-319-17172-2_10

Acquisti, A. (2004). Privacy in electronic commerce and the economics of immediate gratification. In *Proceedings of the 5th ACM conference on Electronic Commerce* (pp. 21–29). EC '04. New York: ACM. doi:10.1145/988772.988777
Acquisti, A., & Grossklags, J. (2005). Privacy and rationality in individual decision making. *IEEE Security Privacy, 3*(1), 26–33. doi:10.1109/MSP.2005.22.
Albareda, L., Lozano, J. M., & Ysa, T. (2007). Public policies on corporate social responsibility: The role of governments in Europe. *Journal of Business Ethics, 74*(4), 391–407. doi:10.1007/s10551-007-9514-1.
Anderson, M. (2008). The gaze of the perfect search engine: Google as an institution of Dataveillance. In A. Spink & M. Zimmer (Ed.), *Web search: Multidisciplinary perspectives* (pp. 77–99). Berlin: Springer-Verlag. In *Web search: Multidisciplinary perspectives*, by Amanda Spink and Michael Zimmer. Berlin.
Anderson, G. (2012). *Just business*. London: Headline.
Arnold, D. G. (2010). Transnational corporations and the duty to respect basic human rights. *Business Ethics Quarterly, 20*(3), 371–399.
Bandura, A. (1999). Social cognitive theory: An agentic perspective. *Asian Journal of Social Psychology, 2*(1), 21–41. doi:10.1111/1467-839X.00024.
Barnes, S. B. (2006). A privacy paradox: Social networking in the United States. *First Monday, 11*(9). doi:10.5210/fm.v11i9.1394.
Barzilai-Nahon, K. (2008). Toward a theory of network gatekeeping: A framework for exploring information control. *Journal of the American Society for Information Science and Technology, 59*(9), 1493–1512. doi:10.1002/asi.20857.
Baym, N. K. (2011). Social networks 2.0. In Mia Consalvo & Charles Ess (Ed.), *The handbook of internet studies* (pp. 384–405). Chichester: Wiley-Blackwell. http://onlinelibrary.wiley.com/doi/10.1002/9781444314861.ch18/summary
Besmer, A., Lipford, H. R., Shehab, M., & Cheek, G. (2009). Social applications: Exploring a more secure framework. In *In Proceedings of the Symposium on Usable Privacy and Security (SOUPS)*.
Black, J. (2001). Decentring regulation: Understanding the role of regulation and self regulation in a 'Post-Regulatory' world. *Current Legal Problems, 54*(1), 103–146.
Blowfield, M., & Murray, A. (2008). *Corporate responsibility: A critical introduction*. Oxford; New York: OUP Oxford.
Blumenthal, M. S., & Clark, D. D. (2001). Rethinking the design of the Internet: the end to end arguments vs. the brave new world. *ACM Transactions on Internet Technology, 1*(1), 70–109.
Boca, P. (2014). *Formal methods: State of the art and new directions*. [Place of publication not identified]: Springer.
Brandimarte, L., Acquisti, A., & Loewenstein, G. (2010). *Misplaced confidences: Privacy and the control paradox*.
Brenkert, G. G. (2009). Google, human rights, and moral compromise. *Journal of Business Ethics, 85*(4), 453–478. doi:10.1007/s10551-008-9783-3.
Burk, D. L. (2011). Toward an epistemology of ISP secondary liability. SSRN Scholarly Paper ID 1920050. Rochester, NY: Social Science Research Network. http://papers.ssrn.com/abstract=1920050
Calhoun, C. J. (Ed.). (2002). *Dictionary of the social sciences*. New York: Oxford University Press.
Cate, F., Cullen, P., & Mayer-Schönberger, V. (2014). Data protection principles for the 21st Century, revising the 1980 OECD Guidelines. http://www.microsoft.com/en-us/download/details.aspx?id=41191
Caudill, E., & Murphy, P. (2000). Consumer Online Privacy: Legal and Ethical Issues. *Journal of Public Policy Marketing, 19*(1), 7–19. doi:10.1509/jppm.19.1.7.16951.
Cavoukian, A. (2014). *The unintended consequences of privacy paternalism /*. Canadian Electronic Library. Canadian Public Policy Collection.
Cerf, V. G. (2011). First, do no harm. *Philosophy & Technology, 24*(4), 463–465. doi:10.1007/s13347-011-0056-1.

Chen, S. (2009). Corporate responsibilities in internet-enabled social networks. *Journal of Business Ethics, 90*(4), 523–536. doi:10.1007/s10551-010-0604-0.
Cragg, W. (2010). Business and human rights: A principle and value-based analysis. In G. G. Brenkert & T. L. Beauchamp (Eds.), *The Oxford handbook of business ethics*. Oxford: Oxford University Press.
Crane, A., Matten, D., McWilliams, A., Moon, J., & Siegel, D. S. (Eds.). (2008). *The Oxford handbook of corporate social responsibility* (1st ed.). Oxford University Press .http://www.oxfordhandbooks.com/view/10.1093/oxfordhb/9780199211593.001.0001/oxfordhb-9780199211593
Crawford, S. P. (2005). *Shortness of vision: Regulatory ambition in the digital age*. http://dash.harvard.edu/handle/1/12933354
Dann, G. E., & Haddow, N. (2007). Just doing business or doing just business: Google, Microsoft, Yahoo! And the business of censoring China's internet. *Journal of Business Ethics, 79*(3), 219–234. doi:10.1007/s10551-007-9373-9.
De Wolf, R., Willaert, K., & Pierson, J. (2014, June). Managing privacy boundaries together: Exploring individual and group privacy management strategies in Facebook. *Computers in Human Behavior, 35*, 444–454. doi:10.1016/j.chb.2014.03.010.
Diamond, L. (2010). Liberation technology. *Journal of Democracy, 21*(3), 69–83. doi:10.1353/jod.0.0190.
Diaz, A. (2008). Through the Google Goggles: Sociopolitical bias in search engine design. In Amanda Spink & Michael Zimmer (Ed.), *Web search*. Information Science and Knowledge Management 14 (pp. 11–34). Berlin/Heidelberg: Springer. http://link.springer.com/chapter/10.1007/978-3-540-75829-7_2
Diller, A. (1994). *Z: An introduction to formal methods* (2nd ed.). Chichester; New York: Wiley.
Donaldson, T. (1992). *The ethics of international business, The Ruffin Series in Business Ethnics*. New York: Oxford Univ. Press.
Donaldson, T., & Dunfee, T. W. (1999). *Ties that bind: A social contracts approach to business ethics*. Boston, Mass: Harvard Business School Press.
Edwards, L. (2011). Role and responsibility of the internet intermediaries in the field of copyright and related rights. Report. Geneva: WIPO. http://strathprints.strath.ac.uk/35492/
Elkin-Koren, N. (2001). Let the crawlers crawl: On virtual gatekeepers and the right to exclude indexing. *University of Dayton Law Review., 26*, 179–188.
Eriksson, J., & Giacomello, G. (2009). Who controls the internet? Beyond the obstinacy or obsolescence of the state. *International Studies Review, 11*(1), 205–230. doi:10.1111/j.1468-2486.2008.01841.x.
Etzioni, A. (1999). *The limits of privacy*. New York: Basic Books.
Floridi, L. (2008). The method of levels of abstraction. *Minds and Machines, 18*(3), 303–329. doi:10.1007/s11023-008-9113-7.
Floridi, L. (2013). *The ethics of information*. Oxford: Oxford University Press.
Floridi, L. (2015a). Free online services: Enabling, disenfranchising, disempowering. *Philosophy & Technology, 28*(2), 163–166. doi:10.1007/s13347-015-0200-4.
Floridi, L. (2015b). Should you have the right to be forgotten on Google? nationally, yes. Globally, no. *New Perspectives Quarterly, 32*(2), 24–29. doi:10.1111/npqu.11510.
Floridi, L., & Taddeo, M. (Eds.). (2014). *The ethics of information warfare*. New York: Springer.
Freeman, J. (1999). Private parties, public functions and the new administrative law. SSRN Scholarly Paper ID 165988. Rochester, NY: Social Science Research Network. http://papers.ssrn.com/abstract=165988
Friedmann, D. (2014). Sinking the safe harbour with the legal certainty of strict liability in sight. *Journal of Intellectual Property Law & Practice, 9*(2), 148–155. doi:10.1093/jiplp/jpt227.
Gerry, F., & Berova, N. (2014). The rule of law online: Treating data like the sale of goods: Lessons for the internet from OECD and CISG and sacking Google as the regulator. *Computer Law Security Review, 30*(5), 465–481.
Goldman, E. (2006). Search engine bias and the demise of search engine Utopianism. SSRN Scholarly Paper ID 893892. Rochester: Social Science Research Network. http://papers.ssrn.com/abstract=893892

Granka, L. A. (2010). The politics of search: A decade retrospective. *The Information Society, 26*(5), 364–374. doi:10.1080/01972243.2010.511560.

Hanel, P. (2006). Intellectual property rights business management practices: A survey of the literature. *Technovation, 26*(8), 895–931. doi:10.1016/j.technovation.2005.12.001.

Hargittai, E. (2007). The social, political, economic, and cultural dimensions of search engines: An introduction. *Journal of Computer-Mediated Communication, 12*(3), 769–777. doi:10.1111/j.1083-6101.2007.00349.x.

Heath, D., Allum, D., & Dunckley, L. (1994). *Introductory logic and formal methods*. Henley-on-Thames: Alfred Waller.

Helgesson, K. S., & Mörth, U. (Eds.). (2013). *The political role of corporate citizens: An interdisciplinary approach, Palgrave studies in citizenship transitions series*. Basingstoke; New York: Palgrave Macmillan.

Hildebrandt, M. (2013). Balance or trade-off? Online security technologies and fundamental rights. *Philosophy & Technology, 26*(4), 357–379. doi:10.1007/s13347-013-0104-0.

Hinman, L. (2005). Esse Est Indicato in Google: Ethical and political issues in search engines. *International Review of Information Ethics, 3*(6), 19–25.

Hinman, L. (2008). Searching ethics: The role of search engines in the construction and distribution of knowledge. In Prof Dr Amanda Spink and Michael Zimmer (Ed.), *Web search*. Information Science and Knowledge Management 14 (pp. 67–76). Berlin/Heidelberg: Springer. http://link.springer.com/chapter/10.1007/978-3-540-75829-7_5

Hoare, C. A. R. (1972). Structured programming. In O. J. Dahl, E. W. Dijkstra, & C. A. R. Hoare (pp. 83–174). London: Academic Press Ltd. http://dl.acm.org/citation.cfm?id=1243380.1243382

Huberman, B. A. (2003). *The laws of the web: Patterns in the ecology of information*. New Ed edition. Cambridge, Mass; London: MIT Press.

Hull, G., Lipford, H. R., & Latulipe, C. (2011). Contextual gaps: Privacy issues on Facebook. *Ethics and Information Technology, 13*(4), 289–302. doi:10.1007/s10676-010-9224-8.

Human Rights Council of the United Nations. (2012). *U.N. Human Rights Council: First Resolution on Internet Free Speech*. http://www.loc.gov/lawweb/servlet/lloc_news?disp3_l205403231_text

Introna, L. D., & Nissenbaum, H. (2006). Shaping the web: Why the politics of search engines matters. SSRN Scholarly Paper ID 222009. Rochester: Social Science Research Network. http://papers.ssrn.com/abstract=222009

Jacky, J. (1997). *The way of Z: Practical programming with formal methods*. Cambridge; New York: Cambridge University Press.

Johnson, D. G. (2009). *Computer ethics* (4 ed.). Upper Saddle River: Pearson.

Jolls, C., & Sunstein, C. R.. (2005). Debiasing through Law. Working Paper 11738. National Bureau of Economic Research. http://www.nber.org/papers/w11738

Karp, D. J. (2009). Transnational corporations in 'bad States': Human rights duties, legitimate authority and the rule of law in international political theory. *International Theory, 1*(01), 87. doi:10.1017/S1752971909000074.

Kerr, I. 2002. Internet users dependence and the duty of loyalty. In T. Mendina and Barbara Rockenbach (Ed.), *Ethics and electronic information* (166–176). North Carolina: McFarland Press.

Kiss, A., & Szőke, G. L. (2015). Evolution or revolution? Steps forward to a new generation of data protection regulation. In S. Gutwirth, R. Leenes, & P. de Hert (Eds.), *Reforming European data protection law, Law, Governance and Technology Series 20* (pp. 311–331). Dordrecht: Springer http://link.springer.com/chapter/10.1007/978-94-017-9385-8_13.

Laidlaw, E. (2008). Private power, public interest: An examination of search engine accountability. *International Journal of Law and Information Technology, 17*(1), 113–145. doi:10.1093/ijlit/ean018.

Laidlaw, E. (2010). A framework for identifying internet information gatekeepers. *International Review of Law, Computers & Technology, 24*(3), 263–276. doi:10.1080/13600869.2010.522334.

Lampinen, A., Lehtinen, V., Lehmuskallio, A., & Tamminen, S. (2011). We'Re in it together: Interpersonal management of disclosure in social network services. In*Proceedings of the SIGCHI Conference on Human Factors in Computing Systems, CHI '11* (pp. 3217–3226). New York: ACM. doi:10.1145/1978942.1979420.

Lanier, J. (2011). *You are not a gadget: A manifesto*. Reprint edition. New York: Vintage.

Lessig, L. (1999). *Code: And other laws of cyberspace*. New York: Basic Books.

Lessig, L. (2007). In support of network neutrality. *I/S: A Journal of Law and Policy for InformationSociety, 3*(1), 185–196.

Lev-On, A. (2009). The democratizing effects of search engine use: On chance exposures and organizational hubs. SSRN Scholarly Paper ID 1481901. Rochester: Social Science Research Network. http://papers.ssrn.com/abstract=1481901

Lev-On, A., & Manin, B. (2007). Happy accidents: Deliberation and online exposure to opposing views. In *Forthcoming* In: T Davies (Ed.), *Online Deliberation: Design Research and Practice*.

Lewandowski, D. (2011). The influence of commercial intent of search results on their perceived relevance. Preprint. February 8. http://eprints.rclis.org/17232/

Lewin, K. (1947). Frontiers in group dynamics. *Human Relations, 1*(2), 143–153.

Madden, M. (2012). Privacy management on social media sites. Pew Internet Report.

Madelin, R. (2011). The evolving social responsibilities of internet corporate actors: Pointers past and present. *Philosophy & Technology, 24*(4), 455–461. doi:10.1007/s13347-011-0049-0.

Matten, D., & Crane, A. (2005). Corporate citizenship: Toward an extended theoretical conceptualization. *Academy of Management Review, 30*(1), 166–179.

Mayer-Schönberger, V. (2011). *Delete: The virtue of forgetting in the digital age*. Princeton: Princeton University Press.

McQuail, D. (1992). *Media performance: Mass communication and the public interest*. London; Newbury Park: Sage Publications.

Metoyer-Duran, C. (1993, January). Information gatekeepers. *Annual Review of Information Science and Technology (ARIST), 28*, 111–150.

Nagenborg, M. (2005). The ethics of search engines (Special Issue). *International Review of Information Ethics* 3.

Negroponte, N. (1996). *Being digital*. New edition. Rydalmere: Coronet Books.

Nissenbaum, H. F. (2010). *Privacy in context: Technology, policy, and the integrity of social life*. Stanford: Stanford Law Books.

Norberg, P. A., Horne, D. R., & Horne, D. A. (2007). The privacy paradox: Personal information disclosure intentions versus behaviors. *Journal of Consumer Affairs, 41*(1), 100–126. doi:10.1111/j.1745-6606.2006.00070.x.

O'Hara, K. (2010). Intimacy 2.0: Privacy rights and privacy responsibilities on the World Wide Web. In Jonathan Zittrain, John Domingue, & Neil Benn. http://eprints.soton.ac.uk/268760/

OECD. (2013). *OECD guidelines on the protection of privacy and transborder flows of personal data* .http://www.oecd.org/sti/ieconomy/oecdguidelinesontheprotectionofprivacyandtransborderflowsofpersonaldata.htm

Okoye, A. (2009). Theorising corporate social responsibility as an essentially contested concept: Is a definition necessary? *Journal of Business Ethics, 89*(4), 613–627. doi:10.1007/s10551-008-0021-9.

Pagallo, U. (2011). ISPs & Rowdy web sites before the law: Should we change today's safe harbour clauses? *Philosophy & Technology, 24*(4), 419–436. doi:10.1007/s13347-011-0031-x.

Palazzo, G., & Scherer, A. G. (2006). Corporate legitimacy as deliberation: A communicative framework. *Journal of Business Ethics, 66*(1), 71–88. doi:10.1007/s10551-006-9044-2.

Pandey, S., Roy, S., Olston, C., Cho, J., & Chakrabarti, S. (2005). Shuffling a stacked deck: The case for partially randomized ranking of search engine results. In *In Proceedings of the 31st international conference on Very Large Databases (VLDB*, 781–792.

Pariser, E. (2012). *The filter bubble: What the internet is hiding from you*. London: Penguin.

Pasquale, F. A. (2006). Rankings, reductionism, and responsibility. SSRN Scholarly Paper ID 888327. Rochester: Social Science Research Network. http://papers.ssrn.com/abstract=888327

Qi, M., & Edgar-Nevill, D.. (2011). Social networking searching and privacy issues. *Information Security Technical Report*, October. doi:10.1016/j.istr.2011.09.005.
Reding, V. (2012). The EU data protection reform 2012: Making Europe the standard setter for modern data protection rules in the digital age. European Commission. http://europa.eu/rapid/pressReleasesAction.do?reference=SPEECH/12/26&format=PDF
Reger, C. M. (2004). Let's swap copyright for code: The computer software disclosure dichotomy. *Loyola of Los Angeles Entertainment Law Review, 24*, 215.
Rosen, J. (2012). The right to be forgotten. *Stanford Law Review Online, 64*, 88.
Rosen, J. (2015). Protecting privacy on the internet is the user's responsibility. *Philly-Archives*. http://articles.philly.com/2012-03-05/news/31124410_1_new-privacy-policy-facebook-search-terms
Santoro, M. A. (1998). Engagement with integrity: What we should expect multinational firms to do about human rights in China.
Schahczenski, C. (2008). Net neutrality, computing and social change. *Acm Sigcas Computers and Society, 38*(2), 27–27.
Scherer, A. G., & Palazzo, G.. 2006. Toward a political conception of corporate responsibility-business and society seen from a Habermasian perspective. SSRN Scholarly Paper ID 952013. Rochester: Social Science Research Network. http://papers.ssrn.com/abstract=952013
Schermer, B. W., Custers, B., & van der Hof, S. (2014). The crisis of consent: How stronger legal protection may lead to weaker consent in data protection. *Ethics and Information Technology, 16*(2), 171–182. doi:10.1007/s10676-014-9343-8.
Schwartz, P. (1999). Privacy and democracy in cyberspace. *Vanderbilt Law Review., 52*(1999), 1607–1071.
Semitsu, J. P. (2011). From Facebook to mug shot: How the dearth of social networking privacy rights revolutionized online government surveillance. SSRN Scholarly Paper ID 1782267. Rochester: Social Science Research Network. http://papers.ssrn.com/abstract=1782267
Shapiro, A. L. (2000). *The control revolution: How the internet is putting individuals in charge and changing the world we know*. 2nd.Printing edition. New York: PublicAffairs.
Smith, H. J., Dinev, T., & Heng, X. (2011). Information privacy research: An interdisciplinary review. *MIS Quarterly., 35*(4), 989–1016.
Solove, D. J. (2008). Understanding privacy. SSRN Scholarly Paper ID 1127888. Rochester: Social Science Research Network. http://papers.ssrn.com/abstract=1127888
Spinello, R. A. (2011). Privacy and social networking technology. *International Review of Information Ethics, 16*, 12.
Spink, A., & Zimmer, M.. (2008). *Web search*. Vol. 14. Information Science and Knowledge Management. Berlin, Heidelberg: Springer Berlin/Heidelberg.
Sunstein, C. R. (2001). *Republic.com*. With a new afterword by the author edition. Princeton: Princeton University Press.
Taddeo, M. (2010). Modelling trust in artificial agents, a first step toward the analysis of E-trust. *Minds and Machines, 20*(2), 243–257. doi:10.1007/s11023-010-9201-3.
Taddeo, M. (2013). Cyber security and individual rights, striking the right balance. *Philosophy & Technology, 26*(4), 353–356. doi:10.1007/s13347-013-0140-9.
Taddeo, M. (2014, September). The struggle between liberties and authorities in the information age. *Science and Engineering Ethics, 21*, 1–14. doi:10.1007/s11948-014-9586-0.
Tavani, H.. (2014). Search engines and ethics. Edited by Edward N. Zalta. *The Stanford encyclopedia of philosophy*. http://plato.stanford.edu/archives/spr2014/entries/ethics-search/
Tavani, H., & Grodzinsky, F. S. (2002). Cyberstalking, personal privacy, and moralresponsibility. *Ethics and Information Technology., 4*(2), 123–132. doi:10.1023/A:1019927824326.
Tavani, H., & Moor, J. H. (2001). Privacy protection, control of information, and privacy-enhancing technologies. *SIGCAS Computer Society., 31*(1), 6–11. doi:10.1145/572277.572278.
Toffler, A., Toffler, H., & Gingrich, N. (1995). *Creating a new civilization: The politics of the third wave*. First Edition edition. Atlanta: Kansas City, Mo: Turner Publishing.

Turilli, M., Vaccaro, A., & Taddeo, M. (2010). The case of online trust. *Knowledge, Technology & Policy, 23*(3-4), 333–345. doi:10.1007/s12130-010-9117-5.
Turilli, M., Vaccaro, A., & Taddeo, M. (2012). Internet neutrality: Ethical issues in the internet environment. *Philosophy & Technology, 25*, 133–151.
Van Couvering, E. (2007). Is relevance relevant? Market, science, and war: Discourses of search engine quality. *Journal of Computer-Mediated Communication, 12*(3), 866–887. doi:10.1111/j.1083-6101.2007.00354.x.
Vedder, A. (2001). Accountability of internet access and service providers – strict liability entering ethics? *Ethics and Information Technology, 3*(1), 67–74. doi:10.1023/A:1011492109277.
Wang, N., Xu H., & Grossklags J. (2011). Third-party apps on Facebook: Privacy and the illusion of control. In *Proceedings of the 5th ACM Symposium on Computer Human Interaction for Management of Information Technology*, 4:1–4:10. CHIMIT '11. New York: ACM. doi:10.1145/2076444.2076448.
Weckert, J. (2005). Trust in cyberspace. In R. J. Cavalier (Ed.), *The impact of the internet on our moral lives* (pp. 95–120). Albany: University of New York Press.
Wettstein, F. (2012). Silence as complicity: Elements of a corporate duty to speak out against the violation of human rights. *Business Ethics Quarterly, 22*(01), 37–61. doi:10.1017/S1052150X00000063.
Wolf, M. J., Miller, K. W., & Grodzinsky, F. S. (2009). On the meaning of free software. *Ethics and Information Technology, 11*(4), 279–286. doi:10.1007/s10676-009-9207-9.
World Economic Forum. (2012). Unlocking the economic value of personal data balancing growth and protection. http://www3.weforum.org/docs/WEF_IT_UnlockingValueData_BalancingGrowthProtection_SessionSummary.pdf
Xu, H. (2012). Reframing privacy 2.0 in online social networks. *University of Pennsylvania Journal of Constitutional Law, 14*(4), 1077.
Ziniti, C. (2008). Optimal liability system for online service providers: How Zeran v. America online got it right and Web 2.0 proves it. *Berkeley Technology Law Journal, 23*, 583.

Chapter 3
The Immunity of Internet Intermediaries Reconsidered?

Georgios N. Yannopoulos

Abstract In Web 2.0 anyone can become a producer of information. As a result, countless 'intermediaries' operate as go-betweens and facilitators to transmit such information. In view of the above, we seek to examine whether they are liable for unlawful or harmful content contained in these transmissions. In 2000 the EU Electronic Commerce Directive (ECD) established a status of immunity for access, cache and host providers. Subsequent legislation regulating electronic communications, data protection, secrecy of communication and enforcement of intellectual property has dramatically increased the obligations of intermediaries and, hence, the degree of their liability, albeit on a cases by case basis and without necessarily challenging the provisions of ECD.

This chapter seeks to identify the exact role of the new power players in the digital world and to discover the boundaries between immunity and liability of Internet intermediaries. Especially their behaviour as *quasi* judges for defending legitimacy in cyberspace is examined in relation to the need for intervention by the judicial authorities. Methods and policies in order to foster an ethos of compliance and social responsibility are proposed for those new gatekeepers in comparison to positive laws and regulations.

3.1 Immunity and Liability in the New Digital Era: A Fine Balancing Act

Following the expansion of Web 2.0, in which anyone can produce and post information on the Internet, a vast amount of unlawful and harmful content has been disseminated: blogs with defamation and insults, fake profiles and blackmails in social networks, downloading of illegal copies of music and movies, 'phishing' frauds and 'denial of service' attacks, cyber bullying, child pornography and offenses committed in cyber games. Countless 'intermediaries' function as

G.N. Yannopoulos (✉)
Law School, National and Kapodistrian University of Athens, Athens, Greece
e-mail: gyannop@law.uoa.gr

M. Taddeo, L. Floridi (eds.), *The Responsibilities of Online Service Providers*,
Law, Governance and Technology Series 31, DOI 10.1007/978-3-319-47852-4_3

go-betweens for the transmission of that information and they include such entities as typical access and host providers (ISPs), search engines, social networks, electronic encyclopaedias, websites for video uploading, blogs, internet games platforms, webpages for short messages etc. Since the early age of the Internet, it has been considered important[1] to review the liability that intermediaries carry for two main reasons: firstly, because the technical contribution of these players is indisputable, as without access and host providers there would be no Internet. Intermediaries are the entities that can be easily traced in order to seek compensation and sometimes play the role of 'scapegoats' in case of extreme hardship endured by subscribers of services, although they might be only the messengers.[2] Secondly, as things stand, it seems that Internet intermediaries are the only entities able to apply an approach to exert *prima faciae* control of Internet content either by (a) intervening during an allegedly illegal or harmful action or (b) by taking preventive or dissuasive measures. Filters, software or similar automated methods of control may be extremely effective in curtailing certain types of illegal behaviour. Therefore, any investigation of the liability of Internet intermediaries is linked with the *stricto sensu* responsibility. Such search would extend to their ability to reasonably control information flows over the Internet, to limit access, to define unilaterally the terms of such access, to block or facilitate users, to impose commercial, political or cultural terms and conditions of managing data and to warrant the security and the integrity of data.

3.1.1 Establishing Links on the Liability of Intermediaries

Under a civil law approach and with a view to as certain indirect liability for intermediaries, the existence of a 'causal link', between services offered to third parties by the Internet intermediary and the carrying out of infringements or unlawful acts must be established. Such legal construction has mainly been influenced by the common law theory underlying secondary liability for infringements in the area of intellectual property and defamation. However, under civil law doctrine, the messenger may not be held liable for the message carried. In theory, it seems like an easy assumption to target internet intermediaries in order to implement regulatory actions, but practice shows that intermediaries, in principle, do not commit unlawful acts related to content themselves, so it would be necessary to prove that they generate and possibly make available such a 'causal link' (i.e. the means) for such acts to the third parties that are the end-users of services offered.

---

[1] See Yannopoulos (2013b), p. 9 and Yannopoulos (2014), p. 787.

[2] Since ancient times the messenger was considered liable cf. verse 277 of Sophocles' Antigone: "*...That no man loves the messenger of ill...*" (transl. R. Jebb).

One possible basis to establish such secondary liability would be to follow the theory of 'adequate cause' and try to prove that the intervention of the intermediary constitutes the indispensable link in the chain of events that has led to the unlawful result. It is obvious that such argumentation could provide adequate grounds to establish liability in cases of copyright infringement, but it seems rather weak in cases of defamation or similar harmful content. It is precisely that particular difficulty[3] to connect the behaviour of intermediaries to the damage caused that has led to the adoption of an immunity approach (e.g. 'safe harbours' according to terminology in US law) for intermediaries. In EU, for the past 16 years a similar system has been set out in the e-Commerce Directive[4] (hereinafter ECD), which established a regime of immunity for Internet intermediaries regarding unlawful and harmful content.

3.1.2 The 'Safe Harbour' of ECD

ECD introduced a horizontal approach governing all types of liability. Analysts agree[5] that Internet liability concerns all types of responsibility both under civil, administrative or criminal law, which leads into a general legal liability. According to certain scholars[6], it has been proposed to introduce a 'special type of liability' reflecting the liability defined in the 4th Section of ECD referred to as: 'liability' (*responsabilité, Verantwortlichkeit*). In view of the consolidation of the internal market such horizontal regulation, no matter whether sanctions are treated as civil, administrative or criminal, it simply means that the safe harbour approach is offered to internet intermediaries without discrimination and without examining the cause of liability.[7] The European system is different in comparison to the system of US, where vertical regulation addresses separately each legislation e.g. for copyright, defamation, trademarks etc. The ECD system is completed with the known prohibition of a general obligation of the intermediaries to permanently monitor content (art. 15 ECD).

[3] G. Yannopoulos (2013b), op. cit. pp. 55–56.

[4] Directive 2000/31/EC of the European Parliament and of the Council of 8.6.2000 on certain legal aspects of information society services, in particular electronic commerce, in the Internal Market (Directive on electronic commerce), OJ L178, 17.7.2000, p. 0001–0016.

[5] See the Report from the Commission to the European Parliament, the Council and the European Economic and Social Committee – First Report on the application of Directive 2000/31/EC of the European Parliament and of the Council of 8 June 2000 on certain legal aspects of information society services, in particular electronic commerce, in the Internal Market (Directive on electronic commerce), COM/2003/0702 final, section 4.6, par. 64.

[6] See for example Ufer (2007), p. 38 and Schmoll (2001), p. 38.

[7] See in that vein the Introductory Report on the Proposal for the German Law for Electronic Commerce (BT-Drs 14/6098 of 17-5-2001, Elektronischer Geschäftsverkehrgesetz – EGG harmonisation of the ECD) stating that immunity concerns also criminal law ("...*die Beschränkungen der Vernatwortlichkeit gelten auch für den Bereich des Strafrechts...*").

Subsequent legislation for electronic communications, data protection, secrecy of communication and intellectual property has dramatically increased the obligations of intermediaries and, hence, the degree of their liability. Such pieces of legislation include Directive 2009/140[8] on better regulation for electronic communications, the e-Privacy Directive 2002/58/EC,[9] the Copyright Directive 2001/29/EC[10] and the Enforcement Directive 2004/48.[11] Each Member State has also imposed very strict rules in order to harmonise with the child pornography Directive,[12] while further restrictions have been imposed by sectoral legislation especially in areas such as online gambling and financial transactions. In view of these developments it seems cumbersome to draw the boundaries between the systems of immunity and newly imposed liabilities.

It can be argued that intensifying liability in the case of intellectual property could be justified by the fact that intermediaries generally derive profits from communicating such data. This approach, fails to fully address other aspects of unlawful and harmful content, and it does not respond to the main dilemma on whether we want intermediaries to act as new 'Gatekeepers'[13] in the digital world. Before addressing that question, it is necessary to underscore that in the era of participative Internet ('Web 2.0') anyone can become a content provider and that the so called 'Internet generation' was born and grown up within that digital environment. In that sense, for that particular e-generation neither intellectual property means something as a right, nor data protection is touching any strings of sensitivity; to date at least.

---

[8] Directive 2009/140/EC, of the European Parliament and of the Council of 25.11.2009 amending Directives 2002/21/EC on a common regulatory framework for electronic communications networks and services, 2002/19/EC on access to, and interconnection of, electronic communications networks and associated facilities, and 2002/20/EC on the authorisation of electronic communications networks and services, OJ L337, 18.12.2009, p. 37–69 (Better regulation directive).

[9] Directive 2002/58/EC of the European Parliament and of the Council of 12.7.2002 concerning the processing of personal data and the protection of privacy in the electronic communications sector (Directive on privacy and electronic communications), OJ L 105, 31.7.2002, p. 37–47.

[10] Directive 2001/29/EC, of the European Parliament and of the Council of 22.5.2001 on the harmonisation of certain aspects of copyright and related rights in the information society, OJ L 167, 22.6.2001, p. 10–19. For the most recent ECJ decision regarding hyperlinks and communication to the public, see Case C-160/15 *GS Media BV v Sanoma Media Netherlands BV, Playboy Enterprises International Inc., Britt Geertruida Dekker.*

[11] Directive 2004/48/EC, of the European Parliament and of the Council of 29.4.2004 on the enforcement of intellectual property rights, OJ L 157, 30.4.2004, p. 45–86. On the enforcement of copyright and the possibility of an injunction against a wi-fi network see the very recent ECJ Case C-484/14 *Tobias Mc Fadden v Sony Music Entertainment Germany GmbH.*

[12] Directive 2011/92/EU of the European Parliament and of the Council of 13.12.2011 on combating the sexual abuse and sexual exploitation of children and child pornography, and replacing Council Framework Decision 2004/68/JHA, OJ L 335, 17.12.2011, p. 1–14.

[13] See, on the contrary, the argument that "...*in a world without fences we do not need gates...*", supported by Bowrey K., *Law and Internet Cultures*, Cambridge University Press, 2005, p. 101.

3.1.3 Liability and the Protection of Other Fundamental Rights

The ECD, reflecting out-dated views, in terms both of the physical and the digital timespan, has established a regime of immunity, while more recent legislation has imposed contradicting obligations to the providers, who are called to observe the rules, while bearing the often demanding compliance costs. Therefore, instead of undertaking an active[14] role for combating the problems, providers are struggling to comply with impracticable rules and regulations. The idea of an International Convention or international custom[15] cannot easily come about either.

Had legislators only focused on prevention, a system of punitive damages, stricter than existing compensation could have been introduced. This could be supported under the theory that the cost of undertaking dangerous activities should be borne by those generating such danger, as for example in environmental disasters, and not by the circumstantial 'victims'. Such civil law oriented analysis argues that certain persons, possessing a source of danger, are better qualified to bear the consequences. However, notwithstanding that such system may not easily be endorsed by certain jurisdictions,[16] this theory does not answer whether Internet intermediaries possess such 'danger generating' sources. While, in the case of hosting providers it could be postulated that they facilitate the posting of harmful or unlawful information, in the case of access providers such capacity may not be attributed, since within their normal everyday business i.e. the provision of access to the Internet, it cannot be argued that they conduct a 'dangerous activity'.

Following that analysis, it seems cumbersome to assign a general degree of liability to Internet intermediaries for unlawful or harmful content; of course, this attitude changes when the intermediary becomes an accomplice to such illegal behaviour. A socially justifiable solution for enterprises acting as sources of 'danger' would be to undertake a portion of that liability. Assuming this 'portion of liability' by the providers may look feasible in theory, but will lead to an increase of the cost of the services provided and it does not take into account the compliance costs already born by the intermediaries. Case law of the ECJ, regarding the imposition of filtering methods[17] shows that the cost factor should not be neglected.

---

[14] See a proposal for active participation and prevention on behalf of intermediaries in de Beer & Clemmer (2009), 375.

[15] For the efforts of international organizations to create and international custom in relation to the Internet see Polanski P., *Customary Law of the Internet, in the search for Supranational Cyberspace Law*, 2007. See also COM (2007) 836, Communication from the Commission to the European Parliament, the Council, the European Economic and Social Committee and the Committee of the Regions on creative content online in the Single Market.

[16] See the attitude of different jurisdictions regarding human rights protection in Stalla-Bourdillon (2010), 492.

[17] See indicatively ECJ cases C-236/08 *Google France v Louis Vuitton Malletier*, C-324/09 *L'Oréal v eBay*, C-70/10 *Scarlet v SABAM*, C-360/10 *SABAM v Netlog*.

Therefore, spilling the 'portion of liability' to intermediaries, need to be implemented keeping in mind all the factors affecting it, including cost.

It should be emphasised that each Member State may attribute a different degree of protection to the fundamental rights of internet users. Furthermore, intermediaries themselves demonstrate different attitudes in balancing the financial interests of the industries against the civil rights of the citizens. For example, countries with a flourishing media industry (such as the US, UK etc.) have been seeking to protect intellectual property. On the contrary, other countries pay more attention to the protection of privacy, freedom of expression, and protection of personality. Therefore, neither a common denominator of 'illegality' nor a common 'culture' for combating infringements is visible.

3.2 Possible Solutions

3.2.1 Signs from the ECJ Case Law

The European Court of Justice in the famous *Promusicae*[18] case has hesitated to reconcile the on-going clash between data protection and copyright and has left to the Member States the initiative to take the necessary measures to strike a balance in relation to the obligations of Internet intermediaries. ECJ, following the principle of proportionality, has tried to discover that point of balance amongst copyright, effective judicial protection and data protection as an element of the protection of personality, without, however, leaning towards any of these aspects. In essence, the ECJ has tried to establish a golden rule between the right to property and the right to privacy. The ambiguity of *Promusicae* does not necessarily indicate bad judgment, but rather shows an internal desire to avoid characterising one of the two fundamental rights as hierarchically superior, leaving, thus the burden of interpretation to judges in the Member State Courts.

In a similar vein, the Court has decided in the *LSG/Tele2*[19] case, that Member States may impose an obligation to disclose 'traffic data' so that those having a legal interest may instigate civil actions for infringements of copyright. Furthermore, the Court has underlined the necessity to interpret national law not just according to EU Directives, but also according to the fundamental rights and principles such as the principle of proportionality. Again, the Court, following *Promusiace* has not favoured any of these rights, emphasising the need to observe the principle of proportionality.

[18] *Productores de Música de España (Promusicae) v Telefónica de España SAU* (Case C-275/06), for comments see Brimsted & Chesney (2008), 275 and Coudert & Werkers (2008), 57.

[19] C-557/07 *LSG – Gesellschaft zur Wahrnehmung von Leistungsschutzrechten GmbH v Te2 Telecommunication GmbH*. Finally, the Supreme Austrian Court has dismissed the claim to disclose 'traffic data' see Decision OGH of 14-7-2009, 4 Ob 41/09x, LSG v. Tele2.

Finally, in the *Scarlet/Sabam*[20] case the Court has moved one step ahead, in order to secure an equilibrium, by stating that intellectual property rights are protected under art 17. par. 2 of the EU Charter of Fundamental Rights *"…[t]here is, however, nothing whatsoever in the wording of that provision or in the Court's case-law to suggest that that right is inviolable and must for that reason be absolutely protected…"*. Under that argument, the ECJ patronises the idea of an *ad hoc* weighting of conflicting fundamental rights, promoting again the principle of proportionality as a guideline.

3.2.2 Notice and Take-Down Procedures

The Notice and Take-Down (N&TD) procedure of the US Digital Millennium Copyright Act (DCMA)[21] has secured acceptance in the US and should be considered by legislators also in other jurisdictions, for other types of offenses beyond copyright.[22]

It is obvious that such methods should draw attention on how to remove content, rather than trying to investigate who is culpable for wrong-doing. The content of the formal notice must be described, conspicuously stating who is asking for what action and for which 'unlawful' content. Details such as the method of authenticating the person requesting the removal, the deadlines for response as well as the ability for defence by the provider (put back content) may also be prescribed by law, while intermediaries should not be held liable in case they remove content, which later has been found not to be 'unlawful'. The decision as to what is harmful or unlawful may be assigned to third parties.[23] In the UK, for example, the Internet Watch Foundation (IWF)[24] undertakes to decide whether certain content falls under child pornography and then informs the authorities, also providing guidelines to the intermediaries for the removal of such content. The IWF handles a large number of complaints which could not have been dealt with by the common police / prosecution authorities.[25] It is a self-regulatory body having as a priority, as per their motto, the protection of minors rather than the freedom of expression.

---

[20] C-70/2010 *Scarelt Extended SA v Société belge des auteurs, compositeurs et éditeurs SCRL (SABAM)*, of 24-11-2011.

[21] Section 512, Title 17, US Code. For a comparative study between EU and US see Ahlert, Marsden & Yung (2004), available at pcmlp.socleg.oxx.ac.uk

[22] See for example the recent Harmful Digital Communications Act (2013) in New Zealand, which has introduced a formal N&TD procedure for defamatory content. For proposals to introduce a similar system for cyberbullying see Areheart (2007), 41. For sexual harassment see Bartow (2009), 383; for safekeeping trade secrets see Rowe (2008).

[23] e.g. in Australia according to the Australian Broadcasting Services Act (1992) notices must be sent to Australian Communications and Media authority (ACMA) see Kleinschmidt (2010), 351.

[24] see Edwards (2005).

[25] In 2008 only, IWF has examined 33.947 complaints of which 80 % concerned sexual exploitation of minors. IWF has come across 1.536 websites of which 99 % were based outside the UK see Reed (2010), 927.

It is questionable, however, whether such a method may be applied for harmful content such as defamation or cyber bullying. While, in the common law system, the opinion of a third party is considered impartial, such notion may not be widely accepted in civil law systems. Therefore, the N&TD method may become just another 'complaints box', having no real impact in combating unlawful content in civil law countries. It is obvious that the decision whether a picture falls under child pornography may be painless. On the contrary, the legal differences between the various degrees of defamation and complicated offences of personality may prove rather cumbersome to resolve. It would be far more difficult for the intermediary to decode legal notions such as 'the principle of proportionality' or 'the basic rules of EU law' or the 'fundamental principles of the ECHR'. Adversaries of the N&TD system claim that intermediaries, fearing the consequences, will act hastily to remove content following any notice. Another radical opinion suggests to introduce preventive censorship before posting any content. As an intermediate approach, we could propose a system of examination before Taking-Down, either by requiring the intervention of a judge or of a third party.

3.2.3 Contractual Clauses

The contract between intermediaries and end-users may include clear clauses regarding the transmission, storage, posting etc. of information or actions, specifying in detail what is to be considered as 'harmful' or 'unlawful' content. In a way, this may be considered as a sort of pressure to the user to accept contractually that restrictions apply to certain content, which may not be posted. Additionally, users could be obliged to accept that the intermediary may remove or block content or even suspend[26] the provision of the service if 'illegal actions' are detected.[27] Moreover, texts of an ethical character could appear on the screen describing the policy for 'permitted uses' or for the 'protection of privacy'. By using similar clauses, intermediaries could try to introduce terms limiting their liability, defining applicable law and jurisdiction and holding the user liable in case of actions of third parties against the provider. It should be noted, however, that provisions prevailing in contracts of adhesion fall under consumer protection legislation, therefore end-users may not be bound by abusive general terms and conditions or clauses imposed without specific negotiation prior to acceptance.

[26] See on that Mifsud Bonnici (2008), 45 and Mifsud Bonnici & de Vey Mestdagh (2005), 133, admitting certain controversies of EU self-regulatory initiatives.

[27] For a similar clause in the Facebook terms and conditions assimilating world-wide users with the citizens of USA, see Solove (2011), 146.

3.2.4 Self-Regulation

Self-regulation is often adopted as a method to reconcile conflicts and comprises any kind of voluntary agreements between intermediaries and right-holders. It has shown positive results in the case of copyright e.g. for the posting of content on YouTube. However, the potential of self-regulatory initiatives may have been overestimated[28] and does not cover any demands for compulsory measures to be taken by the intermediaries, which may be enforced only by means of hard-law or co-regulation. Online Dispute Resolution and Alternative Dispute Resolution (ADR) are also prominent[29], while art. 16 of the ECD encourages the introduction of Codes of Conduct. A similar result may be reached by introducing 'blacklists' of the 'bad' websites e.g. for websites to be avoided by children. Theoretically, such codes of conduct, or other types of soft law, may also introduce common rules of N&TD or, even, a common European standard for N&TD. A practical example is the service provided by Facebook for the prevention of suicides, meaning that the intermediary is entitled to intervene without delay when a user demonstrates such behaviour.

Still, any such intervention touches on the very core of privacy and, in that sense, intermediaries must be accountable for data that they collect and process. Furthermore, providers must inform users for the collection of such data and users should become aware of the exact information collected. Therefore, it is supported that the contractual relationship between users and intermediaries should be transformed to a relationship of confidence. Such an approach seems promising in common law jurisdictions, which assume that a relationship of trust safeguards an asset to a higher degree than a mere contract. However, in continental law, where it seems already difficult to enforce contractual terms, nothing can secure the simple 'keeping of promises'. Additionally, within a 'trust relationship', the parties may not be obliged to act and would adopt a neutral position.

3.2.5 Technical Methods

The application of technical standards may not be ruled out as a regulatory method. It seems far more convenient to adopt certain technical measures (e.g. obligatory use of a filter) than to enforce compulsory rules (e.g. "the transmission of spam is prohibited…"). However, it is not obligatory to impose such method by means of legislation. Certain examples show that providers have taken the necessary

[28] For the myths surrounding self-regulation see Burkert (2006), 245.

[29] See for example in Belgium the Model Code of Conduct Governing On-line Commercial Communications (adopted by the 'Federation des Experts Comptables Europeens' (FEE), available at www.fee.be

measures because they have been coerced implicitly by the authorities.[30] Otherwise, intermediaries could constantly argue that the cost[31] of filtering methods[32] is high.

Another technical solution, without measurable results until know, is the introduction in search engines of certain ratings of the websites according to the respective policy regarding data protection or the protection of minors. It seems that the announcement of 'alliances'[33] against certain threats such as 'phishing' fraud or child pornography, is far more effecting in combating such content. These 'alliances' cooperate with payment providers[34] (such as PayPal) in order to authenticate users, to determine their age and to be able to filter unlawful content. Social networks are also trying to reach a similar arrangement in an effort to verify the age of users.

Readily available help may be provided on a technical cooperation level by means of CERTs (Computer Emergency Response Team). CERTs may provide technical assistance and instructions to the users and suggest methods to tackle complications, could coordinate the defence against large scale attacks (such as viruses or other malicious software), may observe and sometimes obstruct the activity of hackers, could indicate software vulnerabilities and, generally, could provide support for all matters regarding security of information systems, without, however, interception powers. The European Union Agency for Network and Information Security (ENISA) has undertaken to provide guidance for the creation of government CERTs, in the EU Member States. In case of a serious threat, ENISA is entitled to assist, on a theoretical – not operational – level, so as to protect threatened networks, infrastructures and data. The European Commission operates CERT-EU to defend cyberspace attacks against European institutions, organisations and services. Also, the EC3 Centre of Europol, coordinates activities of Member states' law enforcement agencies on cyber-operations.

In the same manner 'safelines' are designed to accept complaints and act as conciliators between users and providers. Normally, they focus on child pornography, child molesting, racist and xenophobic material, but they also examine other cases of illegal content or harassment.

---

[30] For example, in the US, the Child Internet Protection Act does not impose the use of filters. However, libraries not using the filters are not entitled to state funding. See the relevant decision *United States v American Libray Association* (539 US 194, 2003) in Stomberg D., United States v. American Library Association Inc.: The Internet as an inherently Public Forum, *Jurimetrics*, 45, 2004, 59.

[31] For calculating such cost see Schruers (2002), 205.

[32] For a comparison of different systems see Stol W. (2009), 251. In the Netherlands a filtering method has been imposed for child pornography.

[33] See the initiatives in 2012–2013 by Microsoft, Google and other major Internet players.

[34] For the important role of payment providers see MacCarthy (2010), 1037.

3.3 Accepting That Intermediaries May Block or Remove Unlawful Content

The idea of private entities controlling our life is automatically associated with negative connotations. Nevertheless, the role of a 'private' guard is not unknown in the physical world: it has been documented[35] that doctors and pharmacists control the dispense of medicines, small shop owners handle the sale of tobacco products,[36] the proprietors of electronic games parlours administrate admittance of minors,[37] while liquor vendors and bartenders control the consumption of alcohol.[38] In an extreme US case consumers have filed lawsuits against the manufactures of handguns for injuries caused by guns sold.[39]

However, in the first instance the main obstacle is not the acceptance of the model of a private 'judge – guard'. At first, a degree of reconciliation[40] between authorities and intermediaries would be required, the latter being those who control access to the Internet as a medium. It is, hence, far more important to create confidence and social responsibility in a similar manner anticipated by the users from other players of the computer industry. Such confidence should create trust[41] among the users that the providers not only 'live up to their promises', but they are truly validated to conduct their business. The norms for the required level of quality of services may either be based on laws and regulations or on soft law such as ethical rules and codes of conduct that will derive from self-regulatory or co-regulatory schemes, imposing a common platform of performance by the players.

In order to establish a higher degree of liability for Internet intermediaries two suggestions may be brought forward, depending on the type of illegal activity:

(a) *For activities of international criminal demerit*: the catalogue of the Budapest Convention on Cybercrime[42] may help as a common starting point for crimes

---

[35] Goldsmith & Wu (2006), 68.

[36] See for example Directive 2014/40/EU of the European Parliament and of the Council of 3 April 2014 on the approximation of the laws, regulations and administrative provisions of the Member States concerning the manufacture, presentation and sale of tobacco and related products and repealing Directive 2001/37/EC, OJ L 127, 29.4.2014, p. 1–38 [the Tobacco Products Directive]

[37] See an early decision of the Greek Supreme Court (Areios Pagos no. 9/2002) that has awarded compensation to the parents of a minor against the proprietor of a shop with electronic games who had allowed entrance to the minor.

[38] For the sociological aspects of imposing such duty of care in the financial sector see Hamdani (2003), 53.

[39] see Hay & Spier (2005), 1700.

[40] See the Council of Europe 2008 Conference 'Cooperation against Cybercrime', having as a main theme the cooperation of police authorities with intermediaries (proceedings available at www.coe.int). Of the same view see Kioupis (2010), 247.

[41] Of the same opinion see Cohen-Almagor (2010), 381. For an opposite opinion see Nelson (2010), 403, who proposes to control the basic infrastructure i.e payment systems rather than the providers. The same idea has been proposed in Germany in order to control illegal online gambling by cutting the umbilical cord between the '*basisinfrastruktur*' and the provider see Steegmann (2010) and earlier Fox (2003), 187.

[42] Council of Europe Convention on Cybercrime, CETS. No. 185, Budapest, 23.11.2001.

such as child pornography, infringements of IP and activities related to the technological aspects (e.g. hacking, illegal intrusion to systems, data theft, fraud etc.). When faced with such activity intermediaries should be entitled to hinder any further spread, as soon as they are informed and subject to their subjective legal attitude within the escalation degrees of either 'negligence' or 'intent with purpose' of civil law systems.

(b) *For harmful content of international ethical demerit*: this type of activity concerns unlawful / harmful content which is unanimously regulated in each jurisdiction and / or falls within a grey area of illegality, such as hate speech, xenophobia, racism,[43] denial of war crimes, denial of historical events etc. In these instances intermediaries must be held liable (i.e. would be entitled to act) following notification. In both cases (a) and (b), however, immediate action for blocking / removal of content must be tolerated in case of threats to the integrity[44] and security of networks. It is obvious that such effort may be enhanced by technological measures (e.g. filters), while intermediaries may also require informational campaigns targeting groups of users and, especially, vulnerable audiences (e.g. children).

It must be emphasised that Internet intermediaries are not in a 'dependence' relationship against the users, but they derive financial profits. While handling these large volumes of information, intermediaries are best equipped to supervise and to provide help[45] to the users, taking into account their interests. Therefore, the legal relationship[46] between intermediaries and users is not an equitable financial transaction, but presents a weak side, that of the end-user. Hence, an elevated 'duty of care to assist' could be required on behalf of the intermediaries. In that sense, the analysis for the liability of intermediaries must not try to seek the new 'gatekeepers', but rather should try to identify the new co-workers.

3.4 Intermediaries Acting as Quasi Judges

It is a common belief[47] that the ECD has not achieved the goal of promoting electronic commerce. However, it has produced the undesirable side-effect of creating an unofficial regime of umpire for the intermediaries. This unanticipated role has led them to follow two extremely opposed alternatives: either (1) they remove

---

[43] For legislative proposals to combat racism and hate speech see Reed (2010), 79 and Brennan (2009), 123.

[44] It is inconceivable to imagine other networks without the provision of the basic service i.e. a power network without electricity or a water network without water. Therefore, providers must duly supply the envisaged service.

[45] For the introduction of a possible 'duty of help' burdening the intermediaries, as well as the moral dimensions of their liability see Grodzinsky & Tavani (2002).

[46] Kerr (2001), 457.

[47] See First Report, COM(2003)702, op. cit.

content, following precipitate and disputable procedures, contrary to the European tradition of freedom of speech or (2) they demonstrate apathy, maintaining a neutral position not only in view of very serious infringements of individual rights, but also against risks regarding the security and integrity of networks.

Following the immunity regime established by the ECD, subsequent legislation has created a *de jure* enhancement of the degree of liability, without, however, necessarily securing a higher degree of compliance on behalf of intermediaries. Diverging approaches amongst the legislation of the Member States demonstrate that, in the first instance, codes of practice or effective adherence to industry ethics may provide a useful tool to determine a common denominator for the acceptable limits of action by the participants. Another solution can be found in 'hotlines' which may prove far more useful in securing the rights of aggrieved parties, especially because of faster reaction times and lenient procedures.

While discussing the general problems of imposing rules over the new medium, it seems that judges,[48] comprising personal and operational independence, are best suited for balancing legitimacy. Nevertheless, following the European Parliament compromise of 5th November 2009, the Telecommunications Directive 2009/140[49] has adopted in art. 3 an ambitious wording, referring to art. 10[50] of ECHR, without a direct reference to the intervention of judicial authorities, stating that "…*Measures taken by Member States... shall respect the fundamental rights and freedoms of natural persons, as guaranteed by the European Convention for the Protection of Human Rights and Fundamental Freedoms and general principles of Community law. Any of these measures ... liable to restrict those fundamental rights or freedoms may only be imposed if they are appropriate, proportionate and necessary within a democratic society, and their implementation shall be subject to adequate procedural safeguards in conformity with the European Convention for the Protection of Human Rights and Fundamental Freedoms and with general principles of Community law, including effective judicial protection and due process*". The obscure wording of the law, however, does not respond unambiguously to the critical question: are private entities, such as Internet intermediaries allowed to restrict fundamental rights such as the right to access a network? Although the ECHR case-

[48] See the initial text of the famous Amendment 138 of the Telecoms Package as voted by the European Parliament in the 1st reading (24.9.2008): "…*applying the principle that no restriction may be imposed on the fundamental rights and freedoms of end-users, without a prior ruling by the judicial authorities, notably in accordance with Article 11 of the Charter of Fundamental Rights of the European Union on freedom of expression and information, save when public security is threatened in which case the ruling may be subsequent…*".

[49] Te so called 'Freedom Provision' in the 3rd reading of above Amendement 138 (4.11.2009).

[50] See a similar reference to art. 10 ECHR in Internet Recommendation CM/Rec (2008) 6, 26.3.2008 *on measures to promote the respect for freedom of expression and information with regard to Internet filters*, according to which users may object the use of filters. See also CoE document: *Human rights guidelines for Internet Service Providers*, H/Inf (2008) 9, available at www.coe.int

law[51] is clear in several of the above matters, the particular legislation has not been tested in full scale by the ECJ.[52]

Subject to an interpretation that would abolish judges, Internet intermediaries will be charged with quasi-judicial duties: they would be obliged to decide the legality of content being transmitted or hosted in their systems and to take action by blocking, filtering or taking down such content. This is, definitely, a role not to be played by intermediaries and would endanger legal insecurity as of what is legal or not, while camouflaging phenomena of censorship. Having in mind the relatively recently established right of participation to the Information Society,[53] we must evaluate the metamorphosis of Internet intermediaries from 'mere conduits' to 'controllers – guardians' of information. It has been postulated[54] that the modesty of Internet inventors, as far as it concerns the simplicity of the original technical solutions,[55] must also guide the decisions of those now responsible for maintaining the Internet in their capacity either as judges or as law-makers.

Regarding the main problem of establishing liability for Internet intermediaries two guidelines may be followed: on a first level, any self-regulatory or co-regulatory method considered useful should be enhanced and improved in order to achieve reconciliation. On a second level, the existing regulatory model could be transformed in three phases:

(i) certain procedures such as the N&TD method should be formalised and adopted,[56]
(ii) the authorities responsible to control and enforce such procedures should be organised as public and/or private institutions, and
(iii) actual enforcement should take place.

The proposals of the above two levels should be considered as complementary; introducing, for example self-regulation, does not rule out a formal N&TD approach. However, the interference of a judge, at the decision-making stages, is considered an indispensable element inspired by our European legal tradition, especially when civil rights are at stake. Therefore, regardless of the method that we chose and any transitional decisions, the final word lies with the judge.

---

[51] Yannopoulos (2013a), 459 and Yannopoulos (2016), 136.

[52] With the exception of the recent ECJ Decision in the Joined Cases C-293/12 *Digital Rights Ireland* and C-594/12 *Kärntner Landesregierung*, cancelling Directive 2006/24/EC (the Data Retention Directive).

[53] See the Treaty on the Functioning of the European Union (arts 179–180) aiming to promote the development and dissemination of new technologies, the UN initiative for World Summits on the Information Society (www.itu.int), and McKenna A., *A Human Right to Participate in the Information Society*, Hampton Press, 2011.

[54] See Dalal (2011), 1017.

[55] For an overview of the initial technical choices see Zittrain (2003), 32.

[56] See for example the recent Harmful Digital Communications Act (2013) in New Zealand, which has introduced a formal N&TD procedure for defamatory content.

3.5 Concluding Remarks

The on-going clash between the free flows of information and the strategies of certain players, such as the music or movie industry, endorsed by legislators, has led to inefficient outcomes. These actors seek to impede their customers from the benefits of participating to the information society, provided for by the digital revolution, such as the free exchange and streaming of files, the reproduction of information, free postings, public comments and criticism, the creation of content, the participation to social networks. The effort to impose regulations under legal conceptions of the nineteenth and the twentieth century is futile. Laws like the French HADOPI[57] for the restriction of uses of intellectual property, planned laws for the control of blogs and bloggers and for the limitation of freedom of speech are completely anachronistic and mean nothing to the generation that has grown up with You Tube, Facebook and Twitter. On the other hand, increasing the liability of Internet giants, like social networks, search engines etc., who gain enormous profits, seems *de lege ferenda* appropriate and would lead, undoubtedly, to a friendly electronic world.

In a world-wide place of activity efforts for suppressive enforcement of the rule of law are by default condemned to fail. In the digital world users must be convinced[58] that they do not risk something more than in a similar transaction in the analogue universe. Consequently, the demand for freedom to enjoy information society rights must be balanced with the demand for privacy, data protection and security. To cover that vacuum, Internet intermediaries constitute the key factors to find an equilibrium and represent the only reliable partners of the judiciary in order to enforce the law. Therefore, it is not only a matter of increasing their liability, but mainly, a matter of increasing the confidence of the users that the particular players abide by the rule of law.

Those who hold the keys of electronic transactions and who decide about blocking / filtering / taking- down actions, affect fundamental citizen's rights and, therefore, it is crucial that they develop a sense of responsibility. Such duty is not only a matter of statute or case-law, but rather a matter of attitude[59] of the Internet key-players who must seek, in the first place, to create the belief to the simple end-users, who are often aggrieved and offended, that while wandering in the amazing cyberspace they enjoy the same degree of protection like in traditional media, like at home, like during face to face communication.

---

[57] (Loi n° 2009-1311 *du 28 octobre 2009 relative à la protection pénale de la propriété littéraire et artistique sur internet*. The *Haute Autorité pour la Diffusion des Œuvres et La Protection des Droits sur Internet* (HADOPI), is the French authority responisble for the implementation of the law.

[58] For example, in Greece the lack of trust is the second most significant reason for not conducting on-line purchases (by 25 %), the first reason being the lack of credit card, while the third (by 16 %) is the fear concerning the treatment of personal data. See the relevant research of the Athens University of Economics and Business concerning the behaviour of Greek on-line consumers, available at: www.eltrun.gr

[59] See Dworkin's landmark phrase : "…Law's empire is defined by attitude, not territory or power or process…", Dworkin (1986), 413.

References

Ahlert, C., Marsden, C., & Yung, C. (2004). *How 'Liberty' disappeared from cyberspace: The mystery shopper tests internet content self-regulation*. Available at pcmlp.socleg.oxx.ac.uk

Ann, B. (2009). Internet defamation as profit center: The monetization of online harassment, *Harvard Journal of Law & Gender*, Summer, 32, 383.

Areheart, B. A. (2007). Regulating cyberbullies through notice-based liability. *The Yale Law Journal Pocket Part, 117*, 41.

Brennan, F. (2009). Legislating against internet race hate. *Information and Communications Technology Law, 18*(2), 123.

Brimsted, K., & Chesney, G. (2008). The ECJ's judgment in Promusicae: The unintended consequences – music to the ears of copyright owners or a privacy headache for the future? A comment. *Computer Law and Security Report, 24*, 275.

Burkert, H. (2006). Four myths about regulating in the Information Society. In B.-J. Koops et al. (Eds.), *Starting points for ICT regulation*, Asser Press, 245.

Cohen-Almagor, R. (2010). Responsibility of and trust in ISPs. *Knowledge, Technology and Policy, 23*, 381.

Coudert, F., & Werkers, E. (2008). In the aftermath of the promusicae case: How to strike the balance? *International Journal of Law and Information Technology, 18*(1), 57.

Dalal, A. (2011). Protecting hyperlinks and preserving first amendment values on the internet. *The University of Pennsylvania Journal of Constitutional Law, 13*, 1017.

De Beer, J., & Clemmer, C. D. (2009). Global trends in online copyright enforcement: A non-neutral role for network intermediaries? *Jurimetrics Journal, 49*, 375.

Dworkin, R. (1986). *Law's empire*. Cambridge: Harvard University Press.

Edwards, L. (2005). Articles 12–15 ECD: ISP liability. In L. Edwards (Ed.), *The new legal framework for E-Commerce in Europe*.

Fox, M. (2003). Controlling unlawful Internet gambling through the prohibition of bank instruments. *International Company and Commercial Law Review, 14*(5), 187.

Goldsmith, J., & Wu, T. (2006). *Who controls the internet, illusions of a borderless world*. New York: Oxford University Press.

Grodzinsky, F. S., & Tavani, H. T. (2002). Cyberstalking: Moral responsibility, and legal liability issues for internet services providers, in proceedings *IEEE 2002 social implications of information and communication technology*, International Symposium on Technology and Society, ISTAS.

Hamdani, A. (2003). Gatekeeper liability. *Southern California Law Review, 77*, 53.

Hay, B., & Spier, K. (2005). Manufacturer liability for harms caused by consumers. *American Economic Review, 95*(5), 1700.

Kerr, I. (2001). The legal relationship between online service providers and users. *Canadian Business Law Journal, 35*, 457.

Kioupis, D. (2010). Criminal liability on the internet. In I. Stamatoudi (Ed.), *Copyright enforcement and the internet, information law series*, 247.

Kleinschmidt, B. (2010). An international comparison of ISP's liabilities for unlawful third party content. *International Journal of Law and Information Technology, 18*(4), 351.

MacCarthy, M. (2010). What Payment intermediaries are doing about Online Liability and why it matters. *Berkeley Technology Law Journal, 25*, 1037.

McKenna, A. (2011). *A human right to participate in the information society*. Poquoson: Hampton Press.

Mifsud Bonnici, G. P. (2008). *Self-regulation in cyberspace, Information technology and Law Series*. The Hague: Asser Press.

Mifsud Bonnici, G. P., & de Vey Mestdagh, C. N. J. (2005). Right vision, wrong expectations: The european union and self-regulation of harmful internet content. *Information and Communication Technology Law, 14*(2), 133

Nelson, M. (2010). A response to responsibility of and trust in ISPs, by Raphael Cohen-Almagor. *Knowledge, Technology and Policy, 23*, 403.
Reed, C. (2009). The challenge of hate speech online. *Information and Communications Technology Law, 18*(2), 79.
Reed, C. (2010). How to make bad law: Lessons from Cyberspace. *The Modern Law Review, 73*(6), 903.
Rowe, E. (2008). Proposing a mechanism for removing trade secrets from the internet. *Journal of Internet Law, 3*.
Schmoll, A. (2001). *Die deliktische haftung der Internet-Service-Provider*. Frankfurt am Main: Peter Lang.
Schruers, M. (2002, March). The history and economics of ISP liability for third party content. *Virginia Law Review, 88*, 205.
Solove, D. (2011). *Nothing to hide, the false tradeoff between privacy and security* (p. 146). New Haven: Yale University Press.
Stalla-Bourdillon, S. (2010). The flip side of ISP's liability regimes: The ambiguous protection of fundamental rights and liberties in private digital spaces. *Computer Law and Security Report, 26*, 492.
Steegmann, M. (2010). *Die Haftung der Basisinfrastruktur bei rechtswidrigen Internetangeboten: Verantwortlichkeit von Internet- und Finanzdienstleistern im Rahmen des illegalen Online-Glücksspiels*, Nomos Verlag.
Stol, W., et al. (2009). Governmental filtering of websites: The Dutch case. *Computer Law and Security Report, 25*(3), 251.
Stomberg, D. (2004). United States v. American Library Association Inc.: The internet as an inherently public forum. *Jurimetrics, 45*, 59
Ufer, F. (2007). *Die Haftung der internet provider nach dem Telemediengesetz*. Hamburg: Recht der Neuen Medien.
Yannopoulos, G. N. (2013a). Comment on ECHR decision Delfi v. Estonia. *Media & Communications Law (DIMEE), 4*, 459 [in Greek].
Yannopoulos, G. N. (2013b). *The Liability of Internet intermediaries [in Greek]*. Athens: Nomiki Vivliothiki.
Yannopoulos, G. N. (2014). Secondary liability of service providers, report for Greece. *Revue Hellénique de Droit International, 1*, 787.
Yannopoulos, G. N. (2016). Comment on ECHR Decisions Delfi v. Estonia (Grand Chamber) and MTE and Index.HU v Hungary. *Media & Communications Law (DIMEE), 1*, 136 [in Greek].
Zittrain, J. (2003). Internet points of control. *Boston College Law Review, 43*, 32.

Chapter 4
Is Google Responsible for Providing Fair and Unbiased Results?

Dirk Lewandowski

Abstract This chapter discusses the responsibilities of Google as the leading search engine provider to provide fair and unbiased results. In its role, Google has a large influence on what is actually searchable on the Web as well as what results users get to see when they search for information. Google serves billions of queries per month, and users only seldom consider alternatives to this search engine. This market dominance further exacerbates the situation. This leads to questions regarding the responsibility of search engines in general, and Google in particular, for providing fair and balanced results. Areas to consider here are (1) the inclusion of documents in the search engine's databases and (2) results ranking and presentation. I find that, while search engines should at least be held responsible for their practices regarding indexing, results ranking, delivering results from collections built by the search engine provider itself and the presentation of search engine results pages; today's dominant player, Google, argues that there actually is no problem with these issues. Its basic argument here is that "competition is one click away", and, therefore, it should be treated like any other smaller search engine company. I approach the topic from two standpoints: from a technical standpoint, I will discuss techniques and algorithms from information retrieval and how decisions made in the design of the algorithms influence what we as users get to see in search engines. From a societal standpoint, I will discuss what biased search engines mean for knowledge acquisition in society and how we can overcome today's unwanted search monopoly.

4.1 Introduction

In this chapter, I discuss Google's role as the dominant search engine on the market and responsibilities that could derive from this market position. There are many responsibilities that could be discussed in the context of Google (e.g., whether it has

---

D. Lewandowski (✉)
Hamburg University of Applied Sciences, Hamburg, Germany
e-mail: dirk.lewandowski@haw-hamburg.de

M. Taddeo, L. Floridi (eds.), *The Responsibilities of Online Service Providers*,
Law, Governance and Technology Series 31, DOI 10.1007/978-3-319-47852-4_4

responsibilities deriving from it collecting its users' query data), or Online Service Providers in general (see Taddeo & Floridi 2015). I will focus on the results Google provides. I will discuss these results concerning fairness and biases.

First of all, a search engine in the context of this chapter is defined as a computer system that collects content distributed over the Web through crawling, orders the results to a query by machine-determined relevance, and makes these results available to its users through a user interface.

There is a vast body of research on techniques and technologies to improve search engines, on measuring the quality of results of search engines, on the behaviour of search engine users, and on the role search engines play for knowledge acquisition in society. This research is embedded in the wider context of research on the role of algorithms in knowledge acquisition and search engines as socio-technical systems. It is important to understand the decisions made by search engines through their algorithms, as the algorithmic approach to finding information can be seen as "a particular knowledge logic, one built on specific presumptions about what knowledge is and how one should identify its most relevant components. That we are now turning to algorithms to identify what we need to know is as momentous as having relied on credentialed experts, the scientific method, common sense, or the word of God" (Gillespie 2014, p. 168).

The main argument brought forward in this chapter is that every search engine produces *biased results* in some way, resulting from Web crawling, indexing, and results ranking. As there is no perfect or correct results set or ranking, search engine results are always a product of the algorithmic interpretation of the Web's content by the given search engine. Nevertheless, a search engine can still provide *fair results* when there is no preferential treatment of information objects, neither in the process of indexing nor in the process of ranking.

The remainder of this chapter is structured as follows: first, I will elaborate on Google's role as the world's dominant search engine and how Google as a company sees its responsibility for providing fair and unbiased results. Then, I will define the central concepts used in this chapter, namely fair results and unbiased results. I will further discuss the related concepts. I will then focus on the search engines' databases (the indexes) and show how already in building the index, search engines make decisions on which results they will later be able to produce. Then, I will focus on what I call the "algorithmic interpretation of the Web's contents" and how different forms of interpretation shape the results a user gets to see when using a given search engine. After that, I will discuss responsibility issues related to indexing and ranking (or, more general, producing results). I will conclude the chapter with a summary and some suggestions for further research.

4.2 Why Google?

First of all, the importance of search engines for finding information could not be overestimated. Not only are fully-automated search engines like Google the dominant means for finding information on the Web and have made all other approaches to finding content on the Web (like Web directories) nearly obsolete, but more and more information is searched on the Web nowadays instead of other sources outside the Web. While other information sources like social networking sites are sometimes seen as competitors to search engines, as users are directed to information objects through messages displayed there, they do not qualify for *ad hoc* searches, i.e. where a user actually queries an information system in order to find information objects related to his or her information need. Furthermore, when looking at the query volume that search engines process (cf. "Stats: comScore" 2015), we find that search engines not only respond to billions of queries per day, but the query volume is nowhere near declining.

Nearly everybody who uses the Internet also uses search engines (Purcell, Brenner, and Raine 2012). Searching for information is one of the most popular activities on the Internet. On average, European users issue 138 queries per month (comScore 2013). Google's market share is 86% in Europe (comScore 2013) – including eastern European countries, where Yandex has a large market share – with many countries reporting Google's market share well over 90%. Users predominantly relying on one search engine leads to certain problems regarding bias and fairness or at least increases problems resulting from biases inherent in every search engine and from search engine provider's decisions on the fairness of the results and their presentation.

When looking at public statements made by Google, we find that this search engine has a clear view on what its position on the market is and how it should deal with results ranking and transparency related to the rankings. This position can be summarised as follows:

1. *There is competition on the search market, and users can decide to use another search engine without any problem.* In the words of Amit Singhal, Senior Vice President Search at Google: "the competition is only one click away. [...] Using Google is a choice—and there are lots of other choices available to you for getting information: other general-interest search engines, specialized search engines, direct navigation to websites, mobile applications, social networks, and more"[1].
2. *Google generates its results purely through its algorithms, and does not manually interfere with results generated by these*: "No discussion of Google's ranking would be complete without asking the common – but misguided! :) – question: "Does Google manually edit its results?" Let me just answer that with our third philosophy: no manual intervention".[2] And he gives the following reason: "If we

[1] http://googlepublicpolicy.blogspot.de/2011/06/supporting-choice-ensuring-economic.html
[2] https://googleblog.blogspot.de/2008/07/introduction-to-google-ranking.html

messed with results in a way that didn't serve our users' interests, they would and should simply go elsewhere".[3]

3. *Google does not treat its own content preferentially*: "People often ask how we rank our "own" content, like maps, news or images. In the case of images or news, it's not actually Google's content, but rather snippets and links to content offered by publishers. We're merely grouping particular types of content together to make things easier for users. In other cases, we might show you a Google Map when you search for an address. But our users expect that, and we make a point of including competing map services in our search results".[4]
4. *Google is as transparent as possible on how its results are generated*: "Be transparent. We share more information about how our rankings work than any other search engine, through our Webmaster Central site, blog, diagnostic tools, support forum, and YouTube".[5] In another blog post, Singhal says that, "Google's search algorithm is actually one of the world's worst kept secrets".[6] On the other hand, Udi Manber, then Vice President Engineering, Search Quality, said in a blog post: "For something that is used so often by so many people, surprisingly little is known about ranking at Google. This is entirely our fault, and it is by design. We are, to be honest, quite secretive about what we do. There are two reasons for it: competition and abuse".[7]

Much has been written about Google's actual practices, and so I will only summarise some of the findings on Google's practices regarding these statements.

Considering the competition on the search engine market (cf. Lewandowski, 2016), we can firstly see that Google has been the dominant player on the market for years and that it only faces competition in general-interest search engines from Microsoft's Bing. Either other search engines do not have a large enough index to compete with these two engines or they do not provide their own database at all but instead rely on partnerships with either Google or Bing. Regarding vertical search engines, these are to a large degree accessed through general-interest search engines and thus rely on being ranked high in Google.

Google says that it does not manually interfere with its results. While it is true that there is no simple manipulation in the way that Google would manually adjust the organic results for certain queries (although there has been some doubt in the past; see, Edelman 2010), it does exclude certain results due to law or deliberate choice, or it penalizes information objects for not confirming to its self-defined rules. As these information objects are not considered for ranking, they will not have the chance to be discovered through a Google search. Examples for the exclusion of documents from being found include:

[3] http://googlepublicpolicy.blogspot.de/2010/09/competition-in-instant.html

[4] http://googlepublicpolicy.blogspot.de/2010/09/competition-in-instant.html

[5] http://googlepublicpolicy.blogspot.de/2011/06/supporting-choice-ensuring-economic.html

[6] http://googlepublicpolicy.blogspot.de/2010/09/competition-in-instant.html

[7] https://googleblog.blogspot.de/2008/05/introduction-to-google-search-quality.html

- Exclusion of SPAM documents: Some documents are excluded from Google's index due to being irrelevant and classified as SPAM. While it is necessary for any search engine to take action against spamming, the criteria to qualify a document or a website as SPAM are not transparent.
- Penalties for gaming the system: Google reserves the right to penalise certain documents or websites if it finds that the owners of these were trying to "game the system", i.e., trying to achieve better rankings, e.g., by buying links to their documents. Such penalties have nothing to do with the actual quality of the documents' contents.
- Deliberate choice on how to process "Right to Be Forgotten" (RTBF) queries: There is no clearly defined and transparent process on when documents are not shown due to RTBF requests. One has to admit, however, that the RTBF is relatively new and that it may take some time to establish such a process.
- Exclusion of certain sites from vertical searches: Some vertical search engines are built through focussed crawling, i.e., a process where only content from preselected sources is considered for inclusion in the search index. This approach is fundamentally different from Web indexing in general, where a search engine basically crawls all the contents from the Web without humans manually excluding some websites.[8] An example of a vertical search engine that uses focussed crawling is Google News, where humans pre-select news sources that are then regularly crawled for new content. This means that if a website is not considered a news source by Google, its documents do not have a chance to be found through a news search (or in the news box in Google Web search).

A further example where Google does not exactly interfere with the results themselves, but with the process that leads a user to the results, is its interference with the autocomplete function. While Google claims that query suggestions are solely based on other users' past queries and determined automatically[9], there are examples where one can easily see that for certain queries, humans at Google have decided that no suggestions should be made or that suggestions should be filtered (Diakopoulos 2013).

These examples show that Google does not function solely on algorithms and that there are human decisions, not only in the design of the algorithms but also in maintaining the search engine. It is a myth that Google does not manually interfere with the results. However, it is in the interest of Google to prolong this myth because, otherwise, information providers whose content is ranked low within Google's results could argue for them to be ranked higher based on an *assumed* better quality. Google will try not to raise any discussions on the actual quality of its results (apart from it being produced by an algorithm that treats each document the same).

Regarding the question of whether Google gives its own content preferential treatment on its *Web search* results pages (and therefore, using its dominant position

---

[8] It should be noted, however, that there are certain quality thresholds for the inclusion of websites, although whether a website is below such a threshold is determined automatically.

[9] https://support.google.com/websearch/answer/106230?hl=en

in Web search to promote its own content or the results from its vertical search engines, respectively), we can see that giving these results a different (and more attractive) layout than results from competitors alone constitutes preferential treatment. Users are not only attracted by the position of a result but also by its graphical design. For instance, if a news box with a result including an image is presented above the fold on a search engine results page, users will be attracted to it to a large degree (see Lewandowski and Sünkler 2013). Therefore, the if-question may be the wrong one. Instead, one should ask whether search engines have a moral responsibility when it comes to their own content. I will discuss this further below.

This brief comparison of Google's statements with its actual practices shows that Google operates on statements that are at least in part contrary to their actual practices. One could simply qualify these statements as public relations, but the point is that in many cases, even scholars use these arguments when discussing search engines and the role of Google. Moving away from current practices, in the next sections, I will define what fair and unbiased results are and whether search engines are able to provide such results.

4.3 What Are Fair and Unbiased Results?

The Oxford English Dictionary gives several definitions for the term "fair", depending on the context:

> Of conduct, actions, methods, arguments, etc.: free from bias, fraud, or injustice; equitable; legitimate, valid, sound.
>
> Of conditions, circumstances, etc.: providing an equal chance of success to all; not unduly favourable or adverse to anyone.
>
> Of remuneration, reward, or recompense: that adequately reflects the work done, service rendered, or injury received.[10]

For our purposes, we can define fair search results as results that are produced in a way where every document on the Web is treated in the same way by the search engine and, therefore, has the same chance of being found and ranked by that search engine and that there are no human interferences with algorithmic decisions on crawling, indexing and ranking.

Bias, then, is,

> An inclination, leaning, tendency, bent; a preponderating disposition or propensity; predisposition *towards*; predilection; prejudice.[11]

Search engine bias is the tendency of a search engine to prefer certain results through the assumptions inherent in its algorithms. This means that every search engine is biased, as it is impossible to design algorithms without human assump-

[10] http://www.oed.com/view/Entry/67704?rskey=FgCXaJ&result=2#eid

[11] http://www.oed.com/view/Entry/18564?rskey=dJWCZ3&result=1#eid

tions. Therefore, search engine bias does not mean that search results are deliberately manipulated by the search engine vendor but simply that results are ordered in a certain way that is determined by assumptions of what constitutes a good or relevant result in response to queries. It is even at the core of every idea of ranking, based on certain technically mediated assumptions, that certain items are preferred over others.

Yet it should be mentioned that there are other definitions of search engine bias that do not define search engines as biased *per se* but focus on the deliberate preference for certain results (as in the case of Google discussed above, when it favours its own content over content from its competitors. Tavani (2012) summarises the three concerns underlying the definitions of search engine bias:

> (1) search-engine technology is not neutral, but instead has embedded features in its design that favor some values over others; (2) major search engines systematically favor some sites (and some kind of sites) over others in the lists of results they return in response to user search queries; and (3) search algorithms do not use objective criteria in generating their lists of results for search queries. (Tavani 2012)

So, is the question on whether Google is responsible for fair and biased results put the wrong way? If there is no such thing as an unbiased search engine, Google cannot be made responsible for being biased. In my opinion, the bias inherent in search algorithms in fact leads to the severe need for more search engines rather than the demand for Google to reduce or even erase the bias in its search results. I argued for a public search infrastructure (as opposed to alternative search engines) elsewhere (Lewandowski 2014a) and see this as the only solution for dealing with the problem of every search engine being biased by design.

Even if we consider that every search engine is biased by design, Google could still be made responsible for providing fair results. Fair in this sense would be that every information object on the Web has the same chance of being included in Google's database (index) and that every information object in the index has the same chance of being found, solely on the basis of algorithms that treats every information object in the database in the same way.

However, there is an important restriction to this: As search engines not only provide textual documents, but also images, videos etc., they need to treat these different kinds of information objects differently, if only for their different properties. For instance, it is not possible to treat images to be found in the same way by the same algorithm as textual documents. Therefore, it is misleading to speak of *the* index of a search engine, as search engines have multiple indexes for each type of content.

That being said, we must distinguish between fairness in including documents in the indexes and fairness in ranking the results in the indexes in response to a query.

4.4 Providing A Comprehensive and Fresh Index

Search engines make decisions about what to include in their indexes and how often to refresh already known documents. As there is an overwhelming amount of mostly automatically generated content that can be considered SPAM, search engines require technical filters that allow them to not even consider such documents in the crawling/indexing process, which can lead to unwanted exclusion of documents from the search engine. In addition to filtering SPAM content, search engines also apply filters based on – often country-specific – laws (e.g., the "right to be forgotten" in the European Union) and based on the deliberate choices of the search engine provider (mainly in the context of self-defined rules for the protection of children and young persons and takedown notices from copyright holders).

As the Web is of an immense size and continuously changes (Ntoulas, Cho & Olston, 2004), building and maintaining a comprehensive index is a huge challenge (Patterson 2004). Related to these challenges are issues with comprehensiveness, freshness and the deliberate choice of search engines to exclude certain documents from their indexes. The latter can either be deciding not to index certain documents at all or excluding documents after indexing, i.e., not making them available to users in certain countries or regions.

4.4.1 Issues with Comprehensiveness

The first and arguably the most important issue with comprehensive search engine indexes is the size of the Web. While we know that the Web consists of many billions of documents (some years ago, Google even claimed that it knew more than one trillion different URLs[12]), we do not know the exact number, as there is no central registration for URLs on the Web. The best estimates we have are actually from numbers derived through Web crawling (i.e., finding content on the Web through following links), which on a large scale is mainly done by search engines.

Some years ago, search engine companies stopped reporting index sizes. This can be seen as a consequent move, as index sizes do not say much about the quality of a search engine. As there are vast amounts of documents on the Web that a search engine will not want to index (such as copies of the ever-same content and SPAM pages), a search engine having a lot of these documents in its index would surely increase its index size but for no one's benefit.

Then, there is the problem with defining what a document on the Web actually is. One could argue that everything that has an URL should be considered to be a document. However, as such documents can be easily created automatically, and can be built by combining elements from other documents, a lot of documents without any benefit could be (and are) built. This does not have to do with spamming search

[12] https://googleblog.blogspot.de/2008/07/we-knew-web-was-big.html

engines. Consider, for instance, blogs where different kinds of overview pages (such as teasers for all articles from a certain month, teasers for articles tagged with a certain keyword etc.) are generated. We can question whether a search engine should index all these "documents".

However, while this seems to be a purely technical problem, it still comprises decisions about what is worth indexing, and there is no guarantee that no potentially relevant document will slip through. Presumably due to the structure of the Web, certain content, e.g., from certain countries, is not as well represented as content from other countries (cf. Vaughan and Zhang 2007).

Furthermore, search engines have technical and financial limitations regarding index sizes: Even if a search engine wanted to build a complete index of the Web, it would still face limitations due to its technical possibilities and financial resources. No search engine is able to index the Web in its entirety. The problem arising from that, however, may not be the lack of completeness but the lack of transparency regarding the criteria that lead a search engine to index certain documents and exclude others.

4.4.2 Issues with Freshness

Apart from the issues of building a comprehensive index of the Web, search engines must also keep up with the ever-changing Web. New documents are created, existing documents are changed and documents are deleted. The issue related to these Web dynamics is twofold: Firstly, search engines have to make sure to index documents afresh and, secondly, provide fresh results through ranking. It would be a bad idea for a search engine to present a user with a result description that points him or her to a page that no longer exists.

The issue with freshness is that no search engine can keep all the documents in its index current. On the one hand, no search engine could afford to crawl every document every second. On the other hand, even if a search engine would be able to do so, this would account for too much bandwidth and would send too many requests to Web servers. The approach that search engines take is to decide which documents to revisit when based on popularity and on the refreshment rate of each document in the past (Risvik and Michelsen 2002). On the one hand, this leads to a technically feasible solution. On the other hand, decisions about freshness (i.e., which documents to index more frequently) may lead to fairness issues. Preferring popular and/or often refreshed content is a decision that could lead to the oppression of other documents in the results sets.

4.4.3 Issues with Deliberate Choices Made by the Search Engines

Search engine providers also make deliberate choices about documents to exclude from their indexes, sometimes not actually excluding them, but removing them from the results in certain regions or countries. The prime example for this is the "Right to Be Forgotten" (RTBF) in European legislation. Persons can request for certain results to be removed from search results if these results refer to the person's past that is no longer occurring. A problem with the RTBF may lie in it not being precise as to when such data should be removed. However, the issue with search engines and the RTBF lies in that the procedure for having content removed is not transparent. While Google provides information about how many requests it received and how it decided as well as established an advisory council on the topic, there are still no transparent rules on how these requests are treated. So, some documents may have been removed even though they do not fall under the RTBF, while others may have been removed without actually falling under the RTBF.

Very similar is the case of takedown notices by copyright holders. Here also, Google must process a large number of requests, but this time, it mainly processes them automatically. This could lead to documents taken down erroneously, simply due to the sheer volume of these requests and standard procedures to treat them (Karaganis and Urban 2015).

A third area of concern is the protection of children and young persons. At the request of authorities (e.g., the *Bundesprüfstelle für jugendgefährdende Medien* in Germany), Google removes websites from its search results no matter if a child or adult is searching for that content. Furthermore, adjusting rankings in a way that prefers non-offensive content is also a decision on what constitutes a document potentially relevant in response to a query (see the section in rankings below).

From this brief discussion of the RTBF, takedown notices and the protection of minors, we can see how Google and other search engines make decisions about which documents to include in their results sets that are opaque to their users. In contrast to the technical and financial issues related to index comprehensiveness and index freshness, these decisions are, even though they are founded on law, deliberate decisions made by the search engines and, therefore, moral decisions. Mainly due to technical reasons, we cannot expect a search engine to provide a complete and fresh copy of the Web. However, what we could expect from a search engine is to make transparent how its index is built, what is left out and for what reasons.

4.5 Search Engines' Algorithmic Interpretation of the Web's Content

For every query, a search engine – through its algorithms – presents certain results in a certain way. We can call this an "algorithmic interpretation" of the Web's content (Lewandowski 2015a), and users tend to follow this interpretation uncritically (e.g., Pan et al. 2007; Purcell, Brenner, and Raine 2012). Algorithmic interpretation does not only consider the ranking of the results lists (Pan et al. 2007) but also the positioning of elements on the search engine results pages (Lewandowski and Sünkler 2013; Liu et al. 2015), the correctness of results (e.g., (White and Horvitz 2009), the labelling of advertisements and the diversity within the top results (Denecke 2012).

Due to search engines presenting different kinds of results from different collections within search engine results pages (so-called "Universal Search"), the ranking of these results is not only list-based anymore but consists of at least three different ranking functions: (1) Ranking of the results lists from the Web index, (2) ranking of vertical results within collections such as news, images, etc. and (3) ranking of Universal Search containers (i.e., boxes presenting top results from vertical search engines) within the general SERP or the list of ranked Web search results, respectively.

Ranking within collections (whether considering the Web collection or vertical collections) is based on groups of ranking factors as follows (cf. Lewandowski 2015b, p. 91–92):

(1) Text statistics: This is where the query is matched with the representation of the information objects and statistics are applied to rank documents according to their fit with the query. As queries on the Web are usually very short and there is no standard quality for documents found on the Web (opposed to documents in a curated database, like the electronic archive of a newspaper), text statistics alone are not sufficient for ranking Web documents. They must be accompanied by so-called quality factors.
(2) Popularity: Link-based ranking algorithms (e.g., PageRank) as well as click-based algorithms assign popularity scores to documents and use these scores for a ranking based on the assumption that what has been useful to others will also be useful to a given user. Popularity-based algorithms can either take all users into account or only a certain group of users.
(3) Freshness: As new content is produced in large amounts on the Web, the freshness of information objects is also considered in ranking algorithms.
(4) Locality: Information objects are matched to the geographical location of the user.
(5) Personalisation: Information objects are matched to the interests of an individual user, mainly based on his past behaviour (e.g., queries entered, results viewed).

(6) Technical ranking factors: These factors are mainly used to determine how reliable Web servers are in providing results to the user. As a search engine in most cases only links to information objects from external sources, a user clicking on a result on the search engine results page (SERP) will have to wait for the information object to be produced by the server. Search engines take into account how fast a server is able to process requests and how reliable it is (concerning downtimes). Technical ranking factors are an interesting case, as they judge where an information object should be displayed in a results set, not based on the assumed quality of the content of the information object but rather on the convenience for the user to get to the information object.

To understand what types of results search engines prefer, it is important to consider that the popularity group is considered one of the most important to determine the quality of individual documents. This means that while search engines try to measure such things as credibility, this can only be simulated through measuring popularity (cf. Lewandowski 2012).

The ranking of search results is often misinterpreted as either correct or wrong. I argue that this is mainly due to users having success with navigational queries (where they search for a certain website and the aim is to navigate to that website), with transactional queries where they already have a certain website in mind and with informational queries where they either only search for trivia that can be found on a multitude of websites or they already have a website providing that information in mind (such as Wikipedia). Based on their experience of success for these queries, they assume that their favourite search engine will also produce "correct" results for other types of – mainly informational – queries where there may be many relevant results, and, therefore, no single correct ordering of these results (Lewandowski 2014b).

The basis for algorithmic interpretation is the assumptions that search engine engineers put into these algorithms. Little research has been conducted on the motivations, beliefs and assumptions of this group of people. However, the research already done shows that engineers (and other search engine employees) see search engines as rather purely technical systems, and they conform to a capitalistic way of looking at them (van Couvering 2007; Mager 2012).

The effects of algorithmic interpretation can be seen on different levels. Most obviously, algorithmic interpretation affects the ranking of the organic results, i.e., the results produced from the basic Web index. Here, every information object included in the index is treated in the same way by the same ranking algorithms, i.e., the results are ranked in a fair way, although not without bias towards the information objects that fulfil assumptions inherent in the ranking algorithms. The effect of algorithmic interpretation for organic results can best be seen when considering drastic examples like the infamous martinlutherking.org website (Piper 2000) that, even in 2015, still ranks very well in Google, and the results produced by Google in response to queries related to race and gender (Noble 2013; Noble 2012). These examples also show that a good ranking position in Google does not necessarily conform with a result being credible or trustworthy (Lewandowski 2012).

Applying certain algorithms can also lead to search engines presenting one side of an argument or only the results of a certain type or tendency. Some algorithms not only try to rank results according to relevance but also mix different result types within the top results to achieve diversity (Giunchiglia et al. 2009).

Algorithmic interpretation also affects the composition of results pages from different indexes ("Universal Search"). For instance, in addition to results from the Web index, results from vertical indexes like news, images and videos can be included in the results pages. This leads to a manifold ranking: Firstly, the results within each vertical index and the main Web index must be ranked. Then, the top results from these indexes must be incorporated into one search engine results page, where results from the vertical searches are to be positioned, and which is another form of ranking. As the presentation of results on the SERP (and even more importantly, in the area "above the fold" of the SERP) heavily influences users' decision on what results to select, search engines are able to lead users to certain types of results merely through results presentation. An important example for this is Google presenting results from its own vertical search engines (such as Google News, Google Scholar and Google Maps) as attractive boxes within the SERP, which then preferably leads to users clicking on them (Lewandowski and Sünkler 2013).

The personalisation of search results is another form of algorithmic interpretation, this time also related to a user's preferences and interests. Results are then produced according to these assumed preferences, mostly without the user knowing what data about him or her is actually collected and how the use of this data affects his or her results. In extreme cases, personalisation can lead to what Eli Pariser termed the "filter bubble" (Pariser 2011), where information objects presenting contradicting views and beliefs from the users are oppressed, and the users only receive results confirming their already established opinions.

Last but not least, search engines present text-based, contextual advertisements on the SERPs. These can be seen as a distinct type of result, and the often-used term "sponsored link" may describe them best: They are a type of result but different from organic results in that they are paid for. Studies lead to the conclusion that users are not able to properly distinguish between organic results and advertisements (Bundesverband Digitale Wirtschaft 2009; Filistrucchi et al. 2012), and, in the case of Google, ad labelling has not become clearer in recent years (Edelman 2014).

It should also be mentioned that apart from the assumptions underlying search engines' algorithms, there is also external influence on the results, namely in the form of search engine optimisation (SEO). The aim of SEO is to optimise information objects in a way that leads to optimal findability through search engines, mainly Google. Search engine optimisation has grown to be a billion-dollar industry, and, at least for queries assumed to have a commercial intent, it will be difficult to find top results in Google that have not been optimised.

While it is common knowledge in the industry and academia that search results are heavily influenced by search engine optimisers, users' knowledge about these practices seems to be low. Furthermore, we see that users generally know little about search engines' workings in general (see, e.g., Purcell, Brenner, and Raine

2012). They often have misconceptions about how a search engine actually works (e.g., Hendry and Efthimiadis 2008), they are not good at searching and they lack knowledge about search engines' ways of making money. On the other hand, they trust in Google's rankings when it comes to results quality (Keane, O'Brien, and Smyth 2008; Bar-Ilan et al. 2009), sometimes even more than their own judgments (Pan et al. 2007).

4.6 Responsibilities

As we can see from the discussion above, there are multiple areas where we can ask for the responsibility of search engines, especially Google as the dominant player on the market. While we cannot expect Google to provide unbiased results, since we can see that search engine rankings are biased *per se*, we can expect Google to give every information object in its index a fair chance of being ranked in response to a query. "Fair results" here would mean that every information object is treated in the same way. This leads to the conclusion that Universal Search is an unfair treatment to certain results as soon as Google presents results from its own offerings preferentially.

We can also demand for Google to be transparent about its practices, be it the sources its vertical results come from and why they are given preferential treatment or the labelling of its advertisements. While information on both can be found on Google's help pages, we can see in practice that users do not understand – or are not interested in – the workings behind the composition of search engine results pages. This may be seen as the users' own fault, but in its current practices, Google at least accepts that users are deceived about the true reasons for the display of certain results (types).

4.7 Conclusion

We cannot expect a search engine to provide fair and unbiased results. Every search engine is per definition biased in that it is not able to provide sets of correct results, as separated from "incorrect" results. Correct results can only be provided for a subset of queries, mostly navigational queries (Broder 2002; Lewandowski 2014b). With informational queries, search engines can provide *relevant* results. However, as relevance always refers to a given user in a given context (Saracevic 2015), a search engine can only make more or less good guesses to what a user may find relevant in his or her current context.

Before producing a ranked results set in response to a query, a search engine must build an index from content found on the Web. A problem here is the size of the Web and its dynamics. Due to the vast amount of information objects on the Web, search engines produce more results than a user is able to consider for most queries. This means that users must trust the ranking provided by the search engine. However, this does not mean that there may be no additional relevant results (or even results being more relevant to a given user in a given context) on lower positions of the results lists.

So, even if Google treated all information objects in a fair manner, users would still see only a fraction of the relevant results available. And as all ranking algorithms organise results in a certain order based on assumptions about what is relevant to users, results from different search engines could differ considerably (or may not even overlap at all) without the results from one search engine being less relevant than the others.

This leads to the conclusion that to release us from only one (or considering the current competition, from a few) of many possible algorithmic interpretations of the Web's content, we need more search engines. With "more", I do not mean just one or two more search engines but a considerable amount of them. One way to achieve this is to view Web indexing as a public service to be provided for the good of all and then have services built upon that infrastructure (Lewandowski 2014a).

Further research is needed on the actual differences of the algorithmic interpretation by different search engines. While some empirical studies already determined overlaps between search engine results (e.g., Spink, Jansen, Blakely and Koshman, 2006), they do not deal with the actual content found, but from a technical viewpoint with URL overlap only. Furthermore, more research is needed on the types of results and the beliefs reproduced through search engine algorithms.

References

Bar-Ilan, J., Keenoy, K., Levene, M., & Yaari, E. (2009). Presentation bias is significant in determining user preference for search results—A user study. *Journal of the American Society for Information Science and Technology, 60*(1), 135–149.

Broder, A. (2002). A taxonomy of web search. *ACM Sigir Forum, 36*(2), 3–10.

Bundesverband Digitale Wirtschaft. (2009). *Nutzerverhalten Auf Google-Suchergebnisseiten: Eine Eyetracking-Studie Im Auftrag Des Arbeitskreises Suchmaschinen-Marketing Des Bundesverbandes Digitale Wirtschaft (BVDW) e.V.*

comScore. (2013). *Europe digital future in focus: Key insights from 2012 and what they mean for the coming year*. From http://www.comscore.com/content/download/19217/947363/file/2013-uk-future-in-focus.pdf. Accessed 12 May 2016.

Denecke, K. (2012). Diversity-Aware search : New possibilities and challenges for web search. In D. Lewandowski (Ed.), *Web search engine research* (pp. 139–162). Bingley: Emerald Group Publishing Ltd.. doi:10.1108/S1876-0562(2012)002012a008.

Diakopoulos, N. (2013). Sex, violence, and autocomplete algorithms: What words do Bing and google censor from their suggestions? Slate. http://www.slate.com/articles/technology/future_tense/2013/08/words_banned_from_bing_and_google_s_autocomplete_algorithms.single.html Accessed 12 May 2016.

Edelman, B. (2010). *Hard-coding bias in google 'Algorithmic' search results*. Benedelman.org. http://www.benedelman.org/hardcoding/. Accessed 12 May 2016.

Edelman, B. (2014). *Google's Advertisement Labeling in 2014*. Benedelman.org. http://www.benedelman.org/adlabeling/google-colors-oct2014.html. Accessed 12 May 2016.

Filistrucchi, L., Tucker, C., Edelman, B., & Gilchrist, D. S. ((2012). Advertising disclosures: Measuring labeling alternatives in internet search engines. *Information Economics and Policy, 24*(1), 75–89.

Gillespie, T. (2014). The relevance of algorithms. In T. Gillespie, P. Boczkowski, & K. Foot (Eds.), *Media technologies* (pp. 167–193). Cambridge, MA: MIT Press.

Giunchiglia, F., Maltese, V., Madalli, D., Baldry, A., Wallner, C., Lewis, P., Denecke, K., Skoutas, D., & Marenzi, I. (2009). Foundations for the representation of diversity, evolution, opinion and bias.

Hendry, D. G., & Efthimiadis, E. N. (2008). Conceptual models for search engines. In A. Spink & M. Zimmer (Eds.), *Web searching : Multidisciplinary perspectives* (pp. 277–308). Berlin: Springer.

Karaganis, J., & Urban, J. (2015). The rise of the robo notice. *Communications of the ACM, 58*(9), 28–30. doi:10.1145/2804244.

Keane, M. T., O'Brien, M., & Smyth, B. (2008). Are people biased in their use of search engines? *Communications of the ACM, 51*(2), 49–52.

Lewandowski, D. (2012). Credibility in web search engines. In M. Folk & S. Apostel (Eds.), *Online credibility and digital ethos: Evaluating computer-mediated communication* (pp. 131–146). Hershey: IGI Global.

Lewandowski, D. (2014a). Why we need an independent index of the web. Information retrieval; digital libraries. In R. König & M. Rasch (Eds.), *Society of the query reader: Reflections on web search* (pp. 49–58). Amsterdam: Institute of Network Culture.

Lewandowski, D. (2014b). Wie Lässt Sich Die Zufriedenheit Der Suchmaschinennutzer Mit Ihren Suchergebnissen Erklären? In H. Krah & R. Müller-Terpitz (Eds.), *Suchmaschinen (Passauer Schriften Zur Interdisziplinären Medienforschung, Band 4)* (pp. 35–52). Münster: LIT.

Lewandowski, D. (2015a). Living in a world of biased search engines. *Online Information Review, 39*(3), 278–280. doi:10.1108/OIR-03-2015-0089.

Lewandowski, D. (2015b). *Suchmaschinen Verstehen*. Berlin Heidelberg: Springer Vieweg.

Lewandowski, D.(2016). Status Quo und Entwicklungsperspektiven des Suchmaschinenmarkts. In T. Pellegrini and J. Krone (Eds.), *Hanbuch Medienökonomie.* Berlin Heidelberg: Springer. doi: 10.1007/978-3-658-09632-8_38-1.

Lewandowski, D., & Sünkler, S. (2013). *Representative online study to evaluate the revised commitments proposed by Google on 21 October 2013 as Part of EU Competition Investigation AT.39740-Google Report for Germany*. Hamburg.

Liu, Zeyang, Yiqun Liu, Ke Zhou, Min Zhang, and Shaoping Ma. (2015). Influence of vertical result in web search examination. In *Proceedings of SIGIR'15, August 09–13, 2015, Santiago, Chile*. New York: ACM.

Mager, A. (2012). Algorithmic ideology: How capitalist society shapes search engines. *Information, Communication & Society, 15*(5), 769–787. doi:10.1080/1369118X.2012.676056.

Noble, S. U. (2012). Missed connections: What search engines say about women. *Bitch Magazine, 54*, 36–41.

Noble, S.U. (2013). Google search: Hyper-visibility as a means of rendering black women and girls invisible. InVisible culture: An electronic journal for visual culture. http://ivc.lib.rochester.edu/google-search-hyper-visibility-as-a-means-of-rendering-black-women-and-girls-invisible/. Accessed 12 May 2016.

Ntoulas, A., Cho, J., & Olston, C. (2004). What's new on the web?: The evolution of the web from a search engine perspective. In*Proceedings of the 13th international conference on World Wide Web* (pp. 1–12). New York: ACM.

Pan, B., Hembrooke, H., Joachims, T., Lorigo, L., Gay, G., & Granka, L. (2007). In google we trust: Users' decisions on rank, position, and relevance. *Journal of Computer-Mediated Communication, 12*(3), 801–823.

Pariser, E. (2011). *The filter bubble: What the internet is hiding from you*. London: Viking.

Patterson, A. (2004). Why writing your own search engine is hard. *Queue, 2*(2), 49–53.

Piper, P. S. (2000). Better read that again: Web hoaxes and misinformation. Searcher. *Searcher, 8*(8), 40.

Purcell, K., Brenner, J., & Lee, R. (2012). *Search engine use 2012*. Washington, DC: *Pew Internet & American Life Project*.

Risvik, K. M., & Michelsen, R. (2002). Search engines and web dynamics. *Computer Networks, 39*(3), 289–302.

Saracevic, T. (2015). Why is relevance still the basic notion in information science ? In F. Pehar, C. Schlögl, & C. Wolff (Eds.), *Re:inventing Information Science in the Networked Society. Proceedings of the 14th International Symposium on Information Science (ISI 2015), Zadar, Croatia, 19th–21st May 2015* (pp. 26–35). Glückstadt: Verlag Werner Hülsbusch.

Spink, A., Jansen, B. J., Blakely, C., & Koshman, S. (2006). A study of results overlap and uniqueness among major web search engines. *Information Processing & Management, 42*(5), 1379–1391.

Stats: comScore. (2015). Search engine land. http://searchengineland.com/library/stats/stats-comscore. Accessed 12 May 2016.

Taddeo, M., & Floridi, L. (2015). The debate on the moral responsibilities of online service providers. *Science and Engineering Ethics*. doi:10.1007/s11948-015-9734-1.

Tavani, H. (2012). Search engines and ethics.. [Plattformname.]http://plato.stanford.edu/entries/ethics-search/. Accessed 12 May 2016.

van Couvering, E. (2007). Is relevance relevant? Market, science, and war: Discourses of search engine quality. *Journal of Computer-Mediated Communication, 12*(3), 866–887.

Vaughan, L., & Zhang, Y. (2007). Equal representation by search engines? A comparison of websites across countries and domains. *Journal of Computer-Mediated Communication, 12*(3), 888–909.

White, R. W., & Horvitz, E. (2009). Cyberchondria. *ACM Transactions on Information Systems, 27*(4): Article No. 23. doi:10.1145/1629096.1629101.

Chapter 5
Speaking Truth to/as Victims – A Jurisprudential Analysis of Data Breach Notification Laws

Burkhard Schafer

Abstract This paper analyses Data Breach Notification Duties from a jurisprudential perspective. DBNDs impose duties on people who are victims of a crime, duties whose violation in turn can trigger criminal sanctions. To analyze what type of duties a democratic society under the rule of law can impose on victims, we need a conceptual framework that links duties to participate in crime investigation and prosecution to specific roles a person can have in relation to a crime. Duff and Marshall have developed such a theory of the criminal law, which the paper applies to DBNLs, combining their approach with Floridi's concept of the infosphere.

5.1 Introduction

This chapter analyses a comparatively new legal duty for Online Service Providers, the duty to inform customers, and in some cases also one or several government agencies, when the company suspects that its security was breached and customer data has been acquired by unauthorised third parties. An ever-increasing number of jurisdictions has created such a duty through specifically tailored statutes. Despite this success in terms of international adaptation, and despite several high profile cases that led to intensive discussion about their merits and dangers, Data Breach Notification Laws (DBNLs) have not attracted much interest from legal theorist. At first glance, they seem to be just another rather technical compliance burden, something that only concerns those that need to know, (typically the compliance officers in large companies), and which can be addressed best through adequate procedures, or much worse through form filling. It has not given rise to interesting questions of statutory interpretation, and hence to a dearth of litigated cases.

B. Schafer (✉)
School of Law, The University of Edinburgh, Edinburgh, Scotland
e-mail: b.schafer@ed.ac.uk

M. Taddeo, L. Floridi (eds.), *The Responsibilities of Online Service Providers*, Law, Governance and Technology Series 31, DOI 10.1007/978-3-319-47852-4_5

By contrast, this paper argues that DBNLs are a significant addition to the conceptual vocabulary of the law, and challenge deeply ingrained intuitions about criminal justice and procedural fairness. They differ in significant and philosophically interesting ways from the vast majority of other crimes and regulatory offences, which means that their analysis can enable us to rethink much more broadly the role of criminal law in a democratic society, and what conceptual and relational vocabulary we need to describe and account for it.

In particular, I will argue that a fruitful way to understand DBNs is to bring together ideas developed by Antony Duff and Sandra Marshal on the role of criminal law in democratic republics, and the ideas developed by Luciano Floridi on the changing nature of excuses and defences in an age of data abundance. DBNLs can be seen as the manifestation of predictions Floridi has made first in his 1999 paper on Information Ethics as a separate discipline of studies (Floridi 1999), and as legal reflections of changes in the ethical landscape that also feature prominently in his more recent work. If seen through this lens, their conflict with established ideas and intuitions about the role and intrinsic structure of the criminal justice system is not any longer a mere anomaly, inadvertently created by insufficiently thought through ad-hoc reactions by legislators to the latest panic, but rather heralding a much more significant change in the way the criminal law thinks. Their broader significance thus established, we will then turn to the work of Duff and Marshal, who have argued that traditional (criminal) law theory fails to account for some of the most iconic institutions of criminal law, such as the duty to participate in some countries as jurors during trial, the duty of a witness to speak the truth, or the duty (some) victims may have to denounce their attacker. In response, they developed a "role centric" extended vocabulary that allows to discuss and analyze these phenomena within the framework of a much more ambitious communitarian vision of criminal law in democratic societies. We will argue that DBNLs are best understood in this extended theoretical framework. Marshal and Duff's analysis drew its examples mainly from well-established legal institutions. We argue that their approach is even more important for the philosophy of a future criminal law for the information society. Moreover, we will argue that also many, if not all, of their paradigmatic examples are best understood as a type of "communication duty", setting the foundations for an information-theoretic theory of criminal law and procedure.

In the first part, the paper introduces DBNLs, starting with a historical overview, and then contrasts the US approach to that favoured in the EU. The paper concludes with a short discussion of the future of DBNLs in the forthcoming EU Data Protection Regulation. In the second part, the paper explains why DBNLs are indeed a new and unique form of regulation, in a much more radical way than this is normally considered, raising ethical and jurisprudential questions that go beyond the technical particularities of DBNLs. We are then using ideas from Floridi's Information Ethics and Duff and Marshal's communitarian and communicative theory of the criminal trial to outline a possible answer to these questions, or at least advocate for a specific theoretical framework in which they should be addressed.

5.2 Data Breach Notification Duties: Telling Customers the Whole Ugly Truth

5.2.1 Oh, What a Tangled Web We Weave, When First We Practise To Deceive!

In early 2005, the data broker ChoicePoint became aware of a potential security breach, fraudulent access to customer data by some small businesses that were also ChoicePoint clients. The pattern of activity indicated that the data had been used for the purpose of ID fraud. The personal data of 163,000 customers had potentially been compromised.[1] Ten years later, in August 2015, Ashley Madison, the extramarital-affair brokerage website, became the victim of a successful hacking attack by a group calling itself the "Impact Team". Potentially up to 32 million user profiles were affected, though this may exaggerate the number of victims, as one of the stated objectives of the hack was to expose "fake" profiles, which according to hackers were plentiful. The Ashley Madison (AM) data contained names, email addresses, physical addresses, login information, and partial credit card payment information. While the latter exposed victims to the danger of financial loss, login information together with easily searchable personally identifiable data such as names and addresses allowed angry spouses all over the world to see how much time exactly their partner had spend in the pursuit of an affair. The consequences for many of them were devastating, driving some customers into suicide (Segall 2015).

While in both scenarios, customers of online service providers were put at risk when their personal identifiable data was leaked after an attack on their OSP, the legal protection offered to the affected groups of customers differed in one significant aspect. ChoicePoint, being an US company, informed initially the police of the suspected data breach, and soon after in compliance with the Californian Data Breach Notification laws, also its 35,000 Californian customers. This limited release of information resulted in a public outcry, and the company soon afterwards approached all US customers whose records had been improperly accessed. Even then though, Choice Point attempted "minimal compliance". The Californian law, the first of its kind, had only been in force for 15 month when the breach had been noticed. Initially, ChoicePoint restricted the search for leaks of its client database to those records that had been created only after the law had come into force. It was again public outcry and political pressure by investigators and regulators that forced ChoicePoint to extend its analysis to all of its records, increasing the number of customers who needed informing from 145000 to 163.000. While ChoicePoint accepted a duty to tell the truth about the danger it had created for its clients, it dragged its heels in also telling the whole truth.

Ashley Madison customers were even less lucky. Being headquartered in Canada, AM was not subject to a similar legal obligation. When the "Impact Team" hackers

[1] United States v. ChoicePoint, Inc., No. 1:06-CV-0198 (N.D. Ga. Feb. 15, 2006), http://www.ftc.gov/os/caselist/ choicepoint/stipfinaljudgement.pdf; see also Otto et al. (2007).

initially approached AM, threatening the release of the data unless its parent company, Avid Life Media, shut down Ashley Madison and its sister site, "Established Men", AM kept the breach secret, hoping undoubtedly that the hackers would not act on their threat.They miscalculated, and it was through a statement released on the Internet by the hackers that AM customers learned about the risk of exposure that they were now facing:

> Time's Up!
>
> Avid Life Media has failed to take down Ashley Madison and Established Men. We have explained the fraud, deceit, and stupidity of ALM and their members. Now everyone gets to see their data.
>
> Find someone you know in here? Keep in mind the site is a scam with thousands of fake female profiles. See ashley madison fake profile lawsuit; 90–95% of actual users are male. Chances are your man signed up on the world's biggest affair site, but never had one. He just tried to. If that distinction matters.
>
> Find yourself in here? It was ALM that failed you and lied to you. Prosecute them and claim damages. Then move on with your life. Learn your lesson and make amends. Embarrassing now, but you'll get over it.
>
> Any data not signed with key 6E50 3F39 BA6A EAAD D81D ECFF 2437 3CD5 74AB AA38 is fake.

That the world learned of this event through a an Internet announcement from the hackers prevented many AM customers from taking timely mitigating action – from coming clean to their spouse to changing their credit card, to the more evolved strategies to manage online reputations.

The two cases raise interesting philosophical issues. In both of them, we find appeal to "truth telling duties", both moral and legal. Data breach notification laws had imposed on ChoicePoint a duty to tell the truth about data breaches to their customers, even though this was likely going to be a statement against the company's own interest. ChoicePoint complied, hesitantly, telling the truth, but never quite the whole truth. AM, at the danger of sounding puritan, was in the business of facilitating lies, enabling its customers to cheat on their partners. But it was not this that attracted the ire of hackers, but the suspicion that in doing so, they also lied to their own customers, or at least facilitated and benefited from a lie: using inflated numbers of (female) profiles to attract new (male) subscribers, even tough they new, or should have known, that many of the profiles were fake. Finally, AM lied by omission, keeping its customers unaware of the risk they were facing, and encouraging as a result continuous use of AM's services, disclosure of which could later harm them.

This paper is concerned with this last truth telling duties only: a legal duty to tell the truth, potentially against one's own interest, to one's customers, when one has failed to keep their personal data secure.

5.2.2 *When Sorry Is the Hardest Word to Say: Data Breach Notification Duties in Law*

Having seen how data breach notification laws have worked, or failed to work, in two examples, we now have a brief look at the history and structure of data breach notification laws. In 2002, California became the first country to enact a Data Breach Notification law, S.B. 1386, which took effect in July 2003.[2] The law requires

> a state agency, or a person or business that conducts business in California, that owns or licenses computerized data that includes personal information, as defined, to disclose in specified ways, any breach of the security of the data, as defined, to any resident of California whose unencrypted personal information was, or is reasonably believed to have been, acquired by an unauthorized person.[3]

The ChoicePoint security breach discussed above became a main driver behind the wider adoption of this type of law. – too obvious had it become that faced with a serious threat, Californians enjoyed substantially better protection than the residents of other states, even when they were all customers of the same US-based entity. As of February 2016, forty-seven states, the District of Columbia, Guam, Puerto Rico and the Virgin Islands have enacted Data breach notification laws (see Needless 2009; Lee 2006). While the general idea of a duty to notify customers about a data breach has thus been taken up enthusiastically across the US, the specific form that this duty takes differs in its details between these jurisdiction. While e.g. the Californian law was part of consumer protection legislation (Simitian 2009 p. 1014), subsequent proposals in other states emphasized one potential criminal harm, identity theft, as the main rationale that informs a DBN duty (see Skinner 2003; Draper 2006 also for an argument for state wide legislation). This classification matters – the relation between laws that protect privacy and laws that address concerns about identity theft is complex and sometimes antagonistic, so for instance Towle (2003 261–264). If consumer privacy is the main goal, the trigger for a DBN duty will be very low. The mere fact that someone's privacy was breached is the harm in this case, which can also mean that not every notified loss of data will entitle the affected customer to legal recourse against the company. This is particularly true for the US, which does not recognize a general tort of privacy invasion. If a data breach results, at worst, in embarrassment, at best in more targeted online advertising, the only action that receiving a data breach notification can trigger will be to take personal mitigation strategies - if this is appropriate, possible and within the knowledge and means of the customer affected. If by contrast the concern is identity theft, then the trigger can be much higher – companies will have to inform their customers only if there is a reasonable chance that they will become victims of criminal activity and as a result suffer economic harm (Winn 2009 1134):

> The nature of any causal connection between security breaches and concrete harms suffered by consumers such as identity theft remains unclear. Because American consumers are not

[2] Cal. Civ. Code §§ 1798.29, .82, .84.

[3] ibid, Sec 4 1798 .82 a.

> protected by a general right of information privacy, mere notice that a security breach has occurred is not associated with any right to compensation. Attempts to establish a right to damages following receipt of a security breach notice through class action lawsuits have generally only succeeded in clarifying the degree to which no such right exists, although many businesses suffering breaches have chosen on a voluntary basis to provide their customers with credit monitoring services to reduce the risk of harm from identity theft.

In the European context to which we will turn below, the situation is fundamentally different. Here, in particular the upcoming General Data Protection Regulation will not only reaffirm the general right to information privacy, but also provide an enforcement mechanism.

More generally, we can distinguish eight dimensions along which different DBNLs may differ, not just in their US incarnations, but universally. Building on a similar list suggested by Stevens (2005 p 5) DBNLs can differ:

(1) in identifying who must comply with the law (e.g. private, public or both; sector specific or non-specific)
(2) in how they define "personal information"
(3) in how they define "breach of security";
(4) in prescribing what harm to the customer must have occurred or is likely to occur, if any, for notice to be triggered;
(5) in setting the required degree of certainty that a company must have that a breach occurred, and, where applicable, that harm is likely to result from the breach – e.g. a reasonable suspicion or direct knowledge.
(6) in who must be notified (e.g. customers only, customers and a state agency)
(7) in how the notification has to be delivered, and with what minimum content.
(8) if there are exemptions and safe harbors, e.g. non-interference with a police investigation or use of encryption
(7) and in the penalties, enforcement authorities, and remedies that non-compliance brings with it

Orthogonal to these dimensions that determine scope and content of DBNLs, we can distinguish different objectives, or benefits, that these laws may bring to the customers (Goel and Hanay 2014). DBNLs could:

A) assist existing customers of the breached entity who were harmed as a consequence of a data breach to get compensated for their loss. With higher and higher costs of e-discovery, this will assist especially customers with smaller losses in jurisdictions where the sides carry their own costs regardless of outcome of the litigation.
B) assist existing customers in mitigating their risk as a result of a data breach, e.g. by alerting them to change their credit card details.
C) assist existing and potential customers in choosing companies that have a good track record in their security policies.
D) motivate companies that face the costs associated with A-C, to invest more into cybersecurity, to prevent breaches in the first place (see e.g. Romanosky Sasha and Acquisti 2009 p. 1067).
E) assist law enforcement agencies, governments and regulator bodies in getting a more comprehensive picture of the cybersecurity landscape, allowing them to assign recourses or in identifying need for legislative action.
F) assist other companies working in the same field, in getting a more comprehensive picture of the cybersecurity landscape, allowing them to develop more efficient cybersecurity strategies for themselves.

Depending on which of these six aims a DBNL law tries to achieve, different choices in the seven parameters will result. If for instance the main aim is to enable customers to protect themselves, the duty to inform about a breach will often require a format that is deemed to be particularly helpful for the customer, and may include a requirement to suggest appropriate action or offer assistance.[4] Similarly, if harm mitigation rather than compensation is the main goal, triggering conditions will be comparatively low, the sooner the customers knows that socially embarrassing information may become public knowledge, the better they can protect themselves. If compensation is the main objective, triggering conditions can be higher and may require evidence of actual harm. If it is not the individual customer who is at the centre of the law, but the main aim is to incentivise companies to invest in cyber security for everybody's benefit, safe harbours may be created for companies that meet certain security thresholds, and exempt them from DBNL requirement.[5] If the main beneficiaries of DBNLs are law enforcement and other governmental agencies, the law may create a dual notification system with different triggering conditions: A very sensitive trigger to report even only suspected breaches to a designated agency, which can then be combined with a higher threshold for informing also individual customers, e.g. positive knowledge that harm is likely to occur.[6] If the main aim is to improve cyberresiliance in the industry through sharing of information about breaches and the methods that were employed, voluntary information sharing systems and other self-regulatory approaches may be the preferred option. (so a proposal by Nowey and Federrath 2007).

Europe's attitude to DBNLs is a typical example of an "society centric" approach to mandatory breach disclosure. The EU included a breach notification duty in the Directive on Privacy and Electronic Communications in 2009, to be implemented in national legislation by 2011. This initial attempt was specific to personal data held by telecommunication service providers. This Directive has since then been clarified, amended and harmonized by the European Commission Regulation (EU) N° 611/201 from 2013. The Regulation also ensures that both traditional telecom providers such as telephony companies and Internet Service Providers (ISPs) are covered.

[4] So e.g. the US Health Information Technology for Economic and Clinical Health Act (HITECH Act) which in additions to information about the facts of the breach (what data, when and, if known, by whom) also mandates information about the steps individuals should take in response to the breach; steps that are carried out to investigate the breach, and steps individual may want to take to mitigate, and protect against further harm.

[5] In the US, entities that are regulated under the Gramm-Leach-Bliley Act or the Health Insurance Portability and Accountability Act (HIPAA) and meet their more stringent and prescriptive regulations and guidelines are frequently exempted from DBNLs. See Stevens (2005) p. 6. Use of strong encryption can also create a safe harbour. (Burdon et al. 2010a, b).

[6] So in the Interagency Guidance on Response Programs for Unauthorized Access to Customer Information and Customer Notice, 70 Fed. Reg. 15,736 (Mar. 29, 2005), a data breach notification system for the financial sector.

Directive 2009/136/EC created a two-tier framework for the reporting of data breaches. It imposed duties to inform both regulatory authorities and personal data breaches to affected individuals, where a data breach is defined as a defined as:

> [...] a breach of security leading to the accidental or unlawful destruction, loss, alteration, unauthorised disclosure of, or access to, personal data transmitted, stored or otherwise processed in connection with the provision of a publicly available electronic communications service in the Community.

Unlike its state-level counterparts in the US, the notification requirements of the Directive were initially limited to providers of telecommunication services such as ISPs. In marked deviation from the normal pattern, the EU opted here for a sectorial approach, whereas the US state laws have general application to all types of organizations, supplemented by a number of sector-specific, subsidiary legislation. However, for the EU too this soon became superseded in the evolving regulatory landscape. In the same year that saw the passing of the Regulation clarifying and harmonizing the Data Breach notification duties under the e-Privacy directive, the Commission published its proposal for a Directive on Network and Information Security, which contains breach notification duties for certain "market operators", including health providers, financial services and also transportation providers. However, under Chap. 4 Art 14 of the draft directive, this would be a duty to inform the competent national authorities only. These then may "inform the public, or require the public administrations and market operators to do so, where it determines that disclosure of the incident is in the public interest".

We will focus mainly on notification duties under the Data Protection Directive and the Regulation that followed it. Article 3 of the Regulation describes the two-tier approach. Service providers have to report *all* personal data breaches to a competent national authority without undue delay. Only if the data breach is likely to harm the privacy of the data subject will the provider be required to notify also the person affected. Unlike the US laws, this requirement does not only affect subscribers or customers of the service provider, also third parties can benefit from this duty. Notifying affected non-customers can be particularly difficult. As the regulation notes:

> While providers should have contact details of their subscribers, given their direct contractual relationship, such information may not exist for other individuals adversely affected by the personal data breach. In such a case, the provider should be permitted to notify those individuals initially through advertisements in major national or regional media, such as newspapers, to be followed as soon as possible by an individual notification as provided for in this Regulation (Commission Regulation (EU) No 611/2013 at (14))

We note here for the later discussion that the duty of the service provider is not just an expression of a contractual bond between the provider and the affected data subject. Rather, the duty is (also) owed to the wider public.

If in consultation with the national authority, the service provider concludes that the nature of the breach or the data obtained will not adversely affect the data subject, no notification is required. This prevents the often voiced concerns that excessive information could result in a "boy who cried wolf" scenario where customers learn to ignore notifications (so e.g. Schwartz and Janger 2007 p. 955). The term "Adversely affected" is not strictly defined in the Directive. However, Recital 61 lists a number of indicative situations, ranging from identity theft or fraud to physical harm, significant humiliation or damage to reputation. The recital makes it clear that unlike the US approach, the concern of the EU regulation is not solely identity theft and economic loss, but wider social harms, including reputation damage or embarrassment. That means in particular that a person can be adversely affected by a data breach even if the use of the very same data, had it been legally obtained, would have been lawful.

Article 4(3) also contains a safe harbour provision for providers who can demonstrate to the national regulator that technological measures have taken the breached data out of use for unauthorized users.[7] In this EU law follows the US standard. This also means that criticism and concerns regarding the over-reliance on encryption (see e.g. Burdon et al. 2010a) also affects the EU approach.

Notification to both regulators and individuals has to occur "without undue delay", with the Regulation specifying that notification to the regulators should normally be within 24h after a breach was detected. Speed is deemed more important than completeness, so that providers who have become aware of a breach, but have not established all the details that they are required to report, are mandated to amend any initial breach notification as and when new information becomes available.[8] Importantly though, this provision also entails an active duty by the provider to establish "the whole truth" of the breach. They are not just passive reporters of information that they have obtained. This decreases the temptation to be "intentionally ignorant", but as we will discuss in more detail below, also creates a quite significant active duty for the entity whose security was breached. We can already here introduce the comparison to traditional crimes that will concern us in the next section. In most jurisdictions, victims of a burglary do not have a positive duty to report that their home was ransacked. Even where such a duty exists, it will not entail a requirement to carry out independent investigations to identify what was stolen, the method of entry, or an analysis what, if any, crimes the offender will be able to commit with items stolen from the house. This difference in approach between breaking into a physical dwelling and hacking into a digital one is, so we argue, manifestation of a seismic shift in the criminal law that merits theoretical reflection.

The Directive and its clarifying Regulation left considerable leeway to national authorities to specify the contend and format of the notification to customers, but it should at least include "the nature of the breach, relevant contact points to gain further information and the recommendation of further measures to mitigate adverse effects of the breach". Here more than anywhere else in the legislation do we see the "harm mitigation rational" at work. Recital 15 of the Regulation however specifies that the information should be "about the breach only", and not hidden e.g. in an invoice, advertising material or other extraneous information. The whole truth about the breach is required, but also nothing but the truth.

---

[7] Art. 4(3), Directive 2009/136/EC.

[8] Commission Regulation (EU) No 611/2013 Art 2 (2).

Finally, the Directive adds a number of reporting obligations on providers, who will have to maintain an inventory of data breaches. This inventory has to note all the relevant facts that the provider was able to establish, the effects of the breach and any remedial action that was taken. Regulators will use the information gained from these registers to evaluate the efficiency of existing legislation and guidelines, and to develop best practice. We noted above that the European approach extends the scope of notification duties beyond customers to third parties whose data was held. Here we see this idea extended further – the beneficiaries of this aspect of the notification duty is not any individual who suffered harm, but through improved efficiency of the legal system, everybody.

The forthcoming General Data Protection Regulation finally extends the notification duty to all data controllers. It follows otherwise mainly the approach outlined above, but with two important differences. First, the regulatory environment that it creates includes a much improved enforcement mechanism for data protection violations. This also means that companies that report a breach may face in addition to any possible action by the individuals affected also substantive fines by the regulators. We recall that in the US, there is no general tort of privacy violation, though individuals affected by a data breach can sue if they can show that they suffered economic harm through the negligence of the breached entity. The availability of class actions in the US legal system gives this potential bite. In the European context under the Regulation, while class actions are not normally found in European jurisdictions, fines can be levied by the data protection authorities without a need to show concrete loss for individuals.Second, the proposed Amendment to the draft Regulation phrases the required content of the notification in an interesting way (our emphasis) in its Art 32:

> The communication to the data subject referred to in paragraph 1 shall be *comprehensive* and use clear and plain language. It shall describe the nature of the personal data breach and contain at least the information and the recommendations provided for in points (b), (c) and (d) of Article 31(3) and information about the rights of the data subject, including redress.

Following on from the duty to carry out active investigative action discussed above, we find here another expression of the duty to tell "the whole truth".

5.3 Towards a Jurisprudence of Data Breach Notification Duties

In our discussion so far, we developed an analytical framework to describe data breach notification duties. Despite differences in detail, caused by different national cultures, regulatory environments and identified aims of these laws, we could distil a typical expression of the obligation. Entities that control personal data on behalf of their customers must

- inform, in case of a breach of
- sensitive personal data

- a relevant authority AND the affected customers
- immediatly
- but without jeopardising police investigation
- comprehensively and completely
- yet in a useful way and with advice on how to minimise harm.

We indicated above that this type of regulation is not best understood as just another compliance duty. Rather, it is potentially indicative of a substantial change in the way in which we think about the relation between (criminal) law and citizenry in democratic societies. In particular, it impacts on two basic intuitions about the role certain actors play in the justice system, the degree in which citizens can be obligated to play an active role in maintaining the criminal justice system, and the type of excuses they have from performing these duties. This in turn allows us to ask questions about the adequacy of existing jurisprudential theories that try to capture the systemic nature of law, and the degree of differentiation of our analytic vocabulary that is needed to give an account that can legitimately claim to capture not just what law is, but "the whole of the law".

At a first approximation, DBND laws fall into a group of regulatory tools that Cass Sunstein called "regulation through disclosure"(Sunstein 1999). Regulation through disclosure is bidirectional. It gives incentives to the party who owes the duty to minimise the number of triggering events, and it enables in return the affected third parties to make prudent choices. As we saw, DBNLs perform both tasks, making them therefore also part of the "choice architecture" (Leonard, Thaler and Sunstein 2008). Schwarz and Janger (2007 p 915) thus compare them to the duty of hospitals to publicize performance results, labelling of household appliances with energy-efficiency ratings and the duty of factories to disclose information about toxic releases of one's factory.

The last of these examples is of particular interest to us. It is driven by a concern for the biosphere, either because of its instrumental value for humans, or even, in deep ecology, due to its intrinsic value (so e.g. Naess 1973). In either case, a collective responsibility arises to "be concerned about" and "look after" the environment, something that can best be done if information about problems is spread widely. A duty to report oil spills e.g. will give an incentive to companies to invest in security equipment, since reporting a spill harms at the very least the reputation of the company, at worst exposes it to litigation. At the same time, it allows agencies and citizens to take appropriate action to mitigate the consequences of the spill, be it by direct involvement in the clean up, or by reinforcing the deterrent aspect of notification duties through a boycott call or similar use of market tools. Big data, so we are told frequently, is the new oil (for examples see Hirsch 2013). But if the history of oil exploration tells us anything, then it is that there will be oil spills. Just as spill notification duties protect the natural environment, the biosphere, so do DBNLs protect the information environment – or as Floridi terms it, the infosphere (Floridi 2002). Just like Naess conception of the biosphere, in Floridi's conception the infosphere has intrinsic value, which means in particular that we may have duties towards its constituent parts, the "inforgs". These may or may not (also) be humans.

Not only then do we have "information as environmental regulation" (Esty 2004), in Floridi's framework we can now draw the inverse connection as well, and think of DBNL's as "environmental protection regulation *of* information", regulation that forces us to create a "clean", i.e. truthful and transparent information environment.

Spill notification duties not only allow us to rethink more radically the connection between information and regulation. Unlike the energy efficiency warning or the hospital performance indices, they are directly linked to possible sanctions such as fines. Frequently, the polluters will have to inform (also) the very agency that may fine them for the breach. We have seen that in the US incarnation of DBNLs, the connection between legal sanctions and notification remains indirect and, in practice, weak. In the absence of a self-standing privacy tort, legal action will take typically the form of class actions by consumers against the company. However, the company only discloses *that* a breach has taken place, the onus then falls on the pursuer to show that recoverable harm, typically in the form of financial loss, incurred. The difficulty to prove actual harm means that despite an increase in class action litigation due to DBNDs, the number of *successful* class actions remains low (Schneider 2009). For the EU, especially under the forthcoming General Data Proetction Regulation, the situation is different. Here, the entity that (also) receives the DBN is also the agency that has considerable powers to fine companies who allowed a data breach to happen. DBNLs are therefore not just a form of nudging or "tolerant paternalism" (Floridi 2015), they incorporate also a more traditional, sanction based approach.

Because of this direct nexus between the pro-ethical, choice enabling and the punitive aspect of DBNDs, they challenge two deeply held intuitions about the duties that come with certain "roles" people can have regarding a crime: as accused, and as victims. First, and most obviously, a company that is subject to a DBNL plays the role of a accused. Its behaviour at least contributed to a harm that occurred to others, and they may well face legal sanctions as a consequence. The duty to notify about a breach makes the company susceptible to sanctions. Actively disclosing the breach therefore potentially impacts on its right against self-incrimination, the legally protected choice to remain silent.[9] Since the right to remain silent is an archetypical "communication right", it is maybe not unsurprising that it was traded in to achieve wider regulatory goals for Internet governance. Similar trade offs have been done between encryption, key escrow and key-disclosure laws. Here the right to communicate in a specific way, i.e. encrypted, is balanced by a duty to communicate against one's own interest. This can be by disclosing the key if requested to do so in response to a warrant, or proactively by depositing the key with a trusted third party (Desmedt et al 2001).

[9] This problem is explicitly acknowledged in the German implementation of the Directive. § 42a S. 6 BDSG ensures that the information cannot be used in criminal proceedings against the company that reported the breach. (However, German law does not recognize a fruit of the poisonous tree doctrine, so information that the prosecutors found themselves after investigating in response to the notification is probably not affected) . However, German law only awards the right against self incrimination to natural persons, so that legal persons such as companies, the typical data controller, will not be protected by this rule.

Limiting the right to silence however is not restricted to Internet regulation, rather, it this has been a general trend, affecting not just disclosure duties, but criminal procedure in general. For the UK, this has been documented by Zander (1995); in the US, 18 U.S.C. § 3501 aimed to overturn through federal law the Miranda warning and its right to silence, but ultimately was struck down by the Constitutional Court in *Dickerson v. United States*, 530 U.S 428 (2000).

Seeing DBNLs also as an expression of a general trend to limit the right to remain silent opens up the avenue to interpret this development through the conceptual lenses of a philosophy of information. In a world where the default is one of "communication always on", withdrawing from this continuous exchange of information is arguably a more meaningful act than silence was in a society where communication channels were scarce.

If the infosphere is, as Floridi argues, the totality of Being, then the act of withholding information, just as its erasure or destruction, takes on a new significance. Floridi discusses silence but briefly in his 2005 paper on the relation between semantic information and meaningful data (Floridi 2005). He discusses examples where silence carries primary information (silence as tacit approval, e.g.) and situations where it isn't (silence because the person has not heard the question), and also analyses situations where the listener does not know which of the two prevails (Floridi 2005 p. 353). Legal contexts add a new layer of complexity; here we find a normative prohibition to draw certain inferences from silence, even though such an inference would be epistemologically sound (Kurzon 1995). It is this conflict between the epistemologically sound and the deontologically permissible that led Bentham to argue against the very possibility of a right to remain silent (Bentham 1827). On the other hand, we have good reasons to believe that such a right, properly framed, protects on balance the innocent more than the guilty (so Seidman and Stein 2000; critical Bibas 2002). Attempting a jurisprudential analysis of the right not to communicate, including a right to remain silent within a criminal trial, would go beyond the scope of this paper. We note here only that the apparently benevolent pro-ethical and disclosure-centric approach to regulation puts pressure on this concept, at a time when the wider social benefits and costs of a right to withdraw from communication are not yet fully understood.

DBNLs carve an exception into the right against self-incrimination, but as we saw, in this they are but part of a general trend of regulation through disclosure. In this sense, they are indeed very similar to duties to report oil spills or other failures by companies to ensure a safe and clean environment. In another aspect though, they go significantly beyond this type of duty. DBNLs "could" be understood as an expression of the duty to mitigate harm that an actor's negligent actions cause. If seen like this, the role of the entity whose data was breached is that of a more or less active colluder in the harm that befell third parties (typically, the customers), and a result of falling short in their duty of care towards them. Their legal role is based on a specific causal nexus to the crime that has occurred. This is different from a general duty of a witness of criminal activity to report that crime to the police. Western legal systems do not normally have a statutory duty that compels mere witnesses of a planned or completed crime to come forward and assist actively in crime prevention and investigation. Where such duties exist, they are typically reserved for most serious crimes, *and* restricted to situations where the crime can still be prevented. Art 138 of the German Criminal Code is a typical example: It makes it an offence not to inform the authorities about knowledge of a planned war of aggression, high treason, murder, forgery, violent robbery or kidnapping, but only when it is still possible to prevent the criminals from completing their activity. Here, we have a clear harm prevention rationale that creates a duty towards "one's neighbours", similarly to the duty of care in delict law.

Somewhat different in nature is the disclosure duty under Sec 19 of the UK Anti-Terrorism Act (2001), which makes it an offence not to disclose suspicions about the involvement of third parties in terrorist activities, provided they gained this information as part of their "trade, profession, business or employment."[10] Here, specific social roles that link a person to a crime trigger the disclosure duty. We note that this is a very recent addition to UK criminal law, which previously did not include an active duty to come forward with information about criminal activity – another indicator for our theory that "regulation through information duties" is accelerating as a result of the informational environment leaking into or taking over the physical environment.

Finally, we have of course certain information duties of witnesses *as part of* a criminal trial. They differ from the duties described above in that they take place in an environment designed to elicit information about a singular past event, and are in response to the exercise of explicit powers vested into the judiciary.

Unlike the duties of the witness in court, DBN duties apply pro-actively, before a formal investigative process has even started. Unlike the information duties of the German criminal code, they apply even to breaches that may turn out to be of minor concern, and more importantly, typically apply *after* the breach has happened and the unauthorised access to the personal information is completed. The crime prevention rationale that informs the German law is therefore only of limited applicability. DBNLs are closest to the information duties in UK anti-terrorism law: The connection to the crime is created through a specific occupation or professional role (here the similarity to the DBNL under the Directive, restricted to telecommunication providers, is particularly obvious). However, DBNLs are unique in that they impose this duty not only on neutral third parties whose connection to the crime is more or less accidental, they make demands on the *victim* of the crime in question. We might not feel particularly sympathetic to the Ashley Madisons of this world, or ISPs that invest insufficiently in cybersecurity and so expose their customers to risks. Nonetheless, they are the direct victims of a criminal activity, a victim that now faces additional criminalisation if it fails to assist, pro-actively, in the investigation of the crime against them. This even though there can be from the perspective of the company good prudential reasons to abstain from such active support of the police.

Imposing duties on the victim of a crime, potentially against his/her own interests and under the threat of criminal law sanctions, makes DBNLs unusual. They challenge our understanding of what it entails to be a victim of a crime, and the

[10] Terrorism Act (2000) Sec 19 (b).

protective duties and obligations the stakeholders in the criminal justice systems have towards victims. At the same time, as we have seen, they are nonetheless part of a more general trend to impose communication and disclosure duties. A jurisprudential analysis should be able to account therefore for both, the general shift towards imposing disclosure and communication duties on citizens and companies as active collaborators in the criminal justice system, and also for the specific question of the role of victims in the criminal law. When do victims have a moral duty to enable the prosecution of the crimes against them, and under which conditions is it appropriate, for a democratic republic under the rule of law, to turn this moral or civic aspiration into a legal duty, with criminal law sanctions in case of non-compliance? To answer these questions, we need a general theory of the relation between citizens and the criminal law, a theory that asks what actions are the appropriate subject of criminal sanctions, and what duties and obligations can the criminal law rightfully impose on the citizenry. With other words, the question needs to be answered within a general theory of (criminal) law, a theory that outlines the totality of rules and obligations that the criminal law creates.

Grand legal theory in the tradition of Austin, Kelsen or Hart struggles to provide us with an appropriate analytical framework for this task. Their gaze is primarily on the "officials", the legislator as the source of law, the judiciary as its interpreter, and possibly other officials such as police for its enforcement. The citizen is in these systems a largely passive recipient of commands, his/her duties merely to obey the sovereign, or, in Hart's version, obey those officials that they recognise as legitimate. Recast in the language of information, the information flow in these systems is (largely) uni-directional, from the top of the pyramid to its base. The Austinian sovereign directs his demands to the citizens, who obey or face sanctions, but do not otherwise play an active role or communicate with the sovereign. The Kelsenian Grundnorm "informs" or "authorises" the subordinate norms, which in turn inform more and more specific norms right up to the exercise of sovereign power by officials in individual cases, but again, there is no flow of information from the base to the Grundnorm. Hart's model is more differentiated, with its tripartite distinction between sovereign, officials, and citizens. Here, we find a limited degree of reciprocity in the Rule of Recognition. The recipient of information, the legal officials, have to reason "upstream" about the sender of the information, the sovereign, and interpret the orders as having legitimate authority. But even in this model, the citizens, qua citizens, are not playing an active role in maintaining the justice system. The criminal law is something external to them, a body of rules laid down by others for them to obey (Duff 2010a, b p. 300). In a democracy the officials will, of course, also be citizens who are bound by the law, and some citizens will be officials. But the way in which the distinction between officials and citizens, or between official roles and the role of ordinary citizen, is typically drawn gives the ordinary citizen *qua* citizen an essentially passive role in relation to the criminal law. True, in democracies, citizens are also sovereign, and can be seen as the ultimate authors of laws. But as Wintgens (2006) has shown, jurisprudence as general theory of law has never been able to close this communicative loop, for systematic reasons deeply embedded in the history of western thought. Legisprudence, the involvement of the citizen as legislator, is an inevitable blind spot in all those traditions of legal thought that ultimately gave us Kelsen, Hart and Austin.

The shortcomings of this conceptual approach have been recognised amongst others by Duff and Marshall in a series of papers on the role of the criminal law, the criminal trial and criminal punishment in democratic societies (see Duff 2010a, b; Duff et al 2015). The traditional model, so they argue, cannot account for all those characteristics of the criminal trial that do not fall into the neat divide between officials and citizens. A broad range of phenomena, from the duty of citizens to participate, in some countries, as jurors in the fact-finding process, to the specific duties of witnesses as "truth speakers", or the role of criminals in their own rehabilitation, thus remains unaccounted for. It therefore gives at best a partial account of what it means to be a "system of criminal law". Duff and Marshall propose an alternative approach centred around distinct roles, connections that people can have relative to a crime, its investigation and ultimately, its punishment. In particular, they suggest that we can distinguish the following roles a citizen may play in the practical process that is a crime investigation and prosecution (Duff 2015 p3–4):

1) official and professional roles, where the citizen "in uniform" is employed by the polity and in this role given special powers, responsibilities and privileges. These include
   - police officers
   - judges
   - prosecutors
   - civil servants involved in legislative drafting
2) official roles filled by lay participants. Here too the polity endows the citizen with special rights and duties, but they serve not as employed professionals. These include amongst others
   - jurors (the paradigmatic example)
   - lay judges, justices of the peace etc
   - special constables
3) the final type of roles are those that a citizen acquires by his or her connection to a particular crime. These so Duff and Marshal suggest include amongst others
   - victim
   - witness
   - offender
   - suspect/defendant
   - 'ex-offender'

For our discussion, we suggest to add another category: duties a citizen can have as a result of special skills, knowledge or position. A paradigmatic example can be seen in the current legal action between Apple and the FBI (see Kerr 2016). Here, the FBI is using the All Writs Act[11] that allows courts to compel citizens with special skills or knowledge to assist the police. This type of duty is neither fully in category 2 nor category 3, though it shares aspects with both. Justifications for assigning

[11] 28 U.S.C. § 1651.

special duties to people with these skills could e.g. emphasise the communal investment, in terms of education and training, that they enjoyed.

All of these roles carry civic or aspirational duties – things we should be doing to be a good citizen in "our role as X". In many cases, these aspirational or ethical duties will also have a correspondent, but typically more limited legal duty. A model citizen, in their role as witness, will testify the whole truth however painful. In law, we may however allow them to be silent to avoid self-incrimination, or acknowledge possible conflicts with other social roles such as that of a spouse, and give them exemptions or privileges.

This framework then allows us to ask if our vision of the criminal law is compatible with recognising any of these roles in law – *should* we use e.g. lay people as police officers? It also allows us to ask precisely under what conditions we *should* or at least legitimately *can* turn a civic duty into a legal obligation. And finally, it allows us to ask how we can resolve conflicts between these roles while staying true to our vision of criminal law in a democratic society.

As we saw above, being able to talk about specific roles also enabled us to place DBNLs into a first conceptual framework. DBNLs call upon online service providers *in their role as* professional businesses with special expertise, *in their role as* witnesses of a crime, and *in their role as* a victim of a crime. In attaching specific legal duties to these roles, we also saw how they are asked to take on functions that have previously been reserved to certain officials, in particular the investigative duties they incur to be able to inform their customers about the "whole truth" of the breach. Here we see the problematic aspect of their involvement: As experts, they are tasked to investigate, as witnesses, to testify truthfully, while as victims and as enablers, they may be entitled to considerations that limits these duties.

Duff and Marshal integrate their role-based scheme within a wider communitarian theory of criminal law in a democratic society, and also a specific conception of the trial as "public holding to account" (Duff 2001). While the role-based conceptual schema can be separated from these further theoretical commitments, their emphasis on the trial as a communicative, multi-partner performance opens up the possibility to cross-fertilise their analysis with theories of communication and information ethics. For our question, the role and duties of the victim are of particular interest. As noted above, we do not normally in law require active participation of victims of crime in the prosecution of their aggressors. There are however a number of more or less isolated exceptions to this rule, especially when a trial is already under way. One highly controversial example is prosecutions against victims of domestic violence who withdraw their allegations during the court proceedings. Marshall shows how we can make sense of this within a communication centric, communitarian model of the criminal law. Victims are not just witnesses, they are a *specific type of* witness, one that is irreplaceable even when there are other sources to ensure a conviction. In the witness, the harm done to the community is personified, his or her testimony therefore necessary to communicate the wrongness of the action to judge, public and indeed also the perpetrator (Marshall 2004). This latter aspect may seem as the most counter-intuitive, but flows naturally from the conception of the trial as a communicative process. It can fulfil its true function only if all roles meet their communication duties as participants, and that includes the duty *to* the accused to explain to them the nature of their actions *in their wrongness* – setting the trial as a normative practice apart from say a historical investigation. Marshall (2015 p. 299) writes:

> A trial thus calls a defendant to answer not just to an alleged individual victim, though it does indeed do that, but also to the whole polity for the wrong that he allegedly committed; and it constitutes, in part, an expression, articulation, and application of what are purported to be the shared, 'public' values of the polity.

The communication acts that are performed as part of a trial are therefore identity-creating and constitutive. Again Marshall (2015 p. 298):

> To understand the nature of law we have to understand its role as partly constitutive of a political community and therefore as an object for identification, as playing an important role in a people's sense of who they are. [...] Citizens are responsible both *to* the state (insofar as they are responsible to one another in their roles as citizens) and *for* the state.

We can see here one possible way to connect Duff and Marshall's account, which takes as examples legal issues that predate the Internet, to the information ethics of Floridi. The advantage of Floridi's conception of the "Infosphere" is that it extends beyond cyberspace, as a general theory of information, and can thus accommodate pre-Internet institutions, processes and practices. Its sensitivity to the way ITCs "re-ontologise" the world in turn matches the constitutive aspects of information practices during the trial that Marshal emphasizes. If the Infosphere, then, as totality of all informational entities, their properties, interactions, processes and mutual relations is the whole of Being (Floridi 2010 p. 8–10), then we can say that the "whole of the law" becomes the totality of information duties that exist between informational agents.

This cross-fertilization of the communicative model of the trial by Marshall and Duff with the philosophy of information by Floridi is, so we argue, necessary to fully account for and do justice to DBNLs. For all the traction that Duff and Marshall give us in understanding them through the lenses of their role-based analysis of the criminal law, DBNLs differ in important aspects from the paradigmatic cases that they discuss. First, the duty that DBNLs establish precedes the trial – unlike the case of a victim who falsely withdraws her accusation during a formal trial, here the duty is to actively come forward and enable the trial in the first place. Second, as administrative offences, DBNLs seem far away from the "common law of crime" that Duff and Marshall aim to explain, and indeed, Duff has expressed concerns about criminalizing in this way "ancilliary" rule violations, that is violations of rules that only exist to enable the enforcement of "primary" criminal norms. The Terrorism Act discussed above and its duty to inform the police about suspicious behavior is explicitly cited by him (Duff 2010a, b p 294). Finally, Duff and Marshall's account is intimately linked to a notion of citizenship, and ultimately rooted in an anthropological account of human flourishing. For our purpose, this poses two challenges: First, we saw in the historical part how DBNLs became successively more inclusive, protecting first only customers in the state of the breached entity, then all customers, worldwide, and finally, in the EU Directive, all affected parties worldwide, whether customers or not.

Secondly, the duty is owed, typically, by corporations and therefore juridical persons – it is at least subject to discussion in how far Duff and Marshall's approach covers these as "citizens". Within the more general information ethics framework, these issues disappear – inforgs as the main holder of information rights and duties are an all-encompassing concept that includes artificial and natural persons. Secondly, since the infosphere transcends national boundaries, it is at best loosely tied ot a concept of citizenship. DBNLs then really become the new "spill disclosure laws", rules that contribute to the "cleanness" of the informational environment. Just as we owe it to each other to protect our natural environment, so do we owe it to each other to protect the informational environment. As victims, we are not suffering a loss individually, rather, we become mere stand-ins for what is a loss to all of us, and hence, a duty to assist in mitigating the harm that was caused.

5.4 Conclusion

At the centre of our analysis were data breach notification laws as a new duty for online service providers. At first sight, these were mere compliance duties, bereft of theoretical interest. On closer inspection, they turned out to be deeply problematic, challenging sound intuitions about basic notions of procedural fairness and ignoring the special attention that victims of crime should be given in the criminal justice system. Finally, we saw how a communication-centric, role based approach to the criminal law allowed us to make sense of these legal duties, while an information ethical perspective moved them from the periphery of the legal system to an ideal typical and constitutive form of regulation of the Internet age. To achieve this, we had to ask how legal theory needs to respond to the information revolution. Seen through the prism of information theory, traditional accounts of the legal system in the tradition of Hart, Kelsen or Austin, give us a hierarchical, mono-directional information flow from sovereign to officials, and from officials to citizens.

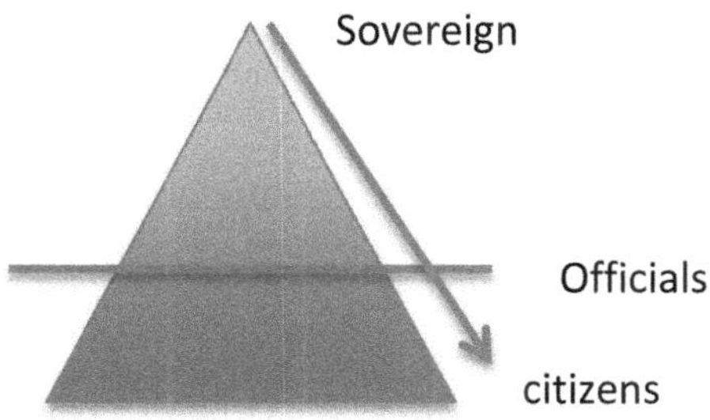

Instead, what we need to account for law in the Information age is a multi-agent system, that may contain locally hierarchical structures, but is globally a network of multi-directional communication rights and duties. Victims owe subtly different communication duties to themselves, to the legal officials, to the jurors and indeed to the perpetrator to give a truthful and complete account of what happened. Similarly, judges have communication duties, in particular the duty to give reasons, that they owe towards victim, accused/convict and the wider public. These can vary in intensity, required communication mode or explicitness between these.

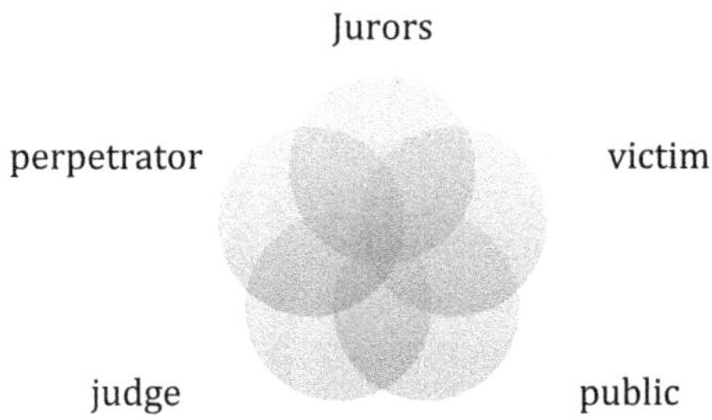

The totality of these duties constitutes "the whole of " the Information law, as a normative extension of the infosphere, which ultimately turns reflexively on itself: creating and curating the infosphere as its "environmental protection law".

Acknowledgement This work was supported by the Arts and Humanities Research Council [grant number AH/M009610/1]

Bibliography

Bentham, J. (1827). *Rationale of judicial evidence 1–8*. New York: Rothman & Co.

Bibas, S. (2002). The right to remain silent helps only the guilty. *Iowa Law Review, 88*, 421–432.

Burdon, M., Low, R. and Reid, J. F.. (2010a). If it's encrypted it's secure! The viability of US state-based encryption exemptions. In: *Proceedings of the 2010 IEEE International Symposium on technology and society: Social implications of emerging technologies*. IEEE http://eprints.qut.edu.au/32781/1/c32781.pdf. Accessed May 2016

Burdon, M., Reid, J., & Low, R. (2010b). Encryption safe harbours and data breach notification laws. *Computer Law & Security Review, 26*, 520–534.

Desmedt, Y., Burmester, M., & Seberry, J. (2001). Equitability in retroactive data confiscation versus proactive key Escrow. In K. Kim (Ed.), *Public key cryptography* (pp. 277–286). Springer: Berlin.

Draper, A. (2006). Identity theft: Plugging the massive data leaks with a stricter nationwide breach-notification law. *Journal Marshall & Law Review, 40*, 681–703.

Duff, A. (2001). *Punishment, communication, and community*. New York: Oxford University Press.

Duff, A. (2010a). A criminal law for citizens. *Theoretical Criminology, 14*(3), 293–309.

Duff, A. (2010b). *The boundaries of the criminal law*. Oxford: Oxford University Press.

Duff, A. (2015). Legal reasoning, good citizens, and the criminal law. *Minnesota legal studies research paper* 15–18. Available at SSRN: http://ssrn.com/abstract=2618684 or http://dx.doi.org/10.2139/ssrn.2618684

Duff, A., Farmer, L., Marshall, S. E., Renzo, M., & Tadros, V. (2015). *Criminalization: the political morality of the criminal law*. Oxford: Oxford University Press.

Esty, D. C. (2004). Environmental protection in the information age. *NYUL Review, 79*, 115–212.

Floridi, L. (1999). Information ethics: On the philosophical foundation of computer ethics. *Ethics and Information Technology, 1*, 33–52.
Floridi, L. (2002). On the intrinsic value of information objects and the infosphere. *Ethics and Information Technology, 4*, 287–304.
Floridi, L. (2005). Is semantic information meaningful data? *Philosophy and Phenomenological Research, 70*(2), 351–370.
Floridi, L. (Ed.). (2010). *The Cambridge handbook of information and computer ethics*. Cambridge: Cambridge University Press.
Floridi, L. (2015). Tolerant paternalism: Pro-ethical design as a resolution of the Dilemma of toleration. *Science and Engineering Ethics, 21*, 1–20.
Goel, S., & Shawky, H. A. (2014). The impact of federal and state notification laws on security breach announcements. *Communications of the Association for Information Systems, 34*, 37–50.
Hirsch, D. D. (2013). The glass house effect: Big data, the new oil, and the power of analogy. *Maine Law Review, 66*, 373–396.
Kerr, O. (2016). Preliminary thoughts on the Apple iPhone order in the San Bernardino case: Part 2, the All Writs Act https://www.washingtonpost.com/news/volokh-conspiracy/wp/2016/02/19/preliminary-thoughts-on-the-apple-iphone-order-in-the-san-bernardino-case-part-2-the-all-writs-act/. Accessed May 2016.
Kurzon, D. (1995). The right of silence: A socio-pragmatic model of interpretation. *Journal of Pragmatics, 23*, 55–69.
Lee, S. (2006). Breach notification laws: Notification requirements and data safeguarding now apply to everyone, including entrepreneurs. *Entrepreneurial Business Law Journal, 1*, 125–153.
Leonard, T. C., Thaler, R. H., & Sunstein, C. R. (2008). Nudge: Improving decisions about health, wealth, and happiness. *Constitutional Political Economy, 19*, 356–360.
Marshall, S. (2004). Victims of crime: Their station and its duties. *Critical Review of International Social and Political Philosophy, 7*, 104–117.
Marshall, S. (2015). It isn't just about you' victims of crime, their associated. Duties, and public wrongs. In A. Duff, et al. (Eds.), *Criminalization: The political morality of the criminal law*. Oxford: Oxford University Press.
Naess, A. (1973). The shallow and the deep, long-range ecology movement. A summary. *Inquiry, 16*, 95–100.
Needles, S. A. (2009). The data game: Learning to love the state-based approach to data breach notification law. *NCL Review, 88*, 267–310.
Nowey, T., & Federrath, H. (2007). Collection of quantitative data on security incidents. In*The second international conference on availability, reliability and security ARES* (pp. 325–334). Vienna: IEEE.
Otto, P. N., Antón, A. I., & Baumer, D. L. (2007). The choicepoint dilemma: How data brokers should handle the privacy of personal information. *IEEE Security and Privacy, 5*, 15–23.
Sasha, R., & Acquisti, A. (2009). Privacy costs and personal data protection: Economic and legal perspectives. *Berkeley Technology Law Journal, 24*, 1061–1101.
Schneider, J. W. (2009). Preventing data breaches: Alternative approaches to Deter Negligent handling of consumer data. *Boston University Journal of Science & Technology Law, 15*, 279–304.
Schwartz, P. M., & Janger, E. J. (2007). Notification of data security breaches. *Michigan Law Review, 105*, 913–984.
Segall, L. (2015, September 8). Pastor outed on Ashley Madison commits suicide. *CNNMoney*.
Seidmann, D. J., & Stein, A. (2000). The right to silence helps the innocent: A game-theoretic analysis of the Fifth Amendment privilege. *Harvard Law Review, 114*, 430–510.
Simitian, J. (2009). UCB security breach notification symposium March 6, 2009: How a bill becomes a law, really. *Berkeley Technology Law Journal, 24*, 1009–1018.

Skinner, T. H. (2003). California's database breach notification security act: The first state breach notification law is not yet a suitable template for national identity theft legislation. *Richmond Journal Law & Technology, 10*, 1–40.
Stevens, G. M. (2005). Data security breach notification laws. *CRS Report for Congress* R42475. https://www.hsdl.org/?view&did=706636. Accessed May 2016
Sunstein, C. R. (1999). Informational regulation and informational standing: Akins and beyond. *University of Pennsylvania Law Review, 147*, 613–675.
Towle, H. K. (2003). Identity theft: Myths, methods, and new law. *Rutgers Computer & Technology Law Journal, 30*, 237–326.
Winn, J. K. (2009). Are "Better" security breach notification laws possible? 2–3. *Berkley Technology Law Journal, 24*, 1133–1165.
Wintgens, L. J. (2006). Legisprudence as a new theory of legislation. *Ratio Juris, 19*, 1–25.
Zander, M. (1995). You have no right to remain silent: Abolition of the privilege against self-incrimination in England. *Louis ULJ, 40*, 659–676.

Chapter 6
Did the Romans Get It Right? What Delfi, Google, eBay, and UPC TeleKabel Wien Have in Common

Peggy Valcke, Aleksandra Kuczerawy, and Pieter-Jan Ombelet

Abstract This chapter provides an analysis of the recent landmark cases of the CJEU and ECtHR on the role and liability of Online Service Providers (OSPs) in policing content on the Internet. The authors assess legal arguments in Google France, eBay, UPC TeleKabel Wien and Delfi to discover how the rulings accommodate the right to freedom of expression and the safe harbours in the EU E-Commerce Directive. They examine how knowledge and awareness of illegal content impact the OSPs responsibilities and duties (of care) to restrict content. Finally, they seek guidance in Roman law and the concepts of 'bonus pater familias' and 'bonus ac diligens scriptor' to fully understand the position and the responsibilities of the OSPs.

6.1 Introduction

Both the EU Court of Justice in Luxembourg (CJEU) and the European Court of Human Rights (ECHR) in Strasbourg have in recent years rendered a number of controversial and much-criticised judgments concerning the role and liability of Online Service Providers (OSPs). Both Courts have been accused of putting unreasonable burdens on OSPs and of sacrificing communication and information freedoms on the altar of intellectual property and personality rights. Some claim that the earlier attempts of national courts to enlist online intermediaries to police content

P. Valcke (✉) • A. Kuczerawy • P.-J. Ombelet
University of Leuven, Leuven, Belgium
e-mail: peggy.valcke@kuleuven.be

M. Taddeo, L. Floridi (eds.), *The Responsibilities of Online Service Providers*,
Law, Governance and Technology Series 31, DOI 10.1007/978-3-319-47852-4_6

on the Internet – leading to the adoption of 'safe harbours' for intermediaries in the E-Commerce Directive in 2000 – are now copied by Europe's highest courts – thereby undermining (at least the spirit of) the E-Commerce Directive. This contribution will offer a snapshot of the legal principles set out in recent landmark cases of the CJEU and ECHR, including the Google France, eBay, UPC TeleKabel Wien and Delfi cases. The legal grounds and arguments put forward by the respective courts in order to hold OSPs liable (or not) will be examined and critically assessed. In particular, it will be studied why, or why not, the Court accepted the OSP's defense based on the liability exemptions in the E-Commerce Directive. Even though the cases dealt with different legal conflicts – involving copyright, privacy, defamation and hate speech – the lecture will contend that there is remarkable consistency in their outcome; a consistency that may go back to a basic principle that even the Romans already understood.

6.2 Safe Harbours in the E-Commerce Directive

Fifteen years ago, the European Union (EU) legislator adopted the E-Commerce Directive with a view of fostering the development of information society services in the internal market. In order to remove obstacles resulting from divergences in Member States' legislation and from legal uncertainty, the Directive introduced harmonized rules for certain legal aspects of electronic commerce. A key section concerns the liability of intermediary service providers for third party content/infringements (Section 4; Articles 12–15) which was introduced in response to existing and emerging disparities in Member States' legislation and case-law concerning liability of service providers acting as intermediaries. According to recital 40 of the Directive such discrepancies prevented the smooth functioning of the internal market, in particular by impairing the development of cross-border services and producing distortions of competition. Legislators across Europe were also concerned about disproportionate burdens on intermediaries (and hence, higher costs) in case courts would hold OSPs – often the most visible and easiest targets for third parties that feel harmed by Internet users' acts – too easily liable for their users' content or actions (Van Eecke, 2011). Other concerns related to possible chilling effects on freedom of expression and risks to privacy rights in case online intermediaries would be forced to take up the role of Internet police and would start monitoring and censoring their users' information to escape liability claims (Kuczerawy and Ausloos 2015).

Section 4 introduces a special protection regime for online intermediaries that covers three types of activities (mere conduit, caching and hosting) and includes two types of protection (against liability and monitoring obligations) (Van Eecke, 2011). Contrary to the 1998 US Digital Millennium Copyright Act (DMCA) the E-Commerce Directive shields intermediaries from different kinds of (civil and criminal) liability for all types of illegal content or activities (infringements on copyright, defamation, content harmful to minors, unfair commercial practices, etc.), and not only copyright violations. The provided immunity applies under certain conditions specified for each type of service. Pursuant to Article 14, for instance, hosting providers will not be held liable for the information they store for third parties on condition that "(a) the provider does not have actual knowledge of illegal activity or information and, as regards claims for damages, is not aware of facts or circumstances from which the illegal activity or information is apparent; or (b) the provider, upon obtaining such knowledge or awareness, acts expeditiously to remove or to disable access to the information."

The safe harbour provisions in Articles 12–14 do not affect the possibility for courts or administrative authorities in the Member States of requiring the provider to terminate or prevent a *specific* infringement. However, Article 15 of the E-Commerce Directive protects online intermediaries against injunctions that would require them to monitor or actively search their systems for illegal content *in general*. In the SABAM v. Scarlet (C-70/10) and SABAM v. Netlog (C-360/10) cases, the EU Court of Justice made it clear that national courts cannot order Internet service providers to implement general filtering systems to prevent the unlawful use of musical and audiovisual works by their users. A 'general filtering system' was understood by the Court as a system which would apply indiscriminately to all the provider's customers, as a preventive measure, exclusively at its expense, and for an unlimited period.

In other words, the E-Commerce Directive reflects the view that it cannot be considered fair to impose on OSPs a general monitoring obligation, on the one hand, and that it is appropriate to limit their liability for third-party content transmitted via their systems, on the other hand. The same idea can also be found in the Council of Europe's Declaration on freedom of communication on the Internet of 2003 (see, more specifically, principle 6).

6.3 Case Law of the European Courts

The intention of the EU legislature was to legally treat online intermediaries as unengaged middlemen who are not responsible for the actual content they transmit or host, similar to the traditional role of postal services and telephone operators. However, court decisions at Member State level following the adoption of the E-Commerce Directive have often deviated from that position. Legal doctrine has identified significant discrepancies in the national case law, not only between Member States, but also within the same Member State (see e.g. Van Eecke, 2011 and Valcke, Lenaerts and Kuczerawy, 2014). Van Eecke states three reasons for such divergences: interpretation issues relating to Section 4 of the E-Commerce Directive (and its corresponding national provisions), the broad wording of EU intellectual property legislation, and the increasing pressure from policy makers and judicial authorities on online intermediaries to take on a policing role towards their users (Van Eecke, 2011).

Although the CJEU was initially – after its *Google France* and *L'Oréal v. eBay* judgments – seen as offering counterweight to national trends towards involving online service providers in speech regulation, it is now criticized for taking a more prudent stance when it comes to shielding OSPs. The same criticism has been voiced with regard to the ECHR, in particular in the context of the recent Delfi ruling, which was the first case in which the Court had been called upon to examine a complaint about liability for user-generated comments on an Internet news portal (Husovec 2016).

6.3.1 Delfi

6.3.1.1 Facts

In 2006, Delfi's online news portal published an article regarding a change in ferry routes delaying the opening of an ice road. The news story elicited several comments, some of which were abusive and defamatory. The Estonian domestic courts ruled that Delfi acted as a publisher and therefore was liable for the comments since it should have deleted them on its own initiative. Delfi was ordered to pay the injured person the equivalent of 320€ in compensation for nonpecuniary damage. On 10 October 2013, the First Chamber ruled that there was no violation of Article 10 of the European Convention on Human Rights (hereafter: Convention). Delfi asked for a referral to the Grand Chamber, claiming wrongful qualification as a publisher and arguing that the Estonian Supreme Court's judgment had had a "chilling effect" on freedom of expression. The judgement of the Grand Chamber confirmed, however, that the imposition of publisher liability on Delfi did not constitute a violation of Article 10 of the Convention.

6.3.1.2 Reasoning of the European Court of Human Rights

The Estonian domestic courts had classified Delfi as a traditional publisher and not as an intermediary. On the basis of this classification, the Estonian domestic courts applied the Civil Code and the Obligations Act, rather than the E-Commerce Directive and the (Estonian) Information Society Services Act. The classification resulted in denying Delfi the protection of the safe harbour foreseen in Article 14 of the Ecommerce Directive.

Interestingly, the Estonian Supreme Court recognized a difference between traditional publishers and Internet media, stating that "it cannot reasonably be required of a portal operator to edit comments before publishing them in the same manner as applies for a printed media publication". The Grand Chamber applauded the distinction made by the Estonian Court and pointed out that the approach is consistent with the international instruments in this field. Specifically, the Grand Chamber referred to the Council of Europe Recommendation on a new notion of media

(Recommendation CM/Rec(2011)7). The Recommendation promotes a 'differentiated and graduated approach' which "requires that each actor whose services are identified as media or as an intermediary or auxiliary activity benefit from both the appropriate form (differentiated) and the appropriate level (graduated) of protection. Similarly, for the purpose of Art. 10 ECHR the "duties and responsibilities" of different media actors may be different, with regard to third-party content (paragraph 113). After making the observation, however, the Court accepted the finding of the Estonian Supreme Court that "[b]ecause of [their] economic interest in the publication of comments, both a publisher of printed media and an Internet portal operator are publishers/disclosers as entrepreneurs" (paragraph 112). Despite its support for a nuanced approach, the Court declared that qualification of Delfi as a publisher was foreseeable (paragraphs 120–129) and proceeded by referring to its own case law on traditional media.

The Grand Chamber agreed with the definition of Delfi as a publisher, but deemed it crucial to delineate the scope of the judgement in the light of the facts of the case (paragraph 111). The Grand Chamber considered the case to only concern the duties and responsibilities of Internet news portals when they provide on a commercial basis a platform for user-generated comments on news articles and when some users on this platform engage in clearly unlawful speech, which infringes the personality rights of others and amounts to hate speech and incitement to violence against them (paragraph 115). The Grand Chamber highlighted that the impugned comments in the present case were not disputed as being hate speech and speech advocating acts of violence and therefore on their face manifestly unlawful. The hateful comments did not enjoy the protection of Article 10 of the ECHR and the freedom of expression of its authors was not at issue (paragraphs 117–118).

The Grand Chamber also took into account the nature of the Delfi news portal. The portal was named in an open letter from the board of an Estonian newspaper to the Minister of Justice, the Chief Public Prosecutor and the Chancellor of Justice as a widely known "source of brutal and arrogant mockery" (paragraph 15). Moreover, Delfi actively called for comments which, after publication, could only be modified or deleted by Delfi but not by the actual authors of the comments (paragraph 145).

In the circumstances at present, the Grand Chamber considered that the rights and interests of others and of society as a whole may entitle Contracting States to impose liability on Internet news portals if they fail to take measures to remove clearly unlawful comments without delay, even without notice from the alleged victim or from third parties (paragraph 159).

6.3.1.3 Fierce Criticism

In the dissenting opinion the judges Sajó and Tsotsoria expressed their scepticism toward the judgement. For example, the judges were highly critical of the foreseeability analysis (paragraph 20 of the dissenting opinion). The judges pointed out that the relevant provisions of the Civil Code (General Principles) Act and the Law of Obligations Act were "quite general and lack[ed] detail" and that "[o]nly divine legal counsel could have been sufficiently certain that a portal operator would be liable" under given circumstances (paragraph 20). Even the nature of the impugned comments was not obvious to the dissenting judges. They argued that "[t]hroughout the the whole judgment the description or characterisation of the comments varies and remains non-specific" (paragraph 12). According to Sajó and Tsotsoria the extent to which the comments amount to a real threat should have been analysed properly (paragraph 14).

The dissenting judges also pointed out the risk of collateral censorship. As a result of the judgement, all comments will have to be monitored from the moment of publication. The judges emphasised that comments sections on the Internet should remain a crucial part of the exchange of ideas among citizens (Ombelet and Kuczerawy, 2015). The judges labelled the delineation of the Grand Chamber as a mere "damage control" and feared it would create an incentive for active intermediaries and blog operators to discontinue offering a comments feature (paragraphs 8–9).

Similar doubts have been expressed by Voorhoof (2015), who questions the pertinence of the delineation performed by the Grand Chamber. Voorhoof argues that the delineation would limit the protection for freedom of expression to social media, personal blogs and websites ran as a hobby. Quoting the dissenting judges, Voorhoof agrees that "freedom of expression cannot be a matter of a hobby". He further criticises the overbroad (pre-) monitoring of all user generated content. Obliging online platforms to filter or monitor users' comments in order to prevent any possible liability for illegal content would constitute a new paradigm for participatory online media. According to Husovec (2016), "Delfi can now serve as a trump card to overcome any threat to freedom of expression".

6.3.2 Delfi Through the Lens of CJEU Case Law

While it is not the role of the ECHR to interpret EU law, the question may be raised whether Delfi would have been decided differently by the CJEU, had it been called upon to decide the case in light of the E-Commerce Directive. Specifically, the case is an excellent opportunity to assess whether the provider of hosting services must be neutral in order to benefit from the liability exemption in the E-Commerce Directive. In the final judgement, the Grand Chamber describes the relevant provisions in the E-Commerce Directive and the case law of the CJEU in the introductory part of its judgment ("The Facts"), but does not come back to it in its actual judgment ("The Law"). Instead, it spends considerable amount of time explaining why the domestic courts could reasonably classify Delfi as a 'publisher'. Apart from the economic interest factor (paragraph 128), the Grand Chamber points out that Delfi "exercised a substantial degree of control over the comments published on its portal" (paragraph 145). Said control was exercised by prohibiting comments with infringing or illegal content through the platform rules, and Delfi's possibility to restrict the commenting option for those who violated the rules. Moreover, Delfi had the sole power to edit or remove the posted comments and installed an automated filtering mechanism, albeit not a particularly effective one (Kuczerawy and Ombelet, 2015). Delfi also relied on a notice-and-take down system and deleted comments upon notice. In the case at hand, Delfi deleted the litigious comments less than 24 h after receiving a notification about their illegal character. The notification, however, was sent only 6 weeks after the comments had been posted. Although measures taken by Delfi were fairly standard, the domestic courts considered them insufficient to prevent causing harm to third parties. They were sufficient, however, to qualify Delfi's activities as being not "merely technical, automatic and passive in nature" (paragraph 146).

6.3.2.1 Active vs. Passive: Neutrality Requirement (Google Adwords)

The requirement that an intermediary's activities are of a mere technical, automatic, and passive nature is based on recital (42) of the E-Commerce Directive. These properties of the service imply that the intermediary has neither knowledge of nor control over the information it transmits or stores. The wording of the recital, however, is problematic.

While it purports to address all of the exemptions of the Directive, some argue that the scope of this recital should be limited to the transmission and access services identified in Articles 12 (mere conduit) and 13 (caching). As is further clarified in recital (43), not being involved in any way with the transmitted information is actually a condition for liability exemption for mere conduit and caching services. The exemption for hosting in Article 14 of the Directive is not limited in scope to either transmission or access services. According to Van Eecke (2011), Article 14 in fact does not require a passive role of the hosting provider in order for the protection regime to apply. A hosting provider can still be protected even if it is not completely passive – as long as it does not have knowledge or control over the data which is being stored. This approach is referred to as 'storage but no knowledge' test. Following this line of reasoning, active intermediaries such as Delfi could still benefit from the safe harbour offered by the E-Commerce Directive, provided that they do not have knowledge or control over the data which is being stored.

The restrictive interpretation of recital (42) is not commonly agreed on. In *Google Adwords*, the CJEU held that recital (42) also applies to hosting services (paragraph 113–114). The CJEU held that, in order to establish whether the liability of a referencing service provider may be limited under Article 14 of Directive 2000/31, it is necessary to examine whether the role played by that service provider is **neutral**, in the sense that its conduct is merely technical, automatic and passive. If this is the case, the referencing service provider cannot be held liable for the data stored at the request of an advertiser and any trademark infringements resulting thereof, unless, after having obtained knowledge of the unlawful nature of those data or of that advertiser's activities, it fails to act expeditiously to remove or to disable access to the data concerned. The CJEU clarified that the mere fact that a referencing service is subject to payment, that the provider sets the payment terms or that it provides general information to its clients cannot have the effect of depriving that provider of the exemptions from liability provided for in the E-Commerce Directive. In the same vein, the Court pointed out that concordance between the keyword selected and the search term entered by an Internet user is not sufficient of itself to justify the view that Google has knowledge of, or control over, the data entered into its system by advertisers and stored in memory on its server. By contrast, if the provider takes up a more active role in the drafting of the commercial message which accompanies the advertising link or in the establishment or selection of keywords, this may trigger liability. The CJEU left it to the national court to assess the actual role played by Google.

6.3.2.2 Actual vs. Constructive Knowledge: Diligent Economic Operator Criterion (eBay)

In *L'Oréal v. eBay*, however, the CJEU seemingly lowered the threshold to benefit from the liability exemption by rephrasing the "**neutrality**" requirement with "**lack of knowledge**". Contrary to what Advocate General Jääskinen had suggested, the Court did not come back on its previous position and reject the neutrality requirement for hosting providers, but converted it to the absence of knowledge about the data. The Court clarified that this entailed two aspects: lack of *general* knowledge and lack of *specific* knowledge.

The first aspect entails that the service provider should not obtain knowledge of what is stored on its systems by virtue of the *general* way in which its service is designed or operated. Similar to *Google Adwords*, the Court held that Article 14 of the Directive applies to hosting providers if they do not play an active role that would allow them to have knowledge or control of the stored data (paragraph 112–116). The main factor is how the service is designed or operated. The fact that the operator of a website sets the terms of its service, is remunerated for that service and provides general information to its customers cannot have the effect of denying it the exemptions from liability provided for by Directive 2000/31 (paragraph 115). These types of activities would not lead, in the CJEU's opinion, to the knowledge or control of the stored information. Such effect could be achieved, however, if the service provider assisted customers in optimising the presentation of certain information, or promoted certain information (paragraph 116).

The second aspect relates to knowledge in *specific* cases. Following Article 14 (1)(a) and (b) of the E-Commerce Directive, the CJEU refuses to exempt a service provider from liability – even in situations where it has confined itself to a merely technical and automatic processing of data – if it has actual knowledge of the illegal activity or information, or, as regards claims for damages, if it is aware of facts or circumstances from which the illegality is apparent, and does not act expeditiously to remove, or disable access to the illegal information in question. 'Apparent' illegality occurs, according to the CJEU in *L'Oréal v. eBay*, when "a **diligent economic operator** should have identified the illegality in question" (paragraph 120). Such 'constructive knowledge' covers every situation in which the provider concerned becomes aware, in one way or another, of such facts or circumstances (paragraph 121). In particular, it covers both the situation in which the operator of an online marketplace uncovers an illegal activity or illegal information as the result of an investigation undertaken on its own initiative, and the situation in which the operator is notified by a third party. Such notification represents, as a general rule, a factor indicating 'awareness', although it could turn out to be insufficiently precise or inadequately substantiated (paragraph 122). Requirements for a valid notification, as well as interpretations of actual and constructive knowledge, differ across the EU countries (Kuczerawy 2015).

The diligent economic operator standard implies a duty of reasonable care, although it remains unclear whether it is only triggered by human knowledge (e.g. if a staff member is notified), or whether computer knowledge would suffice (e.g. if a software filter would find illegal information). According to Van Eecke (2011), the wording used by the CJEU in *L'Oréal v. eBay* (e.g., "the operator of an online marketplace uncovers…an illegal activity", and "the operator should have realized that the offers…were unlawful") seems to indicate the first scenario.

Delfi deleted the defamatory comments immediately upon obtaining actual knowledge of them. The question is whether Delfi's attempts to keep its portal free of harmful comments could be interpreted as leading to constructive knowledge. The measures undertaken by Delfi included filtering mechanisms and the occasional deletion of comments on its own initiative. Despite the voluntary monitoring undertaken by Delfi, the comments remained on the platform for 6 weeks. The Grand Chamber noted that the establishment of the unlawful nature of the comments "did not require any linguistic or legal analysis since the remarks were on their face manifestly unlawful" (paragraph 117). The standard articulated by the Grand Chamber is actually not much different from the CJEU standard of "any diligent economic operator". As a result, the same conclusion with regard to Delfi's liability might be reached under the E-Commerce Directive (Kuczerawy and Ombelet, 2015).

6.3.2.3 Reasonable Measures (UPC Telekabel Wien)

The ruling in Delfi challenged the concept of general monitoring obligations. Article 15 of the E-Commerce Directive prohibits Member States from imposing on providers of mere conduit, caching or hosting services a general obligation to monitor information they transmit or store. The same provision prohibits introducing a general obligation to actively seek for facts or circumstances indicating illegal activity.

Considering the goal of the Directive the prohibition makes sense. An obligation to conduct general monitoring of content would counteract the limited liability paradigm because intermediaries who actively seek for illegal activities would no longer be neutral and passive in nature (OECD 2011).

The prohibition towards monitoring obligations concerns solely monitoring of a general nature. Article 15 does not prohibit monitoring obligations in a specific cases, nor does it affect orders by national authorities in line with national legislation (Recital (47)). The Directive additionally allows Member States to require hosting providers to apply duties of care, which can reasonably be expected from them (Recital (48)). The duties of care should only be introduced to detect and prevent certain types of illegal activities, foreseen by national law. The Directive does not clarify what exactly such duties of care entail. As a result, the boundary between the duties of care and general monitoring obligation is not clear. Some commentators consider recital (48) as contradictory to article 15 (Barceló, Koelman 2000). Similar confusion resurfaced as a result of the Delfi ruling. Is the obligation to prevent the harm by deleting the illegal content after the publication, on its own initiative, as put on Delfi, still merely a duty of care that should be reasonably expected of such actor or does it constitute a general obligation to monitor all the content?

In the previously mentioned Scarlet v. SABAM / SABAM v. Netlog cases (which involved copyright infringements), the CJEU stressed that ISPs cannot be required to install a filtering system preventing the illegal downloading of files which applies indiscriminately to all its customers. However, in its ruling of March 2014 in the *UPC Telekabel Wien v. Constantin Film* case (again dealing with copyright infringements), the CJEU confirmed that "outcome prohibitions", i.e. injunctions ordering ISPs to block users' access to a website infringing copyright without specifying how to achieve that result, are compatible with EU law, including fundamental rights. Such an injunction does not seem to infringe the very substance of the freedom to conduct a business, since, *firstly*, the ISP can choose to put in place measures which are best adapted to the resources and abilities available to him; and *secondly*, the injunction allows its addressee to avoid liability by proving that he has taken *"all reasonable measures"* (paragraph 52–53). According to the Court, this possibility of exoneration has the effect that the addressee of the injunction will not be required to make "unbearable sacrifices". It considers this position justified in the light of the fact that the OSP is not the author of the infringement of the "fundamental right of intellectual property" which had led to the adoption of the injunction concerned.

The Court acknowledges that the principle of legal certainty requires that the OSP knows in advance (i.e. before any decision imposing a penalty on him is adopted) that the measures taken were indeed those which could be expected of him in order to prevent the proscribed result. It therefore elaborates further on what may be considered "reasonable measures" (paragraph 55–63). On the one hand, the measures should be effective; i.e. they should have the effect of preventing unauthorised access to protected content or, at least, of making it difficult to achieve and of seriously discouraging Internet users accessing the unlawful content (although the Court does not expect 100% effectiveness; it explicitly recognizes that the measures may not be capable of putting a complete end to the infringements of the intellectual property right). On the other hand, the measures should be "strictly targeted", in order to ensure compliance with the fundamental right of Internet users to freedom of information; i.e. they should not unnecessarily deprive Internet users of the possibility of lawfully accessing the information available and must allow for Internet users to assert their rights before court.

6.3.2.4 The Special Case of News Publishers (Papasavvas)

Can online newspapers – like Delfi – benefit from the liability limitations in Articles 12–14 of the E-Commerce Directive? This was the question the CJEU had to answer in the *Papasavvas* case, in which a Cypriot citizen had brought an action for damages against a newspaper company and its editor-in-chief claiming that two of their articles, published online, constituted defamation (Case C-291/13).

In its judgment of 11 September 2014 the CJEU considered that the concept of 'information society services' in the E-Commerce Directive covers the provision of online information services for which the service provider is remunerated, not by the recipient, but by income generated by advertisements posted on a website. However, the Court also found that a newspaper publishing company which posted an online version of a newspaper on its website had, in principle, knowledge about the information which it posted and exercised control over that information. Therefore, it could not be considered to be an 'intermediary service provider' within the meaning of Articles 12–14 of the E-Commerce Directive, whether or not access to that website was free of charge.

In other words, in *Papasavvas* the CJEU made it clear that the limitations of liability specified in Articles 12–14 do *not apply* to the case of a newspaper publishing company which operates a website on which the online version of its newspaper is posted, since it has knowledge of the information posted and exercises control over it. The case, however, concerned articles edited and published by the newspaper itself, and not users' comments, like in Delfi. Can an online newspaper still be considered "to have knowledge of the information posted and exercise control over it" when the harm does not result from the newspaper's own articles, but from the comments posted by its readers? Even though the CJEU did not answer that question explicitly, its previous case law on intermediary liability – despite the wide variety of actors involved – points us implicitly into the direction of a legal standard that we have been familiar with for a very long time.

6.4 The Role of Roman Law and Ethics

6.4.1 From the "Bonus Pater Familias"...

Indeed, the terms of "diligent economic operator" (cf. eBay) and "reasonable measures" (cf. UPC Telekabel Wien) in the CJEU case law on intermediary liability reflect the concept of the "bonus pater familias", a legal concept the roots of which go back to ancient Roman law. Both in continental law and in common law, the concept of the "bon père de famille", respectively the "reasonable man", is used as a standard of care in different areas (mainly tort law, but also contract law and criminal negligence).

In the area of journalistic liability, the standard of care can be illuminated by the ethical codes developed by press associations, journalism councils and individual news media. In Belgium, for instance, the courts have developed the concept of "bonus ac diligens scriptor" to evaluate a journalist's civil liability in defamation cases. In doing so, they have explicitly relied on journalist's codes of ethics to provide guidance on how a reasonably competent and diligent journalist should act (Voorhoof 2012). Several journalism councils, as well as individual news media, have in recent years developed ethical guidelines regarding user-generated content. Those ethical guidelines merit our attention when assessing an online news portal's duty of care with regard to users' comments.

6.4.2 …to Press Ethics…

The Code of Practice of the Flemish Council for Journalism contains specific guidelines concerning the handling of user content by the press (Flemish Council for Journalism, 2010). Although these guidelines recognize that, on discussion forums, opinion contributions are firstly the responsibility of the author of the contribution concerned, the medium publishing the contributions is also professionally and ethically responsible for proper moderation of the forum. The guidelines leave some flexibility to the publisher by outlining different scenarios which can qualify as proper moderation: either checking incoming reactions for admissibility prior to inclusion on the forum (i.e. pre-monitoring); or reading submissions prior to publication and publishing only selectively (i.e. active moderation); or (using post-monitoring) utilising the necessary methods for removal of inappropriate material as quickly as possible. They also list several methods that can be implemented for the timely removal of inappropriate content: (1) prior registration of users; (2) clear reference on the site to the terms and conditions of use; (3) use of an electronic filter triggered by certain terms; (4) the option to report inappropriate reactions to a forum moderator; (5) moderation prior to publication and continuous discussion guidance when concerning sensitive topics. Finally, they highlight that anonymous contributions are only to be published as an exception and that the editorial staff must still be in possession of the contributor's personal information.

Similarly, the Finish Council for Mass Media annexed their Guidelines for Journalists in 2011 with provisions related to material generated by the public on a media website (Finnish Council for Mass Media, 2011). The annex declares that (1) the editorial office shall monitor their websites and try to prevent the publication of content that violates privacy and human dignity, (2) the editorial office shall promptly delete content that comes to its attention that violates privacy and human dignity, (3) the public must be given the opportunity to inform editorial offices of inappropriate content in such a way that the informant receives due confirmation and (4) a clear demarcation must be kept on media websites between forums reserved for the public and editorial content.

The Press Ethical Rules of the Danish Press Council on their part state that their rules also cover edited discussion contributions (Danish Press Council). If unedited discussion items are brought, visible and clear guidelines on such items should be published and an effective procedure for handling complaints of such items should be set up by the relevant medium.

As a fourth example, the BBC has adopted editorial guidelines for user generated content on their online platforms (BBC). The guidelines state in section 17 that every online space where user generated content is published must have a moderator who can remove illegal and inappropriate content. There must also be a named individual in the relevant division to take editorial responsibility for the content, ensuring that the space maintains appropriate overall standards of moderation and hosting. The BBC further highlights in its Editorial Policy on User Generated Content that their journalists should aim to apply the same approach to pictures, audio and video supplied by members of the public, as they do to any other material they handle as journalists. The journalists should not automatically assume that the material is accurate and should take reasonable steps where necessary to seek verification. As digital manipulation tools become more accessible, journalists also need to be on their guard against photo manipulation and hoaxing. The journalists should aim to achieve accuracy by (1) the accurate gathering of material using first hand sources wherever possible, (2) checking and cross checking the facts, (3) validating the authenticity of documentary evidence and digital material and (4) corroborating claims and allegations made by contributors wherever possible.

6.4.3 ...and Back to the Ice-Roads

Admittedly, the previous section only highlights a couple of – randomly chosen – ethical codes, but it nevertheless indicates that Delfi's approach towards user-generated content does not seem in line with "best practices" in the press sector. It, for instance, did not register contributor's details, and its automatic word-based filter had failed to filter out odious hate speech. In the CJEU's terms, one could say that Delfi did not act like a "diligent economic operator".

Even if Delfi did not have actual knowledge of the illegal information until it received notice, one could argue that Delfi acted in such way that a trier of fact might impute constructive knowledge. Considering the type of comments in the case (hate speech) and the level of controversy of the topic Delfi could have been expected to take reasonable measures to prevent the harm caused by the publication. So despite the criticism voiced in that sense, the Delfi ruling does not appear to be incompatible with Article 14 of the E-Commerce Directive when applying old concepts like the "bonus pater familias" and press ethics as interpretation criteria.

6.5 Concluding Remarks: A Duty of Care for the Courts

The Delfi ruling should be read with great caution, though, and should not be used to widen news portal's liability for third-party comments beyond what is strictly necessary in a democratic society to protect the rights of others. As we have outlined above, the ruling applies only in the very specific circumstances that were characteristic for the Delfi case, i.e. a professional news portal and hate speech, and it is of crucial importance that this is well kept in mind by courts that will rule on comparable conflicts in the future. The danger that results from unduly stretching the Delfi ruling beyond these limitations (and, according to some, even in the ruling itself) lies in "collateral damage for online freedom of expression" (Voorhoof, 2015).

OSPs fearing possible liability might be overzealous in removing any questionable content. Or, alternatively, the decision could result in website operators shying away from any voluntary monitoring in fear of possible repercussions. Exercising too much control by the means of monitoring content could compromise the neutral status of the OSP and, consequently, deprive them of the safe harbour protection (Kuczerawy, Ausloos 2015). As a result, OSPs might find themselves in a Catch-22 situation. The E-Commerce Directive does not contain a 'Good Samaritan' clause, such as the one offered by the US Communications Decency Act of 1996 (CDA), Section 230 (c)(2). The CDA shields providers or users of an interactive computer service from civil liability in case voluntary screening or blocking measures designed to restrict access to or availability of offensive content prove to be imperfect.[1] Moreover, the CDA ensures that no provider or user of an interactive computer service shall be treated as the publisher or speaker of any information provided by a third party (US CDA, Section 230 (c)(2)). A plea for an introduction of a similar level of protection in the EU has been voiced on several occasions (Stalla-Bourdillon 2015). Some argue, however, that such an absolute immunity of the OSPs would not be compatible with the Convention (Husovec 2016).

In any case, the ECHR seems to live up to its "duty of care" when it comes to confining the Delfi ruling within its limits. On 2 February 2016, it delivered its first post-Delfi judgment on offensive online user-generated content (Voorhoof and Lievens 2016). This time, it held that operators of Internet portals – more specifically of the self-regulatory body of Hungarian Internet content providers, *Magyar Tartalomszolgáltatók Egyesülete (MTE)*, and of the consumer protection column *Index.hu* – were not liable for offensive, though not "clearly unlawful", comments posted by their readers on their respective websites. Anonymous users of *MTE* and *Index.hu* had posted vulgar and offensive online comments following the publication of an opinion on their websites criticizing the misleading business practices of

[1] US CDA, Section 230 (c)(2) provides that: No provider or user of an interactive computer service shall be held liable on account of – (A) any action voluntarily taken in good faith to restrict access to or availability of material that the provider or user considers to be obscene, lewd, lascivious, filthy, excessively violent, harassing, or otherwise objectionable, whether or not such material is constitutionally protected; (B) any action taken to enable or make available to information content providers or others the technical means to restrict access to material described in paragraph (1).

two real estate websites. Using the criteria established in its Delfi ruling (Kuczerawy and Ombelet, 2016), the ECHR this time found that by holding *MTE* and *Index.hu* liable for the third-party comments, the Hungarian courts had violated the right to freedom of expression as guaranteed by Article 10 of the European Convention of Human Rights. Since the incriminated comments in this case did not amount to hate speech or incitement to violence, it is to be accepted that the operators took sufficient precautions and acted as responsible and diligent intermediaries by installing an effective notice-and-take-down system. In other words, online portals are not expected to pre-monitor all third-party comments. The Hungarian courts had held that, by allowing unfiltered comments, the applicants should have expected that some of those might be in breach of the law. For the ECHR, however, such objective duty of care would amount to *"requiring excessive and impracticable forethought capable of undermining freedom of the right to impart information on the Internet"* (paragraph 82). The *bonus pater familias* should not become the Internet's censor...

References

Case Law

CJEU Google France SARL v. Centre national de recherche en relations humaines (CNRRH) SARL (Joined cases C-236/08 to C-238/08) ECLI:EU:C:2010:159

CJEU L'Oréal SA v. Ebay International AG (C-324/09) ECLI:EU:C:2011:474

CJEU Scarlet Extended SA v. Société belge des auteurs, compositeurs et éditeurs SCRL (SABAM) (C-70/10) ECLI:EU:C:2011:771

CJEU Belgische Vereniging van Auteurs, Componisten en Uitgevers CVBA (SABAM) v. Netlog NV (C-360/10) ECLI:EU:C:2012:85

CJEU UPC Telekabel Wien GmbH v. Constantin Film Verleih GmbH (C-314/12) ECLI:EU:C:2014:192

CJEU Sotiris Papasavvas v. O Fileleftheros Dimosia Etairia Ltd (C-291/13) ECLI:EU:C:2014:2209

ECtHR (Grand Chamber) 16 June 2015, Delfi AS v. Estonia (no. 65469/09)

ECtHR 2 February 2016, Magyar Tartalomszolgaltatok Egyesülete and Index.hu ZRT v. Hungary (no. 22947/13)

Legislation

Directive 2000/31/EC of the European Parliament and of the Council of 8 June 2000 on certain legal aspects of information society services, in particular electronic commerce, in the Internal Market (Directive on electronic commerce). *O.J.* 2000, L 178/1

Ethical Codes

BBC. Editorial Guidelines – Section 17: Interacting with our audiences. http://downloads.bbc.co.uk/guidelines/editorialguidelines/pdfs/Section_17_Interacting.pdf

Danish Press Council. Press Ethical Rules. http://www.pressenaevnet.dk/Information-in-English/The-Press-Ethical-Rules.aspx

Finnish Council for Mass Media. 2011. Guidelines for Journalists – Annex: Material generated by the public on a media website. http://www.jsn.fi/en/guidelines_for_journalists/

Flemish Council for Journalism. 2010. Code of Practice. http://rvdj.be/sites/default/files/pdf/code_of_practice.pdf

Literature

Barceló, R.-J., & Koelman, K. (2000). Intermediary liability in the E-commerce directive: So far so good, but it's not enough. *Computer Law & Security Report, 4*, 231–239.

Husovec, M. (2016). General monitoring of third-party content: compatible with freedom of expression? *Journal of Intellectual Property Law & Practice, 11*(1), 17–20.

Kuczerawy, A. (2015). Intermediary liability & freedom of expression: Recent developments in the EU notice & action initiative. *Computer Law and Security Review, 31*(1), 46–56.

Kuczerawy, A., & Ausloos, J. (2015). *NoC online intermediaries case studies series: European Union and Google Spain*. http://dx.doi.org/10.2139/ssrn.2567183

Kuczerawy, A., & Ombelet, P.-J.. (2015). Not so different after all? Reconciling Delfi vs. Estonia with EU rules on intermediary liability. *LSE Media Policy Project Blog*. http://blogs.lse.ac.uk/mediapolicyproject/2015/07/01/not-so-different-after-all-reconciling-delfi-vs-estonia-with-eu-rules-on-intermediary-liability/

OECD. (2011). *Directorate for science, technology and industry, committee for information, computer and communication policy. The role of Internet intermediaries in advancing public policy objectives*. Part II: Forging partnerships for advancing public policy objectives for the Internet economy.

Ombelet, P.-J., & Kuczerawy, A. (2015). Delfi AS vs Estonia. *Reveal Blog*. http://revealproject.eu/delfi-as-vs-estonia/

Ombelet, P.-J., & Kuczerawy, A. (2016). Delfi revisited: the MTE & Index.hu v. Hungary case. *LSE Media Policy Project Blog*. http://blogs.lse.ac.uk/mediapolicyproject/2016/02/19/delfi-revisited-the-mte-index-hu-v-hungary-case/

Stalla-Bourdillon, S. (2015). On Delfi v Estonia… Is it time to adopt a good-Samaritan style exemption? *Peep beep!* https://peepbeep.wordpress.com/2015/06/19/on-delfi-v-estonia-is-it-time-to-adopt-a-good-samaritan-style-exemption/

Valcke, P., Lenaerts, M., & Kuczerawy, A. (2014). Chapt. 7: User generated content. In P. Lambert (Ed.), *Social networking: law, rights and policy* (pp. 83–99). Dublin: Clarus Press.

Van Eecke, P. (2011). Online service providers and liability: A plea for a balanced approach. *Common Market Law Review, 48*, 1455–1502.

Voorhoof, D. (2012). Rechtspraak integreert Code van de Raad voor de Journalistiek bij beoordeling van "fout" in toepassing van artikel 1382 BW – Noot onder Rb. Brugge 30 april 2012, Rb. Brussel 13 december 2011 en Rb. Brussel 27 maart 2012. *Auteurs & Media, 6*, 596–597.

Voorhoof, D. (2015). Delfi AS v. Estonia: Grand chamber confirms liability of online news portal for offensive comments posted by its readers. *Strasbourg Observers*. http://strasbourgobservers.com/2015/06/18/delfi-as-v-estonia-grand-chamber-confirms-liability-of-online-news-portal-for-offensive-comments-posted-by-its-readers/

Voorhoof, D., & Lievens, E. (2016). Offensive online comments – New ECtHR judgment. *Strasbourg Observers*. http://echrblog.blogspot.be/2016/02/offensive-online-comments-new-ecthr.html

Part II
Business Ethics & Corporate Social Responsibilities

Chapter 7
Responsibilities of OSPs from a Business Ethics Point of View

Christoph Luetge

Abstract Certain parts of the debate on the ethical responsibilities of OSPs strike one with their heavily moralising and accusing tone. However, from a business ethics point of view, any assessment of these responsibilities should keep the pivotal role of economic competition in mind. Its exact degree or intensity may vary, but in general, economic competition limits the options of actors, which has consequences for their ethical responsibilities as well. I argue, first, that the OSP discussion should rest less on purely individual concepts of responsibility, but more on concepts of group agency. Second, there are still ways of ascribing responsibility to companies compatible with competition, as much of the debates on Corporate Social Responsibility show.

7.1 Introduction

The responsibilities of online service providers can be viewed from a number of different ethical standpoints. In this chapter, I adopt a business ethics point of view, which I believe is often neglected in the discussion. It is however an important aspect of Online Service Providers' (OSP) responsibilities, since they are companies. In many cases, the companies in question are even quite large ones, such as Google, Facebook or Yahoo. While the ethical principles guiding companies – as well as other organisations – are not different in principle from those in other areas of ethics, the difference lies in the *implementation* of ethical norms. It is safe to expect ethical norms to be greatly more difficult to implement in large organisations, companies and institutions than in smaller entities, communities or groups.

In order to elaborate on this point, I will – after giving some general observations on the debate – proceed in two steps: I will first discuss a key *economic* aspect of norm implementation in companies and organisations, before turning, second, to a

C. Luetge (✉)
Peter Löscher Chair of Business Ethics, Technical University of Munich, Munich, Germany
e-mail: Luetge@tum.de

M. Taddeo, L. Floridi (eds.), *The Responsibilities of Online Service Providers*, Law, Governance and Technology Series 31, DOI 10.1007/978-3-319-47852-4_7

key *philosophical* aspect. The economic aspect deals with the role of competition in the process of norm implementation, while the philosophical one deals with the notion of responsibility in corporations as well as responsibility of collective entities in general. I will end with a conclusion and some possible practical implications.

7.2 The Use of Moralising in the OSP Debate

Certain parts of the debate on the ethical responsibilities of OSPs strike one with their heavily moralising and accusing tone. It is interesting to see how moralising concepts like "greed" or similar ones are used frequently in this debate. Here are some examples:

- "Google is a gigantic parasite that makes a fortune from exploiting the creativity and entrepreneurship of others" – Daily Mail, 24 Mar 2013 (Johnson 2013)
- "No Good Can Come Of Microsoft's Greed" – Daily Press, 11 Feb 2008 (Rau 2008)
- "Google's Greed: How the Search Giant's Lust For Information Is Impacting Your Search Results" – SEO Blog, 22 Mar 2015 (Bailyn 2015)
- "Googles Gier kennt keine Grenzen" – Berliner Zeitung, 06 May 2014 (Haufler 2014)
- "I wonder what Steve Jobs is thinking. Doesn't he see, like, he's the guy in that old Apple commercial who's the mean dictator" – Paul Graham on Bloomberg, 10 Mar 2011 (Schofield 2011)
- "Everybody wants to rule the world" – The Economist, 29 Nov 2014 (The Economist 2014)
- "Googles Gier opfert die Netzneutralität" – netzpolitik.org, 13 Aug 2010 (Pohl 2010)
- "Why Facebook Is Almost as Greedy as Bernie Madoff" – Wall Streeet Journal Video, 24 May 2012 (WSJ.com 2012)
- "Google est-il dangereux pour la démocratie? " – lesaffaires.com, 26 Feb 2013 (Brault 2013)

This type of moralising in the OSP responsibilities debate is not limited to certain countries, but widely used all over the world. It is frequent especially among activists dealing with Internet ethics, data security and related issues. It should be interesting to explore the theoretical or philosophical background of these opinions and ideas, which often are, as I suspect, quite simple anti-capitalist ones. For the purposes of this chapter, it suffices to keep in mind that when dealing with OSPs, we are dealing with mechanisms that belong both to the world of economics as well as philosophy. Considering OSP responsibilities in particular, it should be taken into account that there are crucial conditions of competition for those companies. I will turn to this aspect now.

7.3 Economic Aspect: The Concept and Role of Competition

Anti-capitalist critics[1] tend to believe that criticising capitalism is the same as criticising competition. Yet competition is not an exclusively economic concept. True, economists define it and try to develop conditions for good and bad competition. But we find forms of competition quite different from the ones usually discussed in economics, forms that might be more acceptable to critics of capitalism: in sports, in chivalry, in auctions, in other cultures like ancient China, and even in 'socialist competition'. I will elaborate on the last two ones in turn.

Competition: An Intercultural View According to Harvard historian Niall Ferguson (2010), competition is one of the killer applications that "the West" developed but other cultures (originally) lacked. (The others are science, the rule of law, modern medicine, consumer rule and the work ethic.) Globalisation, according to Ferguson, is a process of others appropriating these killer applications gradually.

However, from an intercultural perspective, competition has always been present in non-European cultures such as the Chinese one, albeit in different forms: While ancient China has traditionally been regarded as anti-competitive and stuck in traditions (Jones 1981), some facts contradict this:

For example, China was conducting highly competitive exams for civil servants. Starting in the seventh century, these exams required the knowledge of canonical Confucian texts. The candidates had to know central books of Confucianism, like the "Analects", the books of Mencius and five others like the "Book of Songs", the "Book of History" and so on. While in the beginning, some elements of these exams still favoured certain groups, especially the aristocracy, these were eventually abolished completely, leaving only a purely knowledge-based exam (Miyazaki 1981). This was a highly competitive system, which in Europe at that time could be compared only to some of the religious orders. It was however not a case of classical economic competition on 'ordinary' markets.

Competition: A Socialist View Capitalist critics who try to uphold the "communist hypothesis" (Badiou 2010) overlook that Lenin himself was not against competition: There is a 1917 speech in which he goes as far as saying that capitalism *destroys* competition and that socialism is in favour of competition by breaking up monopolies:

> [...] capitalism long ago replaced small, independent commodity production, under which competition could develop enterprise, energy and bold initiative to any considerable extent, by large- and very large-scale factory production, joint stock companies, syndicates and other monopolies. [...] competition is replaced by financial fraud, nepotism, servility on the upper rungs of the social ladder. (Lenin 1917)

[1] For example, Chomsky 1999; Chossudovsky 2003; Comte-Sponville 2004; Badiou 2010.

Socialism, according to Lenin, does *not* aim at doing away with competition:

> Far from extinguishing competition, socialism, on the contrary, for the first time creates the opportunity for employing it on a really *wide* and on a really *mass* scale, for actually drawing the majority of working people into a field of labour in which they can display their abilities, develop the capacities, and reveal those talents, so abundant among the people whom capitalism crushed, suppressed and strangled in thousands and millions. (Lenin 1917)

The task of a socialist government is to organise competition:

> Now that a socialist government is in power our task is to organise competition" – and the elimination of monopolies "is the opportunity created for the truly mass display of enterprise, *competition* and bold initiative (my italics).

It is certainly true that Lenin wanted competition mainly to take place within the organisation and administration, but he advocates still competition of some sort:

> (...) we must organise the accounting and control of the amount of work done and of production and distribution by the entire people, by millions and millions of workers and peasants. (...) And in order to organise this accounting and control, which is *fully within the ability* of every honest, intelligent and efficient worker and peasant, we must rouse their organising talent, the talent that is to be found in their midst; we must rouse among them – and organise on a national scale – *competition* in the sphere of organisational achievement ... Competition must be arranged between practical organisers from among the workers and peasants. Every attempt to establish stereotyped forms and to impose uniformity from above, as intellectuals are so inclined to do, must be combated.

Competition and Capitalism: A Different View Thus one of the most important theorists of Marxism is in favour of competition – and this is very much reminiscent of what former UK Business Secretary Vince Cable said in 2010: "Capitalism takes no prisoners and kills competition where it can." (22nd Sept 2010) This is actually a criticism of capitalism that even fierce advocates of the market economy could subscribe to: It is immoral for a company to actively fight competition. It is not immoral to fight competitors within the boundaries allowed by the rules of competition, but from the ethical point of view outlined here, turning against the system of competition itself should be seen as unethical. For example, a former monopolist that envisages competition in the future should not actively engage in preventing competition, lobbying against it or taking steps to discourage potential competitors from taking part in it. It would be ethical, however, to prepare one's company for the future market and taking steps to increase the efficiency of production processes and so on. This is something that the conception of order ethics I am advocating aims at.

Order Ethics: The Philosophical Background The conception of order ethics (cf. Homann/Lütge 2004/2013, Luetge 2015a, 2006, Luetge/Armbrüster/Müller 2006 and Luetge/Mukerji 2016) is based on the social contract tradition that includes authors like Hobbes, Locke, Spinoza, and Kant. The idea of embedding business ethics in a contract-related context has been explored by a number of authors over the past decades, such as Donaldson and Dunfee (1999, 2000), Rowan (1997, 2001) and Wempe (2004, 2008a, b).

Many authors (such as Heugens et al. 2004, 2006 or Wempe 2004) distinguish between contractarianism and contractualism:

Contractarianism is a philosophical position which regards the consent of actors as its only normative criterion. In particular, it assumes that actors are self-interested and give their consent to norms or rules only if they regard this as beneficial for themselves. This is different from a position that takes the consent of actors after deliberation as its main starting point (i.e., discourse ethics).

Contractualism presupposes an internal morality of contracting: it assumes that contracting parties must have certain moral capabilities in order for the contracts to work. For example, Rawls (1993) presupposes a sense of justice (cf. Luetge 2015a, 2016). Moreover, contractualism usually regards actors not as being predominantly self-interested, but rather as being embedded in a more general frame of commitments. In this sense, contractualism is at least in some regards closer to discourse theory and discourse ethics.

Order ethics ("Ordnungsethik") draws mostly on the contractarianism of James M. Buchanan[2], and can be regarded as the complement of the German conception of 'Ordnungspolitik' which stresses the importance of a regulatory framework ("Ordnung") for the economy. Order ethics relies heavily on the Buchanan-type contractarianism, taking up its basic idea that society is a cooperation for mutual self-interest. The rules of a society, and of its economy, are agreed upon by the participants, in a situation like Rawls' *original position*. Here, Order ethics also relies on Rawls: In accordance with his principles of justice, it will be in the mutual interest of all to devise rules that will in principle allow to improve everyone's position, and in particular, that of the least well-off. This idea is taken up in order ethics: the order framework of a society is regarded as a means for implementing ethics.

In the following, I will present the basic logic of the order ethics approach, starting with its account of modern, competitive societies.

Competition as a Social Condition Order ethics takes its start not from an aim to achieve, but rather from an account of the social conditions within which ethical norms are to be implemented: modern societies differ strongly from pre-modern ones. Pre-modern societies were "zero-sum societies", in which people could only gain significantly at the expense of others. This view is concisely expressed in the words of the successful fifteenth century Florentine merchant Giovanni Rucellai (1772), written around 1450: "by being rich, I make others (which I might not even know) poor".

Modern societies, by contrast, are societies with continuous growth, made possible by the modern competitive market economy. In this system, positive sum games are played. Many types of ethics, however, are still stuck with the conditions of pre-modern societies and lag behind: They ignore win-win-situations, for example, when distinguishing sharply between self-interest and altruistic motivation, and when requiring people to be moderate, to share unconditionally and to sacrifice (Luetge

[2] Buchanan 1975/2000; Brennan and Buchanan 1985; for the ethical dimension of Buchanan's work see Luetge 2006.

2005). These attitudes, which make self-interest something evil, would have been functional in a zero-sum society, but they are inadequate for modern societies.

In a situation of zero-sum games, it was necessary to call for temperance, for *moderate* profits, or even for banning interest. But in a modern society, self-interest in combination with the order framework promotes morality in a much more efficient way: Morality is implemented on the level of the order framework which governs the market. Via competition on the market, the position of each individual can be improved, resulting in win-win situations. These are of economic as well as of ethical value, resulting in innovative products at good value for money, of jobs, of income, of taxes – or in the promotion of diversity and pluralism (McCloskey 2006, 2010). Within the positive-sum games of modern societies, the individual pursuit of self-interest is compatible with traditional ethical ideas like universal solidarity.

Competition is central to this conception (cf. Luetge 2014, 2015b): Order ethics emphasises the importance of competition in a modern society. In a key article, Hayek (1978) lists the main advantages of competition. In doing so, however, he is not in favour of competition per se. Rather, he puts forward a functional argument: He insists that competition creates broadly distributed wealth, because

1. it sets incentives for creativity and fosters innovations by pioneers,
2. it disciplines the suppliers and
3. it enables the quick spreading of new ideas and problem solutions. This is Hayek's well-known argument against the possibility of a central planning authority, which, he argues, could never amass nor manage the amount of information necessary to produce and distribute the goods.

In this way, competition, however, mainly creates *pressure*, not freedom. The idea of the market as primarily an expression of freedom (Friedman 1962) is misleading.

Another point that has been added mainly by German economists Walter Eucken (1949) and Franz Böhm of the Freiburg School is that competition is an efficient tool for destroying and eroding temporary positions of power.

Competition, however, has negative aspects, too: in competitive situations, morality is constantly in danger of getting crowded out. The prisoners' dilemma or the stag hunt game are classic models for such situations which can be detrimental to morality if the incentives set by the rules thwart what is deemed ethical.[3] As an example, if corruption is seen as unethical, then rules which allow for corruption (for example, allowing bribes to be deducted from tax) will promote unethical behaviour – no matter how many public calls for morality are being launched.[4] Therefore, order ethics aims at changing the order framework of a society rather than at appealing to moral behaviour. This does not imply that people cannot behave ethically, but rather that ethical behaviour should not get punished by (counter-productive) incentives.

---

[3] Axelrod 1984; for the systematic use of the stag hunt game cf. Binmore 1994, 1998, 2005.

[4] This behaviour has been confirmed in experimental studies such as Gürerk et al. 2006; Binmore 2010 or already Andreoni 1988.

For the order ethics approach, it is important that rules and actions do not conflict with one another. Ethical behaviour on the level of actions can be expected only if there are no counteracting incentives on the level of rules. In the classical model of the prisoners' dilemma, the prisoners cannot be expected to cooperate because the conditions of the situation (the 'rules of the game') are such that cooperation is punished by defection on the part of the other player. Morality thus gets crowded out – and moralizing conceptions will not work.

A similar idea has been sketched by Boatright (1999): He distinguishes between a 'moral manager model' and a 'moral market model' of business ethics. While the former focuses on individuals and calls for managers to become more ethical as individuals, the latter intends to make markets more moral, by improving their rules and order framework.

So as a consequence for the matter at hand, I believe that under conditions of competition, it is futile to expect ethical appeals to directly change intentions. Apart from the problems mentioned above, there would be even further unresolved problems here: *Whose* intentions? The company's CEO's intentions? The board's intentions? Who are we addressing, if we address an organisation as complex as a company of significant size? Are there intentions of the organisation 'as such'? This is futile. I would therefore suggest looking at the rules, formal and informal, rather than at the intentions – and toning down the moralising rherotic.

Competition plays a pivotal role for OSPs. For most of them directly, but even for the bigger ones like Google, competition is always lurking in the background: At least indirectly they compete against other (sometimes possible) competitors and thus need to keep being innovative, to keep being on the forefront of technology etc. – because others might get bigger or better 1 day.

7.4 Philosophical Aspect: The Role of Corporate Responsibility

So far, I have been arguing that there are certain actions or measures we can *not* expect from OSPSs. Now, I will turn to a couple of other measures we can indeed expect from them – bearing in mind that there are mechanisms of self-interest behind them. But these mechanisms can be more complex than some economists may think.

In the field of Corporate Responsibility (CR) or Corporate Social Responsibility (CSR)[5], there are lots of examples of how companies actually are doing many things that traditionally would not have been expected from them. In 1970, it may have been safe for Milton Friedman to state that the only social responsibility of companies is to maximise their profits (Friedman 1970). But obviously, during the last decades, the situation on markets worldwide has developed further massively: big multi-national corporations, but also small and medium-sized enterprises (SME)

[5]And in related concepts like Corporate Citizenship, Creating Shared Value or Corporate Social Accounting.

are active in fields like human rights, diversity, ecological problems and related ones. I will mention just a few examples from the international oil industry, a branch that is mostly only associated with scandals and less with social responsibility (cf. Pies et al. 2009):

- In 1998, Shell built a much longer pipeline than necessary in the Philippines to both avoid ecological damage to the region as well as staying away from the holy ground of the local inhabitants. At that time, Shell was under a lot of pressure resulting from the Brent Spar case. However, 10 years later, independent calculations showed that the initially more expensive solution actually was less expensive in the longer run: production delays and penalties for delayed completion were avoided.
- Other projects include ExxonMobil helping in developing women's economic potential in Angola, Colombia, Egypt and several other countries ("ExxonMobil Educating Women and Girls Initiative", with a budget of 8 Mio. US-$ in 2008), Shell conducting a programme in community building and capacity developing to improve health status and education of residents in Venezuela, or OMV which have similar programmes in Pakistan, Yemen and Iraq. This goes down as far as the "base-of-the-pyramid projects", which aim at simply providing basic relief to poor people in those countries the companies invest in, like providing mobile cookers that can be used with biomass.
- BP and Shell have been setting up independent panels and committees to work as a kind of devil's advocate against themselves. Prominent independent members (like former Dutch Prime Minister Wim Kok) seek to guarantee that these panels are not just used just as 'fig leafs'.
- Reports to external review committees with the aim to enhance transparency, like the "Global Reporting Initiative" and "Publish what you Pay". In the EITI initiative, the extractive industry has agreed to disclose payment flows to government officials.
- There are cases where companies are actually helping governments or officials in complying with standards, as in the case of the Baku-Tbilisi-Ceyhan pipeline. Before building this pipeline, BP, in 2003, signed an agreement with Georgia, Azerbaijan and Turkey, stating that the company waives their rights to compensation payments if the states involved choose to raise their standards in areas like human rights, health care or social security. This causes additional costs for BP, but encourages states to improve their standards. For better monitoring, non-governmental organizations (NGOs) are involved too.

While each of these projects might leave some points open to criticism in detail, in total they are too many to be totally disregarded or neglected, even by outspoken critics (like Chomsky or Badiou). These cases show how genuine interests of local people and inhabitants are taken seriously, not (or not only) by the state but by companies, who nevertheless act in their own interest as well. This can only work if this is not just question of ethical discourse, but if there are economic mechanisms behind it as well.

The economic mechanisms mostly are based on questions of risk. In the globalised world, multinational corporations face a number of risks: traditional business risks like financial risks, production risks, weather risks (in some industries), political risks like the introduction of new tariffs or fundamental changes in trade relations. But there are moral risks as well: corruption, discrimination, child labour, job safety problems and many others. These factors have always been ethically problematic but, in the era of globalisation, they develop into serious economic issues, mainly for two reasons:

First, there have been important changes in legal regulation, such as the U.S. Sarbanes Oxley Act (2002) or the Dodd-Frank Act (2010). Companies can be fined much more severely than before, not only in the US or UK, but in countries with quite different legal frameworks like Germany too.[6] Second, corporations are increasingly being watched over by NGOs – who are starting to become a new element of democratic control in the globalised era.

These kinds of risk become increasingly important for many different kinds of companies – and require them to invest in their ethical capital. From an order ethics point of view, improvements in the legal framework are important, but not sufficient. The legal framework can never be refined enough to cover all possible situations. It leaves companies with not just one, but a number of possibilities to pursue their own interest.

So if rules are incomplete or if there are no rules for a specific situation, contractarianism suggests relying on substitutes: Corporations, as partners in interactions, have the opportunity to commit themselves to certain policies, to mechanisms of trust and fairness, for example. This commitment has to be made credible through organizational measures and must be signalled to others. In this way, actors create by themselves the very reliability that would normally be expected from formal rules. They create a *reputation*, which especially under conditions of globalization is a necessary prerequisite for success in the long run. This commitment must be signalled, and thus becomes an asset for the company. Corporate responsibility rankings have become increasingly visible in the past few years.

From a theoretical perspective, order ethics provides an integrative contractarian view on both situations, those with well-established and those with incomplete rules. In both cases, incentives and sanctions are key issues. In the first case, incentives are set by formal rules, while in the second case, this role is taken up by informal rules in the shape of 'soft' factors like ethics and reputation.

Of course, there are bad examples as well: Some large companies like Volkswagen have, apparently for a long time, ignored risks that at first were only *moral* risks. In dealing with the question whether to use 'cheating' software in their company's cars, Volkswagen officials failed to recognise the possibility that this behaviour would eventually become public – and turn into a major *economic* issue. They have been neglecting warning signals for a long time. Business ethics, in this sense, can be seen as an "early warning system" for moral risks which might become (very expensive) economic risks later on.

[6]Especially important in Germany was the reformulation of the §30 Ordnungswidrigkeitengesetz in 2002.

In the case of OSPs, reputation is obviously an important thing: the brands of some OSPs are among the mostly highly rated ones in the world, like Google or Apple. And putting these brands in danger, risking their devaluation, is something these companies obviously should want to avoid.

Thus, ethical advancements can be had not just against the market, but with the market and through the market – provided adequate rules are in place (on whatever level). Measures taken by companies can now be called ethical, if they have more than narrow instrumental value, but value for others too, in a win-win sense. The creation of win-win situations is the key point in calling something ethical. Some philosophical approaches would probably deny it, but for business ethics, win-win is the framework that makes most sense.

In a win-win framework, order ethics focuses not on changing intentions but on changing the legal rules as well as others, e.g., on the respective branch level (maybe the most relevant one in globalisation) and also on the individual corporate level: especially multinational corporations have considerable systems of rules of their own.

7.5 Collective Responsibility: Philosophical and Economic Aspects

Both the economic as well as the philosophical point are tied together in one question: can the notion of responsibility only be applied to individuals or to collective agents as well? A number of moral philosophers stick to individual responsibility (e.g. Narveson 2002). However, a notion of collective responsibility is indispensable for questions of business ethics as well as of ethics of OSPs. Without it, there is no way to find companies (including OSPs) ethically responsible. Such a notion of collective responsibility has been put forward by authors like List and Pettit (2011) who advocate a group agency approach. As a final point, I will turn to this approach now.[7]

Group attitudes can be derived from individual attitudes by using aggregation rules. Both the unanimity rule as well as the majority rule can be derived in this way, for example: The unanimity rule says that a group believes that the world matches proposition p if and only if all members of the group believe p. The majority rule says that the group believes that the world matches proposition p if and only if a majority of its members believe p. However, some of these aggregation rules run into problems when trying to derive group attitudes. List and Pettit give the following examples:

A group of legislators is considering a budget proposal which currently is running a deficit. Legislators 1, 2 and 3 consider whether they should take measures to

[7] For the following, cf. Mukerji and Luetge 2014.

even out the budget which, as they know, means either increasing taxes or cutting spending. They have the following preferences:

	Increase taxes?	Reduce spending?	Even out budget?
Legislator 1	Preferred	Dispreferred	Preferred
Legislator 2	Dispreferred	Preferred	Preferred
Legislator 3	Dispreferred	Dispreferred	Dispreferred
Majority	Dispreferred	Dispreferred	Preferred

Note that every legislator has consistent preferences as an individual. However, the preferences of group are inconsistent when aggregated by majority rule: The group would prefer a budget without deficit, without wanting to increase taxes or cutting spending – which is inconsistent. This means that if group members' preferences are to be aggregated, this cannot be done in a straightforward way, but is a more complicated process. The group's preference on budget questions has to be determined on the basis of the members' stances regarding tax increases and spending cuts. Or, otherwise, the majority for a balanced budget must be taken as given, implying that the group has to adopt a stance on tax increases or spending cuts which will not necessarily be equal to the view of the majority.

An analogous problem of List and Pettit concerns the aggregation of beliefs. This one refers to global warming. The authors introduce a panel on climate change, which consists of three individuals.

Each member holds beliefs regarding a number of related issues. The first one is whether current emissions are above a given threshold. The second one is whether temperatures will increase if emissions are above this threshold. The third one is whether temperatures will, in fact, increase. The decision process is as follows:

	Emissions above threshold?	If Emissions above threshold, then temperature increase?	Temperature increase?
Individual 1	True	True	True
Individual 2	True	False	False
Individual 3	False	True	False
Majority	True	True	False

A majority judges that emissions are above the threshold and that the temperature will increase if emissions are above the threshold. However, a majority judges that the temperature will not increase. Obviously, this is inconsistent. To resolve this inconsistency, the group has to adopt a different method of aggregation. It cannot aggregate the attitudes of its members proposition-wise. It can either, first, determine its attitude towards the third proposition on the basis of its members' attitudes on that proposition. In that case it cannot adopt the majority view on the first two propositions. Or, second, it can determine its attitude towards the third proposition based on the majority views regarding the first and second proposition. In that case, the group must inevitably reject the majority attitude towards the third proposition.

As these two examples make clear, groups can form beliefs and desires which, though supervening on the beliefs and desires of their individual members, are nevertheless largely independent from them. The group agency notion, thus, is very much in line with ideas from order ethics as sketched above, including its conception of responsibility.

In the light of the above, it makes little sense to argue that a collective entity has no consciousness and therefore cannot be held responsible. This is just too narrow a notion of agency and responsibility – because there might be functional equivalents to consciousness which result in ascribing ethical responsibility to such a collective entity: For example, it must have certain kinds of organisational mechanisms in place for choosing concerted action in a rational way, it must have standards of conduct which are actually being enforced, and defined roles for its individuals to exercise powers. If these conditions are met, then creative decisions are made possible which go beyond mere routine behaviour – and the organisation can be held ethically responsible as a collective entity.[8]

Therefore, the OSP responsibilities discussion should rest less on purely individual concepts of responsibility, but include collective concepts like group agency. In this line, there are ways of ascribing responsibility to companies compatible with competition, as many of the debates on CSR show, some of which I have mentioned here. In the case of Volkswagen, one cannot sensibly say that it was only one individual technician or a small group who made a mistake. It is a question of company structures and mechanisms, of certain rules which a company has or has not set up in an adequate way.

7.6 Conclusion

Certainly, an exhaustive list of responsibilities cannot be given here, but rather is the task of this volume. From my part, I hope to have made it clear that some responsibilities can be expected from OSP (as companies) and some cannot. In general, we can rely on economic mechanisms and expect not only people, but also organisations to improve their calculations: calculate in the longer run, take long term risks into account (reputation risks and legal penalty risks, for example), and consider interests of others – because you need them for your own well-being, your own profit-maximisation in the longer run. From a business ethics point of view, these arguments should be taken into account when considering OSPs' ethical responsibilities.

[8] Cf. already French 1984, 13f.

Bibliography

Andreoni, J. (1988). Why free ride? Strategies and learning in public goods experiments. *Journal of Public Economics, 37*, 291–304.
Axelrod, R. (1984). *The evolution of cooperation*. New York: Basic Books.
Badiou, A. (2010). *The communist hypothesis*. London: Verso.
Bailyn, E. (2015). *Google's greed: How the search giant's lust for information is impacting your search results*. http://firstpagesage.com/seo-blog/thought-leadership/googles-greed-search-giants-lust-information-impacting-search-results/#sthash.tEiEE1Yy.dpuf. Accessed 12 May 2016.
Binmore, K. (1994). *Game theory and the social contract: Vol. 1: Playing fair*. Cambridge, MA/London: MIT Press.
Binmore, K. (1998). *Game theory and the social contract: Vol. 2: Just playing*. Cambridge, MA/London: MIT Press.
Binmore, K. (2005). *Natural justice*. Oxford: Oxford University Press.
Binmore, K. (2010). Social norms or social preferences? *Mind and Society, 9*, 139–158.
Boatright, J. (1999). Presidential address: Does business ethics rest on a mistake? *Business Ethics Quarterly, 9*, 583–591.
Brault, J. (2013). *Google est-il dangereux pour la démocratie?*. https://www.lesaffaires.com/blogues/julien-brault/google-est-il-dangereux-pour-la-democratie/554735. Accessed 12 May 2016.
Brennan, G., & Buchanan, J. M. (1985). *The reason of rules: Constitutional political economy*. Cambridge: Cambridge University Press.
Buchanan, J. M. (1975). *The limits of liberty: Between Anarchy and Leviathan*. Indianapolis: Liberty Press.
Chomsky, N. (1999). *Profit over people: Neoliberalism and global order*. New York: Seven Stories Press.
Chossudovsky, M. (2003). *The globalization of poverty and the new world order*. Oro: Global Outlook.
Comte-Sponville, A. (2004). *Le capitalisme est-il moral?* Paris: Albin Michel.
Donaldson, T. D., & Dunfee, T. W. (1999). *Ties that bind: A social contracts approach to business ethics*. Boston: Harvard Business School Press.
Donaldson, T. D., & Dunfee, T. W. (2000). Book review dialogue: Tightening the ties that bind – Defending a contractarian approach to business ethics. *American Business Law Journal, 37*, 579–585.
Eucken, W. (1949). Die Wettbewerbsordnung und ihre Verwirklichung. *ORDO, 2*, 1–99.
Ferguson, N. (2010, November 18). In China's orbit. *Wall Street Journal*.
French, P. (1984). *Collective and corporate responsibility*. Chicago: University of Chicago Press.
Friedman, M. (1962). *Capitalism and freedom*. Chicago: Chicago University Press.
Friedman, M. (1970, September 13). The social responsibility of business is to increase its profits. *The New York Times Magazine*, p 32f., pp. 122–126.
Gürerk, Ö., Irlenbusch, B., & Rockenbach, B. (2006). The competitive advantage of sanctioning institutions. *Science, 312*, 108–111.
Haufler, D. (2014). *Suchmaschine Googles Gier kennt keine Grenzen*. http://www.berliner-zeitung.de/digital/-suchmaschine-googles-gier-kennt-keine-grenzen-1455016. Accessed 12 May 2016.
Hayek, F. A. (1978). Competition as a discovery procedure. In F. A. Hayek (Ed.), *New studies in philosophy, politics and economics*. Chicago: University of Chicago Press.
Heugens, P., Kaptein, M., & Oosterhout, H. (2004). Ties that grind? Corroborating a typology of social contracting problems. *Journal of Business Ethics, 49*(3), 235–252.
Heugens, P., Oosterhout, H., & Kaptein, M. (2006). Foundations and applications for contractualist business ethics. *Journal of Business Ethics, 68*(3), 211–228.
Homann, K., & Lütge, C. (2004/2013). *Einführung in die Wirtschaftsethik* (3rd ed.). Münster: LIT.

Johnson, L. (2013). *A gigantic parasite: Forget the touchy feely posturing*. Google's tiny tax bill shows how greedy and ruthless it really is. http://www.dailymail.co.uk/debate/article-2314375/A-gigantic-parasite-Forget-touchy-feely-posturing-Googles-tiny-tax-shows-greedy-ruthless-really-is.html. Accessed 12 May 2016.

Jones, E. L. (1981). *The European miracle: Environments, economies, and geopolitics in the history of Europe and Asia*. Cambridge: Cambridge University Press.

Lenin, V. I. (1917). How to organise competition. In: *Collected works*, (4th English Edition translated by Yuri Sdobnikov and George Hanna and edited by George Hanna) (vol 26, pp. 404–415). Moscow: Progress Publishers.

List, C., & Pettit, P. (2011). *Group agency: The possibility, design, and status of corporate agents*. Oxford: Oxford University Press.

Luetge, C. (2005). Economic ethics, business ethics and the idea of mutual advantages. *Business Ethics: A European Review, 14*(2), 108–118.

Luetge, C. (2006). An economic rationale for a work and savings ethic? J. Buchanan's Late Works and Business Ethics. *Journal of Business Ethics, 66*, 43–51.

Luetge, C. (2014). *Ethik des Wettbewerbs: Über Konkurrenz und Moral*. München: Beck.

Luetge, C. (2015a). *Order ethics vs. moral surplus: What holds a society together?* Lanham: Lexington.

Luetge, C. (2015b). Ethique et concurrence dans l'économie de marché. In D. de Courcelles (Ed.), *Actes pour une économie juste* (pp. 211–220). Paris: Lémieux Editeur.

Luetge, C. (2016). Order ethics and the problem of social glue. *University of St. Thomas Law Journal, 12*(2), 339–359.

Luetge, C., Armbrüster, T., & Müller, J. (2016). Order ethics: Bridging the gap between contractarianism and business ethics. *Journal of Business Ethics, 136*(4), 687–697.

Luetge, C., & Mukerji, N. (Eds.). (2016). *Order ethics: An ethical framework for the social market economy*. Heidelberg/New York: Springer.

McCloskey, D. (2006). *The bourgeois virtues: Ethics for an age of commerce*. Chicago: University of Chicago Press.

McCloskey, D. (2010). *Bourgeois dignity: Why economics can't explain the modern world*. Chicago: University of Chicago Press.

Miyazaki, I. (1981). *China's examination hell: The civil service examinations of Imperial China*. New Haven: Yale University Press.

Mukerji, N., & Luetge, C. (2014). Responsibility, order ethics, and group agency. *Archiv für Rechts- und Sozialphilosophie, 100*(2), 176–186.

Narveson, J. (2002). Collective responsibility. *Journal of Ethics, 6*(2), 179–198.

Pies, I., Winning, Av., Sardison, M., & Girlich, K. (2009). *Nachhaltigkeit in der Mineralölindustrie: Theorie und Praxis freiwilliger Selbstverpflichtungen*. Wirtschaftsethik-Studie Nr. 2009-1 des Lehrstuhls für Wirtschaftsethik an der Martin-Luther-Universität Halle-Wittenberg, Halle.

Pohl, A. (2010). *Googles Gier opfert die Netzneutralität*. https://netzpolitik.org/2010/googles-gier-opfert-die-netzneutralitat/. Accessed 12 May 2016.

Rau, M. (2008). *No good can come of Microsoft's greed*. http://articles.dailypress.com/2008-02-11/features/0802110006_1_windows-media-digital-production-yahoo. Accessed 12 May 2016.

Rawls, J. (1993). *Political liberalism*. New York : Columbia University Press.

Rowan, J. R. (1997). Grounding hypernorms: Towards a contractarian theory of business ethics. *Economics and Philosophy, 13*, 107–112.

Rowan, J. R. (2001). How binding the ties? Business ethics as integrative social contracts. *Business Ethics Quarterly, 11*, 379–390.

Rucellai, G. (1772). Ricordanze, Padua

Schofield, J. (2011). *Steve jobs, the mean dictator – Paul Graham*. http://www.zdnet.com/article/steve-jobs-the-mean-dictator-paul-graham/. Accessed 12 May 2016.

The Economist. (2014). *Internet monopolies: Everybody wants to rule the world.* http://www.economist.com/news/briefing/21635077-online-businesses-can-grow-very-large-very-fast-it-what-makes-them-exciting-does-it-also-make. Accessed 12 May 2016.

Wempe, B. (2004). On the use of the social contract model in business ethics. *Business Ethics: A European Review, 13*, 332–341.

Wempe, B. (2008a). Contractarian business ethics: Credentials and design criteria. *Organization Studies, 29*, 1337–1355.

Wempe, B. (2008b). Four design criteria for any future contractarian theory of business ethics. *Journal of Business Ethics, 81*(3), 697–714.

WSJ.com. (2012). *Why facebook is almost as greedy as Bernie Madoff.* http://www.wsj.com/video/why-facebook-is-almost-as-greedy-as-bernie-madoff/5945C59A-102F-4027-8AFE-4F127EDC6589.html. Accessed 12 May 2016.

Chapter 8
Myth or Promise? The Corporate Social Responsibilities of Online Service Providers for Human Rights

Emily B. Laidlaw

Abstract The story of technology and law has often been about the law lagging behind innovation. Corporate Social Responsibility (CSR) instruments, because of their transnational amenability, responsiveness to change and non-binding nature, have been increasingly deployed to fill the perceived governance gap. This chapter examines the various CSR frameworks that guide or govern online service providers (OSP), such as search engine providers, social networking providers, payment providers, and e-commerce platforms. It will focus on three influential initiatives: the United Nations Guiding Principles, the European Commission ICT Sector Guidance and the Global Network Initiative. The core question for OSPs, as with any company, is how do we make them accountable for their human rights impact? This entails analysis of some of the following questions. What value, if any, do such CSR frameworks offer to the protection of human rights online and what are their drawbacks? Relatedly, how do we know if these CSR frameworks are a success? What factors lead to the conclusion that CSR, in a given situation or sector, is an ill-suited device where more traditional legal measures are needed?

8.1 Introduction

Corporate Social Responsibility (CSR) creates a visceral reaction in many, whether it is seen as having promise to embed ethics into a company's operations or is seen as an exercise in public relations that impedes or delays the development of much needed legal reform (Manokha 2004). In the technology sector, debates about the responsibilities of companies, legally or socially, for human rights is now at the forefront of regulatory discussions. 10 years ago that was not the case. Three related developments, I suggest, shifted the landscape such that an information and

E.B. Laidlaw (✉)
Faculty of Law, University of Calgary, Calgary, Canada
e-mail: emily.laidlaw@ucalgary.ca

M. Taddeo, L. Floridi (eds.), *The Responsibilities of Online Service Providers*,
Law, Governance and Technology Series 31, DOI 10.1007/978-3-319-47852-4_8

communication technology (ICT) business's responsibilities for human rights is now central to their governance.

First, a series of high profile human rights abuses drew attention to the responsibilities of ICTs. Notable early examples include the revelation that Yahoo! had provided information to Chinese authorities concerning journalist Shi Tao, which helped lead to his arrest and imprisonment. At the same time, Google launched its search service in China using censored search results (Kopytoff 2011). Google has since stopped this practice (Google 2010). Microsoft shut down a blog by a Chinese journalist who worked for the New York Times (Maclay 2010). These three incidents prompted creation of the Global Network Initiative (GNI), discussed below, one of the leading multi-stakeholder initiatives (MSI) guiding CSR for technology companies.

Since then the scenarios have multiplied. Some involve ICTs causing or facilitating the abuse. Certain British internet service providers (ISP), for example, unbeknownst to their clients, allowed a third party, Phorm, to track users activity online (Bernal 2011). A French company, Amesys, sold surveillance technology to the Libyan government, which allegedly was used to spy on dissidents, who were later arrested and tortured (Nieuwenkamp 2015). Facebook applications shared user information with advertisers without their permission (Steel and Fowler 2010). During the Arab uprising in 2011 Vodafone shut down mobile telephone networks at the direction of the Egyptian government (Garside 2011).

In other incidents, ICTs are involved as intermediaries; while they might not be the perpetrators of the abuse, by virtue of their gatekeeping function they are in a position to regulate it. This is evident concerning content regulation, such as requests to hosts to remove copyright infringing material, or requests to social networking providers (SNP) to remove abusive or threatening posts. Caroline Criado-Perez, for example, was threatened with death and rape for campaigning for a woman to be on the British £10 note. Two of her attackers were convicted for sending menacing tweets, but it also prompted assessment of Twitter's responsibility to control this kind of abuse (the Guardian 2014). Perhaps most affecting of them all was the revelations by former National Security Agency contractor Edward Snowden of massive online surveillance by the Five Eyes Alliance of the United States of America, United Kingdom, Canada, Australia and New Zealand, including online companies collecting and sharing user data with governments pursuant to secret court orders.

Second, at the same time, and partly as a result of, such incidents the importance of human rights in cyberspace started receiving increased scrutiny by the United Nations (UN). Frank La Rue, the former UN Special Rapporteur on the promotion and protection of the right to freedom of opinion and expression, stated in a report that access to the internet is a fundamental right, noting the growing influence of companies: "[g]iven that Internet services are run and maintained by private companies, the private sector has gained unprecedented influence over individuals' right to freedom of expression and access to information" (La Rue 2011, para 44). In 2012 the UN Human Rights Council (HRC) passed a resolution affirming internet freedom as a basic human right (A/HRC/20/L.13). In 2015, partly in response to the Snowden revelations, the HRC affirmed the human right to privacy, including company responsibilities thereto, and appointed a special rapporteur on the right to privacy (A/HRC/RES/28/16). Joseph Cannataci has been appointed to the role and has thus far taken a hard line, advocating for an international convention on privacy (Alexander 2015).

Third, outside of the technology sector a paradigmatic shift took place concerning the business and human rights relationship, most importantly with the drafting of the UN Guiding Principles (GP). The GPs dissected a polarizing issue: how are businesses human rights obligations the same or different from that of states? (Ruggie 2011) The GPs, to be discussed in detail in this chapter, are at once widely accepted and controversial. They were endorsed by the UN in 2011 (A/HRC/RES/17/4) and have inspired many agendas on CSR, as seen in Europe (European Commission 2011) and the UK (HM Government 2013). For the technology sector, the GPs have influenced sector CSR codes and guides, such as the European Commission Guidance for ICTs (Guidance), the GNI and the Technology Industry Dialogue (TID). The are also viewed by some as destructive to the agenda of advancing human rights protection (Deva and Bilchitz 2013).

In this chapter I will discuss the various CSR frameworks at work that guide or govern online service providers (OSP), such as search engine providers, SNPs, payment providers, and e-commerce platforms. I will focus on three influential initiatives: the GPs, the EC Guidance and the GNI. The core question for OSPs, as with any company, is how do we make them accountable for their human rights impact? This involves analysis of some of the following questions. What value, if any, do such CSR frameworks offer to the protection of human rights online and what are their drawbacks? Relatedly, how do we know if these CSR frameworks are a success? What role can the law play in helping embed corporate responsibility for human rights?

8.1.1 Contextualizing CSR

There is a reason why we have seen the rise of CSR-style instruments to address internet law issues, whether at a company level or through MSIs. Top-down legal controls are hard to enforce outside territorial boundaries and internet communications are transnational in nature. These CSR instruments are argued to fill a governance gap and have the advantage, due to their non-binding nature, to bring to the table a wide variety of stakeholders. They also can be responsive to change in a way that command-and-control laws cannot (McBarnet 2007). The story of technology and law has often been about the law lagging behind innovation (Reed 2012). Such instruments often exist, as Dorothée Baumann-Pauly et al assert, 'in part because of the lack of better alternatives' (2015, p 29). The drawbacks of such initiatives is that they are vulnerable to issue selection, can be overly-broad, lack accountability, and only attract those already committed to human rights resulting in free-riding by other companies or lack of standardization across industry (Webb and Morrison 2004).

Further, the kinds of disputes that arise for OSPs online tend to relate to human rights such as free speech and privacy, where there is little consensus even in Western states about their scope and meaning. Given the American base of many of the key internet companies, concerns have been raised in the past that they are exporting the First Amendment (Vick 2005). Thus we turn to CSR frameworks to bridge the governance gap, but equally criticize any instruments that are American situated.

CSR has no one meaning. Indeed, it conceivably has hundreds of meanings (Blowfield and Murray 2008), such that Adaeze Okoye has argued it should be treated as an essentially contested term (2009). It has been used to describe, among others, a business's charitable activities, ethical responsibilities, and duties to its stakeholders, employees or communities. At its broadest, CSR can be described as an umbrella term for the business and society relationship (Blowfield and Murray 2008). Narrower, it might be defined according to a particular issue in a community or industry, such as an oil and gas company's responsibilities concerning the environment or engagement with First Nations, or a security companies responsibilities concerning its personnel operating in conflict zones. The European Commission defines CSR as:

> Corporate social responsibility concerns actions by companies over and above their legal obligations towards society and the environment. Certain regulatory measures create an environment more conducive to enterprises voluntarily meeting their social responsibility. (European Commission 2011)

This definition focuses on the voluntary commitments of businesses concerning their social and environmental impact rather than legal commitments, but envisions government in a meta-regulatory capacity setting the expectation of what these commitments entail. This expectation might be legal in nature, or normative where good behavior is pressured through a system of rewarding or shaming. An example is the United Kingdom's Companies Act (2006), which requires Directors to have regard to their company's impact on the community and on the environment, and requires the company to discuss its CSR policies in its annual report (ss 172 and 417 respectively).

While the European Commission definition blurs the line between voluntariness and legal regulation, it is helpful in identifying that the voluntary commitments of a business for social responsibility, in this case, responsibility for their human rights impact, exists in a wider regulatory and political context. CSR, therefore, should be considered on four levels: international, state, industry and company. International and state frameworks tend to be more formalized, such as the UN Global Compact or OECD Guidelines for Multinationals Enterprises (OECD Guidelines), or strategic, such as *a renewed EU strategy* 2011–2014 *for Corporate Social Responsibility* (EU CSR Strategy), which seeks to unify approaches to CSR across European Union (EU) member states. Industry level initiatives tend to be more regulatory in nature, even if largely of normative force. Company-level CSR is where all initiatives trickle-down to, because they guide a company on the operational changes it must make to meet its responsibilities.

The voluntary responsibilities a business undertakes I describe elsewhere as pure-CSR (Laidlaw 2015), but CSR otherwise, and in the sense used by the European Commission, captures the wider business and society relationship. CSR is outward-looking, and when CSR is about human rights, it arguably has aspects to it of mobilizing or responding to social change. This is different than the targeted and instrumental nature of regulation, which tends to be focused on achieving a specific policy objective. CSR might be deployed through regulation, whether through self-regulatory codes or industry regulation, and indeed I suggest this is CSR in its best form, but this is not its baseline. The struggle with CSR is moving from aspirational-mode to embedding this aspiration into a company's operational structure. Thus the concept of CSR constantly struggles with criticisms that it is an exercise in public relations, or, when it becomes more regulatory in nature, criticisms that it is not CSR at all. The work of John Ruggie in drafting the GPs has helped lift the debate from its polarized past.

8.1.2 International: The Guiding Principles

The GPs were borne out of the failures of the UN *draft Norms on the responsibilities of transnational corporation and other business enterprises with regard to human rights* (2003) (Norms). The Norms would have applied state-like human rights obligations on businesses. As a result of the controversy of this proposal and other issues with the Norms (Kinley 2007), John Ruggie was appointed by the UN Secretary General to be the Special Representative on the Issue of Human Rights and Transnational Corporations and Other Business Enterprises (Bilchitz and Deva 2013). He served in this role from 2005–2011.

Ruggie's job was to investigate ways to improve the human rights performance of companies. He carried out his work in three stages. In the first stage he identified the existing standards that apply to companies concerning human rights. In the second stage he published his Protect, Respect and Remedy framework (Framework) where he sought to create a "common conceptual and policy framework" (Ruggie 2008, para 8). The Framework was accepted by the HRC, the main UN body for the protection of human rights. The HRC then renewed his mandate and he embarked on the third stage his project, which was to recommend how businesses can operationalize the Framework. This culminated in the publication of the GPs in 2011.

It is the GPs that are the global reference point to the work of Ruggie as the Special Representative. The GPs were endorsed by the HRC in June 2011, which gave the GPs a weight of authority that elevated them from mere guidance to something more closely akin to a system of public governance (Ruggie 2013). They have been positively received in the international community. They have influenced countless global standards, such as the OECD Guidelines, ISO 26000 Guidance on social responsibility, the Performance Standards on Environmental and Social Sustainability of the International Finance Corporation and the UN Global Compact. At a regional level, they have influenced, for example, the EU CSR Strategy, the EC

Guidance and the UK's renewed CSR strategy. They also influence key technology sector standards we have today, such as the GNI, TID and the Electronic Industry Code of Conduct.

The GPs identify three pillars: a state's duty to protect human rights, a business's duty to respect human rights, and the need for access to a forum of remediation for those harmed. What this means is that states have a duty to protect, respect and fulfil human rights reflecting traditional legal obligations on states to ensure the human rights of its citizens are protected. The unique aspect of the GPs, for the purposes of this chapter, are the second and third pillars. The second pillar imposes a duty on a business to respect human rights. By human rights it is drawing from internationally recognized human rights, at minimum those set out in the International Bill of Rights and the International Labour Organization's Declaration of Fundamental Principles and Rights at Work 1988. The duty to respect requires that companies do the following:

(a) Avoid causing or contributing to adverse human rights impacts through their own activities, and address such impacts when they occur;
(b) Seek to prevent or mitigate adverse human rights impacts that are directly linked to their operations, products or services by their business relationships, even if they have not contributed to those impacts. (Ruggie 2011)

The GPs root the duty in a company's social license to operate – that the drive for a company to respect human rights is based on social expectation (Ruggie 2008). The duty can therefore be seen to be non-legal in nature, particularly as the external accountability options largely involve reputational costs. It is not, however, simply encouragement of corporate nicety as it tasks companies with operationalizing the duty to respect through a system of due diligence. This requires a company to assess and act on their human rights impact, actual and potential, then monitor their performance and communicate concerning their performance with the public. A basic version of this would be a human rights impact assessment. The goal is for a company to know and show they are respecting human rights in practice (Ruggie 2011, 2013).

The third pillar requires that anyone whose rights have been impacted by a business should have access to a grievance mechanism where they can make a complaint and seek a remedy. The GPs identify three types of remedial mechanisms: judicial, state-based non-judicial and non-state-based. Judicial mechanisms refer to traditional actions before a court. State-based non-judicial mechanisms are complementary to the court system and include, for example, National Human Rights Institutions (NHRIs). Of particular relevance to this chapter are non-state-based mechanisms to resolve disputes, which encompass those run by an industry group, a multi-stakeholder organization or alone by the business. Ruggie identified such mechanisms as being the most underdeveloped of the three types of remedial mechanisms (2013).

Non-state-based mechanisms are valuable, because, among other things, they are cheaper, faster and potentially transnational (Ruggie 2011). However, at what point can a company be deemed successful on this front? Does Facebook's report abuse button, to flag content that user thinks is inappropriate, qualify as the provision of a legitimate non-judicial process? What if there is no opportunity to follow-up, make a case, hear reasons, appeal or speak directly with anyone? It is certainly innovative, and is a model for other SNPs, such as Twitter, which has adopted a similar tool (Wang 2013). However, in the end it is just a tool: when a report is made it is evaluated by a team and a decision is made whether to remove the content. There is no formalized process where the complainant or accused have an opportunity to be heard. The GPs note the importance of states in facilitating access to such mechanisms. Therefore, while the grievance mechanism might be housed and run by an industry group or within a business, what responsibility does the state have to make it happen?

Since the GPs were published it is this second pillar that has been attempted to be operationalized the most. The third pillar, requiring a forum for remediation, has been acknowledged but not yet extensively engaged. The volume of online disputes makes the third pillar a difficult hurdle for many companies. However, the high volume, transnational nature of internet disputes and low value of most disputes makes such non-judicial mechanisms all the more important. We have seen mechanisms developed in the e-commerce sector, such as eBay's Resolution centre, and to resolve domain name disputes, with mechanisms through Nominet.[1] However, we have not seen similar mechanisms developed for content related disputes. The GPs identify seven criteria for effective non-judicial grievance mechanisms: legitimate, accessible, predictable, equitable, transparent, rights-compatible and a source of continuous learning. Operational-level mechanisms should also be based on consultation with stakeholders. The key is that complainants should know about the mechanism, trust it and use it (Ruggie 2011).

The GPs bridge the gap between proponents of CSR as something purely voluntary and proponents of direct application of human rights laws to companies. Despite its positive reception, however, the GPs struggle in areas that have a significant impact on their usefulness to OSPs. The GPs are criticized as inconsistent with human rights logic, because by setting a benchmark of success based on social expectation, human rights duties are arguably weakened and confused (Bilchitz 2013). This lack of grounding in legal duties was also criticized by Justine Nolan, where she questioned the power of authority of the GPs, because most soft laws are effective as complements to hard laws (2013). While the GPs have added authority because of the HRC endorsement, they are not complemented by hard laws. As a common reference point the GPs are powerful. No longer is an argument that a business as *no* human rights responsibilities tenable, although that would have been the case 10 years ago.

The GPs are potentially most effective when implemented at an industry-level where more specific guidance is possible. Baumann-Pauly et al. argue in favour of

[1] Country-level and generic top-level domain names are registered through the International Corporation for Assigned Names and Numbers (ICANN), which spearheaded dispute resolution for domain names with its Uniform Dispute Resolution Policy (UDRP). This is the model for state-level mechanisms such as Nominet.

industry-specific MSIs, such as the GNI, stating that they have a greater chance of success in filling perceived governance gaps than broader-based MSIs, such as the UN Global Compact (2015). Some of the MSIs, in particular at an international or state level, are generalised. For example, the OECD Guidelines were revamped in light of the GPs. The only provision on free speech is:

> Support, as appropriate to their circumstances, cooperative efforts in the appropriate fora to promote Internet Freedom through respect of freedom of expression, assembly and association online. (section II.B.I)

The difficulty with blanket commitments, such as this, is that they do not capture the struggles of many OSPs, nor do they provide practical advice on how to handle a problem. The vice-president of trust and safety at Twitter discussed in an interview the struggle to regulate online abuse (Hudson 2015). It is difficult to assess whether a comment is abusive without context. In such a scenario what does the corporate responsibility to respect require of a company? Would being a member of the GNI provide the kind of advice and support needed to embed this responsibility into a company such as Twitter?

8.1.2.1 European-Level: European Commission ICT Sector Guidance

The EU, as a result of the GPs and the EC's new CSR strategy, led an initiative to develop sector-specific guidance concerning businesses human rights responsibilities in the areas of ICT, employment and recruitment, and oil and gas. The guides were developed by the Institute for Human Rights and Business and Shift. The ICT Sector Guidance (2013) seeks to be practical, thus does not engage the way that the GPs do with wider debates about how to frame a business's human rights responsibilities. The EC's task was much narrower: how to give practical advice to companies in certain sectors concerning how to fulfill the corporate responsibility to respect. The Guidance mainly engages with the second pillar of the GPs, although it does address the third pillar requiring remedial mechanisms.

The Guidance targets the entire ICT industry, 'from telecommunications and Web-based services through software, and electronic device and component manufacturing' (European Commission 2013, p 6). It identifies six core elements to the duty to respect: a human rights policy commitment, a process of due diligence requiring assessment of potential and actual impacts, integration of findings regarding impacts, tracking of impacts, communication of performance to the public and a remedial process. As I have written elsewhere, the greatest strength of the Guidance is as an ideas-bank for how to approach various situations (Laidlaw 2015), in particular for start-ups and small and medium enterprises (SME) that need basic guidance in the form of a checklist. In this respect, the Guidance is a success. Where the Guidance is weak is in providing specific guidance for sub-sectors of the ICT industry in handling difficult scenarios. The human rights issues for the ICT sector are vast and changeable. By pitching the Guidance to the entire industry, it fell short on the kind of detail companies in the trenches of content regulation need. The GNI noted this criticism, stating (in reference to a draft of the Guidance), "by grouping together issues with fundamentally different dynamics, the usefulness of the guidance diminishes" (GNI 2013).

For example, OSPs are at the centre of legal and policy discussions concerning the liability that should be imposed on such intermediaries for enabling communication of third party content that is alleged to be unlawful. Common accusations are that this third party content is defamatory, harassing, offensive or terrorism-related, violates copyright law or infringes privacy. The Guidance acknowledges the critical role of intermediaries, and explores a few of the scenarios where their roles are challenged, but goes no further. For example, the Guidance notes that in the area of government requests for content removal or blocking "[t]his is an area in which discussions about the most appropriate approaches are ongoing." (p 44)

European Digital Rights (EDRi) criticized a draft of the Guidance, because government pressure on businesses to undertake voluntary measures 'leads to actions being taken in a legal grey zone between direct state obligations under international legal obligations, positive obligations of states under relevant legal instruments and the inapplicability of those same instruments to non-state actors.' (2013, p 1) Corporate action, therefore, might be more about pacifying governments who are threatening regulation, rather than as an effort to achieve a policy goal such as privacy (EDRi 2013). This might impact to the effectiveness of an action to protect human rights.

This can be seen with the creation of the United Kingdom's Internet Watch Foundation (IWF) by the ISP industry to address the problem of sharing child sexual abuse images. It was created as a direct reaction to government threats of regulation. The body that emerged failed to have in place features of good governance such as accountability, transparency, consistency and proportionality that might have been featured in a structure that was developed with clearer policy goals, although the IWF is a good example of robust and targeted industry regulation of child sexual abuse images. The failure to engaged more broadly with its role and impact meant that other rights such as the right to free speech or privacy, which were historically not seen as a focal point (or impact) of its regulatory role, didn't receive as much attention (Laidlaw 2015). The reality, however, is that many of these industry bodies, including the GNI (see the Global Internet Freedom and the Rule of Law hearings), exist because of government pushing from the shadows.

8.1.2.2 Multi-stakeholder Industry Initiatives: The GNI

The GNI was publicly launched in December, 2008 after 2 years of multi-stakeholder consultations among academics, civil society organizations, investors and companies. It was created as a response to the increasingly complex issues facing ICT businesses concerning their potential and actual impact on free speech and privacy, thus unlike the broad scope of the Guidance, the GNI is narrowly tailored to two human rights issues. As noted earlier, the creation of the GNI was spurred by incidents where its founding companies acceded to government demands. Thus a foundational element of the GNI is helping corporations navigate government requests that are not consistent with international human rights. The GNI describes its purpose as follows:

> GNI exists to provide guidance to the ICT industry and its stakeholders on how to protect and advance the human rights of freedom of expression and privacy when faced with pressures from governments to take actions that infringe upon these rights. (GNI FAQ)

The focus on government is significant, because some free speech and privacy issues that face OSPs might not be directly related to government pressure. Certainly the Snowden revelations and Arab Spring highlight the precarious situations these companies navigate in dealing with governments. The GNI audit revealed that these companies defended user rights by seeking clarification/modification of government requests, denying them or generally interpreting requests as narrowly as possible to meet human rights standards (GNI 2014). However, OSPs handle other issues (some legal and some ethical) such as data protection concerning collection and sharing of data with third parties (whether government or advertisers), regulating buyer to buyer disputes, facilitating or blocking payments for certain online activities and regulating disputes concerning abusive content. For example, how should Twitter address online shaming? Should Facebook allow pro eating disorder groups? How should Facebook address private requests for removal of content? Should PayPal block payments to copyright infringing sites? The GNI Core Commitments are framed broadly to include all potential impacts, and the GNI has engaged research in this area (ie Newland 2011). However, companies relationships with governments undergirds the organization's reason for being and risks sidelining or minimizing other, relevant human rights issues.

The original company members are Yahoo!, Microsoft, and Google. It has since grown to include Facebook (2012), Procera Networks (2013) and LinkedIn (2014). Evoca and Websense joined in 2011 and left in 2014 (GNI 2014). A glaring absence is Twitter.[2] There are notably no payment providers or e-commerce providers as members, such as Paypal and Amazon. Non-company members include, for example, the Berkman Center for Internet and Society, the Centre for Democracy and Technology and Human Rights Watch (GNI list of participants). Amnesty International, one of participants in drafting the GNI, pulled out just before it launched, because while it saw the GNI as progress concerning human rights issues, it did not see it as enough progress (Johnson 2008). The GNI has promoted the initiative to the entire ICT sector, but it has struggled to recruit telecommunications companies. In 2013, the telecommunications sector-specific initiative TID was launched, which is currently in collaboration with the GNI to share 'best practices, learning, and tools' (TID). This indicates potential for a certain amount of standardization across ICT sectors while allowing room for sector-specific practices.

---

[2] In a letter to Senator Durbin Twitter advised 'it is our sense that GNI's draft policies, processes and fees are better suited to bigger companies who have actual operations in sensitive regions.' (Letter, Alexander Macgillvray, General Counsel, Twitter to Senator Richard Durbin, p 2)

The GNI states it does four things:

1. Provides a **framework for companies**, rooted in international standards
2. Ensures **accountability** of ICT sector companies through independent assessment
3. Enables opportunities for **policy engagement**
4. Creates **shared learning opportunities** across stakeholder boundaries. (GNI About Us)

Its Core Commitments are set out in four documents: Principles, Implementation Guidelines, Accountability Policy and Learning Framework, and Governance Charter. Its Principles uses international human rights principles and bases members obligations on the GPs duty to respect. Concerning free speech it advises, among others, that companies should avoid or minimize government impacts that are inconsistent with international human rights principles. Concerning privacy, companies should protect privacy rights of users in all countries in which it operates. Note here the language operate, meaning the GNI applies to companies' activities in countries in which it has operational control, which is a narrower concept than a company's impact.

Three things are notable about the Core Commitments. First, the GNI emphasizes accountability. The Principles state, 'Participants will be held accountable through a system of (a) transparency with the public and (b) independent assessment and evaluation of the implementation of these Principles.' (GNI Principles) The Implementation Guidelines provide further detail on how to implement the Principles. Namely, member Boards must review the business concerning its impact on free speech and privacy; members must perform human rights impact assessments; the Principles must be internalized to a company through senior management, procedures and communication. (GNI Implementation Guidelines) This helps answer the question: how does a company know whether it has been successful in living up to the the GNI Principles?

The Implementation Guidelines also offer guidance on specific free speech and privacy issues. For example, the Implementation Guidelines emphasize the need to be as transparent as possible with users about demands that might be being made for user data or blocking of content, in effect "regulation by information" as the concept has been described elsewhere (Baumann-Pauly 2015, p 23). Innovative implementation of this is Google's Transparency Report where it details government requests to remove content or obtain information about users, copyright notice and takedown requests, European privacy requests for search removals, security, encryption and traffic data. This model has been replicated by other ICT companies, such as Vodafone (Law Enforcement Disclosure Report 2015) and Microsoft (Transparency Hub).

Second, it is significant that the Principles and Implementation Guidelines require members to engage with public policy. Issues of CSR push businesses into a potentially ill-suited position of being advocates. This is certainly the case for OSPs. The GNI embraces companies in this role making it a condition of membership, stating, '[p]articipants will encourage governments and international institutions to adopt policies, practices and actions that are consistent with and advance the Principles.' (GNI Implementation Guidelines) There is value to companies participating in these discussions. They can speak to the issues that they face and are in a better position to innovate responses. They are also self-serving and we cannot confuse a company with other organizations or individuals that exist to serve the public. Many OSPs business models are built on collection and sharing of their users data, precisely the human rights issues a body such as the GNI exists to address. A body like the GNI works within the boundaries of what these businesses are.[3] It is these types of compromises that concern human rights advocates that MSIs and CSR are not enough to protect and advance human rights.

Third, one of the GNI's greatest values is that it offers 'a safe space' (GNI Who We Are, p 3) for companies to work work through a complicated issue they are facing. The Implementation Guidelines states that the GNI offers an internal multi-stakeholder advisory forum to 'provide guidance to participating companies on emerging challenges and opportunities for the advancement of freedom of expression and privacy.' For example, it encourages shared learning by hosting a Policy and Learning Committee and email list for GNI participants to initiate discussions. Professionals in other fields have such a safe place, such as law society helplines for lawyers struggling with an ethical issue, and this is something I have advocated for in past research (Laidlaw 2015).

The system of accountability invites elaboration. The GNI requires member companies to self-report yearly and undergo an independent assessment every 2 years. The GNI has thus far completed one such assessment of its founding members, Google, Yahoo! and Microsoft (GNI 2014). The audit assessed a series of cases faced by its members between July 2011 and July 2013.[4] The Board concluded that the members were compliant with the GNI Principles. The audit focused largely on government restrictions (47 of the 59 cases reviewed involved a government request), although it did ask broader questions in the context of company-decision making, namely concerning how free speech and privacy was integrated into board leadership, the supply chain, the use of human rights impact assessments and integration of the GNI Principles into business operations.

The assessors selected the cases to review based on consultation with the the company and GNI stakeholders, in the range of 12–20 cases per assessment. In the context of free speech, the GNI identified the following as key: blocking/filtering of content, takedown requests, criminalization of speech protected under international human rights principles, intermediary liability, and selective enforcement of laws as against political dissidents. Concerning privacy, the GNI focused on the following issues: content surveillance, and state requests for user data, in particular, real name registration requirements, formal and informal requests for user information and data retention laws combined with government access requirements.

[3] The GNI demand of companies is to promote the rule of law and be consistent with the Principles (Implementation Guidelines).

[4] It was the third phase in assessment, the first being self-reporting and the second a review of implementation of GNI policies and procedures.

The Assessment appears like a robust engagement with the auditing process as envisioned by the GPs. Three issues, however, were evident from a review of the audit. First, a system of accountability is central to the GPs and the weaknesses of non-judicial mechanisms were evident in its deployment by the GNI in the audit. Snowden revealed that companies such as GNI members Google, Yahoo! and Microsoft had collected and shared customer data with the government at the direction of secret court orders. The GNI does not appear to have been consulted, and the GNI independent assessment did not unearth this information.

This seems to have had two effects. On the one hand it legitimized the importance of a body such as the GNI in holding businesses to account. On the other hand, it made it appear that this accountability was toothless. As I stated in earlier work, '[t]he secret nature of these companies data sharing with governments made a mockery of the human rights audit process, so much so that GNI founder, the Electronic Frontier Foundation (EFF), resigned from the organisation.' (Laidlaw 2015, p 106) The GNI responded in the independent assessment by advocating for law reform:

> Protecting the free expression and privacy rights of Internet users around the world—the goal behind the creation of GNI—has never been so vital. It was not possible, however, to assess the way in which GNI companies respond to U.S. national security requests because of the restrictions under U.S. law that prohibit the companies from disclosing any information related to such requests. This strengthens our belief that legal and policy reform is necessary and advocacy for increased transparency and other changes will be a greater part of our work in future. (GNI 2014, p 3)

Second, the GNI public report on the independent assessments is provided in aggregate form to allow for public disclosure of information that the GNI might not otherwise be able to reveal because of limits on disclosure due to legal prohibitions (such as court orders), user privacy rights and their terms of service, attorney-client privilege and trade secrets. The assessment rightfully notes that:

> [T]he assessments cover cases in which companies use a variety of means to push back on overbroad requests and, in some cases, challenge governments. Making the cases and strategies public could compromise the ability of companies to respond to government requests in a manner that maximizes respect for users' rights consistent with international human rights standards. (GNI 2014, p 15)

This openness of companies to help and the need for confidentiality would be key for the functioning of the GNI as envisioned, but equally the cases studies became generalized reassurance devices rather than public accountability tools. Public accountability is a key part to the success of CSR, because of the normative effect of social expectation. How to balance these conflicting needs is an ongoing dilemma.

Third, the audit reflects a criticism I have made about the GNI in general and that is the narrow focus on government interferences. OSPs also grapple with private interferences with rights through requests for content removal and the like. The project spearheaded by Rebecca McKinnon, Ranking Digital Rights, published a 2015 Corporate Accountability Index (Index), which assessed 16 companies, eight of them internet companies, including GNI members Google, Yahoo!, Microsoft and Facebook. The companies that rated the highest for their commitment to human rights were members of either the GNI or TID (McKinnon 2015a). Interestingly, where all evaluated companies struggled, including GNI members, was matters outside the GNI mandate, largely transparency concerning enforcement of Terms and Conditions and private interferences with rights. This indicates, perhaps, that whatever work the GNI is doing concerning government interference is having a positive effect, but is not stimulating reflection on the part of the companies concerning their wider impact. One way that the GNI can help this aspect is to embrace a wider mandate. The audit performed does not appear representative of these companies' potential and actual human rights impacts, potentially distorting views of their success in respecting human rights.

The GNI notes the importance of access to a remedial mechanism in the GPs and has committed to establishing it in its Governance Documents (GNI Governance Charter). The GNI engaged the nonprofit Shift (one of the drafters of the EC Guidance) to pilot an engagement and grievance mechanism (GNI Annual Report 2012). The concern is with the numerous complaints the GNI might have to field and the complexity of assessing free speech and privacy issues in a global environment. It states,

> The GNI recognizes that it may receive complaints and grievances from users concerning company compliance with the Principles. Due to the complexity of the global landscape regarding online freedom of expression and privacy, and the potential scale of complaints, the GNI will develop an appropriate complaints procedure consistent with its size and available resources. (GNI Governance Charter, cl 6)

At the time of writing, Shift reports that it has provided the GNI with a confidential report on the design of such a public engagement mechanism (Shift). One role the GNI should embrace is helping its members develop appropriate operational-level grievance mechanisms. Facebook, in its self-reporting, details how it approaches requests for data from governments (ie it interprets requests narrowly, challenges some requests in court, seeks transparency). There was no discussion in its self-reporting of how it can animate the grievance mechanism advocated in the GPs. This leaves a swatch of human rights impacts unattended.

As of 2016, the future of the GNI is unclear. It has evolved since its launch.[5] In 2008 it was focused largely on providing shared standards for the ICT sector. It continues in this role (GNI 2014). Its evolution has been in its role as advocate, most evident for surveillance reform. In some ways the Snowden revelations helped push the GNI through early growing pains to a more mature organization. By embracing the role of advocate it is understanding the two sides to CSR as it relates to human rights: the role of regulation in setting standards and accountability and the opportunity to be leaders in social change.

[5] It underwent a strategic review in 2014 to determine whether it was making progress. It revised its Governance Charter and Accountability Framework ramping up its advocacy work, increasing and diversifying fundraising, streamlining its communications strategy with the public, and simplifying its independent assessment process.

The GNI has been criticized for being too narrowly focused on internet providers, or for failing to do enough for the protection of human rights (see Business and Human Rights Resource Centre). Most of the criticisms are with its scope and accountability, and the GNI has made changes to address some of the criticisms. Its audits, for example, have been argued to be biased because companies pick their assessors (EFF 2013), although the Governance Charter states these auditors are vetted for independence (GNI, cl 5). Other criticisms are that the GNI does not sufficiently address issues concerning data collection, sharing and retention practices of its members (EFF 2013). The revisions to the Governance Charter and Accountability Framework and the GNI's advocacy work have gone some way to address these concerns (GNI 2014).

An important assessment of the GNI is whether it is successful, both internally in achieving its objectives and externally as filling a governance gap (Baumann-Pauly 2015). A comprehensive assessment is beyond the scope of this chapter. However, a few points can be made. At its core the criticisms are that the GNI is not accountable enough (See Maclay 2010). As a body, it is an excellent example of an effort to build clarity in the governance gap where these businesses operate. Much of the criticisms of CSR are that it can be a distraction from real movement in the area or real protection of rights. The GNI has shown itself to be responsive to criticism, open to change and able to be more than a public relations exercise. That is likely reflective of the Board leadership, and like all things, the culture and attitude can change with membership. This is a risk with any voluntary code, whether industry or at a company-level. The GNI shows that CSR does not stand in opposition to law reform, but can complement law reform efforts.

An enduring problem for the GNI is the lack of take-up of the regime. 7 years into its launch it has six members, gaining three companies beyond its founding members. Why have so few ICTs signed on? Even if the GNI is considered more narrowly as a framework for internet providers, the lack of take-up is concerning. The risk at the moment is that the GNI fails to evolve. By focusing so narrowly on government impacts it does not engage the wider human rights impacts of its members.

8.1.2.3 The Future

The question is whether CSR can bring about lasting change for ICTs. Can the GNI, for example, prompt governance by its members of their human rights impact in line with international human rights standards? The Fair Labor Association (FLA), which has conducted audits of its members since 2002, analysed the question of whether these audits had been effective. It concluded that 'while audits are reasonably effective in identifying violations they are less so in effecting lasting change.' (see discussion Baumann-Pauly 2015, p 31) In the case of the FLA, despite the audits, the number of human rights violations for factories remained high. Achieving lasting change is a key task for a body like the GNI in a world with few alternatives to address the types of transnational issues that face OSPs.

There continue to be calls for a legally binding treaty for business and human rights (see the establishment of a working group to explore such a treaty: A/HRC/26/L.22), but consensus is unlikely in the areas of free speech and privacy.[6] Given the transnational nature of OSPs and user communications, industry becomes the best contender to embed human rights protection, although government always retains the duty to lead in setting the expectation of companies.[7] As Baumann-Pauly et al argue, industry level initiatives are advantageous compared to broader efforts, because they can lie comfortably between voluntary and mandatory regimes, and companies can influence each other in creating and maintaining industry standards (2015).

These initiatives show that simple categories of CSR as voluntary and law as binding fail to appreciate the network of governance structures that regulate business. For example, transparency was a common theme in the initiatives discussed in this chapter. It is a foundational element of the GNI. Governments are joining the action and California, for example, requires certain retailers and manufacturers to report how they address supply chain issues involving trafficking and slavery (California Transparency in Supply Chains Act (2010), §1714.43). The UK has passed similar legislation (Modern Slavery Act 2015 c. 30).

Our focus should be on how to support CSR, whether MSIs or company level, but not as an alternative to other forms of governance. The better question to ask is: how can CSR be used to complement other efforts to achieve a desired objective? And, what factors lead to the conclusion that CSR, in a given situation or sector, is an ill-suited device where more traditional legal measures are needed? Revenge pornography is a good example here. Most of the OSPs that host such content would never join a body like the GNI. Law reform is a more appropriate route to achieving the desired objective here.[8] How a business should handle requests to take down content in different countries with varying laws more readily involves various stakeholders and regulatory approaches. Are the right people tasked with assessing such content? Does the public know about the Terms and Conditions and other processes the company uses? These are questions where CSR frameworks, whether industry-wide or organization-level, can help a company be more reflective in its practices. British Telecom, for example, implemented a sliding scale tool bar to inform users about its cookies policy (http://www.btexpedite.com/cookie-policy/).

Is CSR enough to protect and advance human rights as it relates to OSPs? The Snowden revelations indicate that the use of audits by the GNI is flawed. It is hard to hold companies accountable when you don't know what they've done. The

---

[6] See the history surrounding the failed negotiations for a hate speech provision in the Convention on Cybercrime (2001).

[7] Recall the first pillar of the GPs of a duty on states to protect human rights. This relationship for internet related issues in explored in more detailed in Laidlaw 2015, chapter 6.

[8] New laws have been passed in some jurisdictions to specifically address revenge pornography. For example, see ss. 33-35 of the United Kingdom's Criminal Justice and Courts Act (2015), 2015 c. 2, and s. 162.1 of Canada's Criminal Code. For discussion of developments in America see Citron 2014.

Sullivan Principles, on which the GNI is based, were created in the 1980s to guide companies in doing business in apartheid South Africa. Sullivan later rejected the Code as not doing enough to protect human rights, rather asserting it had the opposite effect of legitimizing the regime (Downes 2011). It is the same story with the Kimberley Process, an international diamond certification scheme designed to prevent the supply of bloody diamonds. One of its drafters, Global Witness, later resigned from the Process, stating, '[the] past decade has proven that the Kimberley Process cannot clean up the diamond sector on its own.' (Global Witness 2013) 20 years from now will it be the same story for technology sector initiatives such as the GNI?

As we saw, GNI and TID members were ranked higher in the Corporate Accountability Index for commitment to human rights than non-members. Twitter, for example, which is not a member of the GNI, was described by Rebecca McKinnon as showing 'very little evidence of institutionalization of [human rights] commitments.' (Corporate Accountability Index Launch Event 2015) However, all companies evaluated, whether GNI members or not, ranked low on the index concerning communication to users about their practices affecting privacy and free speech (McKinnon 2015b). Thus, while a company might have practices in place concerning how it restricts content, this information is not made public. As the Index explains, '[w]ithout such information it is difficult to hold companies, governments, and other actors accountable when users' rights are undermined.' (McKinnon 2015, p 4) On the other hand, nine of 16 companies evaluated publish transparency reports. Thus, such reports are becoming standard practice, but what is deemed a matter requiring transparency should evolve.

The most pressing task for any CSR mechanism, whether company or industry level, is to tackle the development of remedial mechanisms. Access to a forum of remediation to hear a grievance and provide a remedy is the critical third pillar of the GPs and has thus far been sidestepped by most initiatives and company-level CSR. In the Index, Indian and South Korean-based companies had stronger remedial mechanisms, but such companies are legally required to have such mechanisms in place (McKinnon 2015). This indicates that legal regulation might be important to undergird the development of such mechanisms.

Transparency and access to a remedy go hand-in-hand. CSR at its core is about public accountability, which necessitates the public be informed. Such transparency is how users learn about corporate practices, learn whether their rights have been impacted and learn about grievance mechanisms, if any, and how to use them. In this sense, the essence of the GPs - the requirement that companies must know and show respect for human rights - has only been partially engaged. OSPs are acquiring knowledge of their human rights impact but as of yet are not sufficiently showing respect for human rights to the extent needed to effect lasting change.

References

Alexander, A. (2015, August 24). Digital Surveillance "worse than Orwell", says new UN privacy chief. http://www.theguardian.com/world/2015/aug/24/we-need-geneva-convention-for-the-internet-says-new-un-privacy-chief. Accessed 6 Nov 2015.

Baumann-Pauly, D. et al. (2015). Industry-Specific Multi-Stakeholder Initiatives That Govern Corporate Human Rights Standards – Legitimacy Assessments of the Fair Labor Association and the Global Network Initiative. UNSW Law Research Paper No. 2015-12.

Bernal, P. (2011). Rise and Phall: lessons from the phorm saga. In S. Gutwirth et al. (Eds.), *Computers, privacy and data protection: An element of choice* (pp. 269–283). Dordrecht/New York: Springer.

Bilchitz, D. (2013). A Chasm between "Is" and "Ought"? A critique of the normative foundations of the SRSG's framework and guiding principles. In S. Deva & D. Bilchitz (Eds.), *Human rights obligations of business: Beyond the corporate responsibility to respect* (pp. 107–137). Cambridge: Cambridge University Press.

Blowfield, M., & Murray, A. (2008). *Corporate responsibility: A critical introduction*. Oxford/New York: Oxford University Press.

British Telecom. http://www.btexpedite.com/cookie-policy/. Accessed 6 Nov 2015.

Business and Human Rights Resource Centre. www.business-humanrights.org/Documents/GlobalNetworkInitiative-responses. Accessed 6 Nov 2015.

California Transparency in Supply Chains Act. (2010), Cal. Civ. Code

Citron, D. K. (2014). *Hate crimes in cyberspace*. Cambridge, MA: Harvard University Press.

Companies Act. 2006 c. 46

Convention on Cybercrime. (2001), 23.XI.2001

Criminal Code RSC 1985, c C-46

Criminal Justice and Courts Act, 2015 c. 2

Deva, S., & Bilchitz, D. (Eds.). (2013). *Human rights obligations of business: Beyond the corporate responsibility to respect*. New Delhi: Cambridge University Press.

Downes, L. (2011, March 30). Why no one will join the Global Network Initiative. http://blogs.forbes.com/larrydownes/2011/03/30/why-no-one-will-join-the-global-network-initiative/. Accessed 6 Nov 2015.

Electronic Frontier Foundation. (2013, October 10) GNI resignation letter. www.eff.org/document/gni-resignation-letter. Accessed 6 Nov 2015.

European Commission. (2013). ICT Sector Guidance on Implementing the UN Guiding Principles on Business and Human Rights

European Commission. A renewed EU strategy 2011–2014 for Corporate Social Responsibility, COM (2011) 681.

European Digital Rights. (2013). Comments on the Guidance for the Information and Communication Technologies ("ICT") Sector on Implementing the UN Guiding Principles on Business and Human Rights. http://www.ihrb.org/pdf/eu-sector-guidance/ICT/European-Digital-Rights.pdf. Accessed 6 Nov 2015.

Garside, J (2011, July 26). Vodafone under fire for bowing to Egyptian pressure. http://www.theguardian.com/business/2011/jul/26/vodafone-access-egypt-shutdown. Accessed 6 Nov 2015.

GNI. (2014). Public Report on the Independent Assessment Process for Google, Microsoft, and Yahoo. http://globalnetworkinitiative.org/sites/default/files/GNI%20Assessments%20Public%20Report.pdf. Accessed 26 May 2016.

GNI explanation. (2014, May 26). GNI Presents 2013 Annual Report. https://www.globalnetworkinitiative.org/news/gni-presents-2013-annual-report. Accessed Nov 2015.

GNI. (2013, February) Comments from the GNI on the Draft Guidance for the ICT Sector on the Corporate Responsibility to Respect Human Rights. http://globalnetworkinitiative.org/sites/default/files/GNI%20comments%20on%20EC%20draft%20ICT%20guidance.pdf. Accessed 26 May 2016.

GNI. (2012). Annual Report. http://globalnetworkinitiative.org/sites/default/files/GNI%20 Annual%20Report%202012.pdf. Accessed 6 Nov 2015.

GNI. Frequently asked questions. https://globalnetworkinitiative.org/content/frequently-asked-questions. Accessed 6 Nov 2015a.

GNI list of participants. http://globalnetworkinitiative.org/participants/index.php. Accessed 6 Nov 2015.

GNI. About Us. https://globalnetworkinitiative.org/about/index.php. Last visited 6 Nov 2015b.

GNI Principles. http://globalnetworkinitiative.org/principles/index.php. Accessed 6 Nov 2015.

GNI, Who We are. What we do. Why it matters. https://globalnetworkinitiative.org/sites/default/files/GNI_brochure.pdf. Accessed 26 May 2016.

GNI Core Commitments. https://globalnetworkinitiative.org/corecommitments/index.php. Accessed visited 6 Nov 2015.

GNI Implementation Guidelines. http://globalnetworkinitiative.org/implementationguidelines/index.php. Accessed 5 Nov 2015.

GNI Governance Charter. https://globalnetworkinitiative.org/charter/index.php. Accessed 6 Nov 2015.

Global Witness. (2013, April 1). The Kimberley Process. https://www.globalwitness.org/campaigns/conflict-diamonds/kimberley-process/. Accessed 6 Nov 2015.

Google. (2010, January 12). A new approach to China. http://googleblog.blogspot.com/2010/01/new-approach-to-china.html. Accessed 6 Nov 2015.

HM Government. (2013). Good business: Implementing the UN guiding principles on business and human rights.

Hudson, L. (2015, October 22). 6 Experts on how Silicon Valley can Solve Online Harassment. http://www.wired.com/2015/10/how-silicon-valley-can-solve-online-diversity-and-harassment/. Accessed 6 Nov 2015.

Human Rights Council. A/HRC/26/L.22

Human Rights Council. A/HRC/20/L.13

Human Rights Council. A/HRC/RES/28/16

Human Rights Council. A/HRC/RES/17/4

ICANN, Uniform Domain-Name Dispute Resolution Policy.

The International Bill of Human Rights comprised of the Universal Declaration of Human Rights. 1948, the International Covenant of Civil and Political Rights 1966 and the International Covenant on Economic, Social, and Cultural Rights 1966

The International Labour Organization's Declaration of Fundamental Principles and Rights at Work. 1988

ISO 26000 Guidance on social responsibility.

Manokha, I. (2004). Corporate social responsibility: A new signifier? An analysis of business ethics and good business practice. *Politics, 24*(1), 56–64.

Johnson, B. (2008, October 30) Amnesty criticises Global Network Initiative for online freedom of speech. www.guardian.co.uk/technology/2008/oct/30/amnesty-global-network-initiative. Accessed 6 Nov 2015

Kinley D. et al. (2007). The Norms are dead! Long life the Norms! The politics behind the UN Human Rights Norms for corporations. In: D. McBarnet et al (Eds.), The new corporate accountability: Corporate social responsibility and the law (pp. 459–475). Cambridge: Cambridge University Press

Kopytoff VC (2011, March 6). Sites like Twitter absent from Free Speech Pact. http://www.nytimes.com/2011/03/07/technology/07rights.html?_r=1. Accessed 6 Nov 2015.

Laidlaw, E. (2015). *Regulating speech in cyberspace: gatekeepers, human rights and corporate responsibility*. Cambridge/United Kingdom: Cambridge University Press.

Maclay, C. (2010). Protecting privacy and expression online: Can the global network initiative embrace the character of the net? In R. Deibert et al. (Eds.), *Access controlled: The shaping of power, rights, and rule in cyberspace* (pp. 87–108). Cambridge, MA: MIT Press.

McBarnet, D., et al. (Eds.). (2007). *The new corporate accountability: Corporate social responsibility and the law*. New York: Cambridge University Press.
McBarnet, D. (2007). Corporate social responsibility beyond law, through law, for law: the new corporate accountability. In D. McBarnet et al. (Eds.), *The new corporate accountability: Corporate social responsibility and the law* (pp. 9–58). New York: Cambridge University Press.
McKinnon, (2015a). Corporate accountability index. https://rankingdigitalrights.org. Accessed 6 Nov 2015.
McKinnon, (2015b). Corporate accountability index launch event, 3 Nov 2015.
Microsoft, Transparency Hub. https://www.microsoft.com/about/corporatecitizenship/en-us/transparencyhub/. Accessed 6 Nov 2015.
Modern Slavery Act. 2015 c 30
Newland E et al Account Deactivation and Content Removal: Guiding Principles and Practices for Companies and Users. The Berkman Center for Internet & Society Research Publication Series No. 2011-2009.
Nieuwenkamp, R. (2015, April 30). Responsible business conduct in cyberspace. http://www.ihrb.org/commentary/responsible-business-conduct-in-cyberspace.html. Accessed 6 Nov 2015.
Nolan, J. (2013). The corporate responsibility to respect human rights: Soft law or not law? In S. Deva & D. Bilchitz (Eds.), *Human rights obligations of business: Beyond the corporate responsibility to respect* (pp. 138–161). New York: Cambridge University Press.
Nominet, Dispute Resolution Service Policy.
OECD Guidelines for Multinationals Enterprises.
Okoye, A. (2009). Theorising corporate social responsibility as an essentially contested concept: Is a definition necessary? *Journal of Business Ethics, 89*, 613–627.
Reed, C. (2012). *Making laws for cyberspace*. Oxford: Oxford University Press.
Report of the special rapporteur on the promotion and protection of the right to freedom of expression, Frank La Rue to the United Nations General Assembly. (2011). http://www2.ohchr.org/english/bodies/hrcouncil/docs/17session/A.HRC.17.27_en.pdf. Accessed 20 May 2016.
Resolution centre. http://resolutioncenter.ebay.com. Accessed 6 Nov 2015.
Ruggie, J. (2011). Guiding principles on business and human rights: Implementing the United Nations "Protect, respect and remedy" framework. www.business-humanrights.org/media/documents/ruggie/ruggie-guiding-principles-21-mar-2011.pdf. Accessed 6 Nov 2015.
Ruggie, J. (2008). Protect, Respect and Remedy: A Framework for Business and Human Rights, Report of the Special Representative of the Secretary General on the issue of human rights and transnational corporations and other business enterprises. www.reports-and-materials.org/Ruggie-report-7-Apr-2008.pdf. Accessed 6 Nov 2015.
Ruggie, J. (2013). Just business: Multinational corporations and human rights. New York: W.W. Norton & Company.
Shift, Advising the Global Network Initiative on a Public Engagement Mechanism. http://shiftproject.org/project/advising-global-network-initiative-public-engagement-mechanism. Accessed 6 Nov 2015.
Steel, E., Fowler, G. A. (2010, October 18). *Facebook in privacy breach*. http://www.wsj.com/articles/SB10001424052702304772804575558484075236968. Accessed 6 Nov 2015.
Technology Industry Dialogue.
The Electronic Industry Code of Conduct.
The Guardian. (2014, January 24). Two jailed for Twitter abuse of feminist campaigner. http://www.theguardian.com/uk-news/2014/jan/24/two-jailed-twitter-abuse-feminist-campaigner. Accessed 26 May 2016.
The Global Internet Freedom and the Rule of Law hearings led by Senator Richard. Durbin. http://durbin.senate.gov/public/index.cfm/pressreleases?ID=c3078a7d-bfd9-4186-ba86-2571e0e05ec8. Accessed 6 Nov 2015.
The Performance Standards on Environmental and Social Sustainability of the International Finance Corporation.

United Nations draft Norms on the responsibilities of transnational corporation and other business enterprises with regard to human rights. (2003), E/CN.4/Sub.2/2003/12

United Nations Global Compact.

Vick, D. (2005). Regulating hatred. In M. Klang & A. Murray (Eds.), *Human rights in the digital age*. London: Cavendish Publishing.

Vodafone. (2015). Law Enforcement Disclosure Report. http://www.vodafone.com/content/index/about/sustainability/law_enforcement.html#. Accessed 6 Nov 2015

Wang, T. (2013, August 3). Our Commitment. https://blog.twitter.com/en-gb/2013/our-commitment. Accessed 6 Nov 2015.

Webb, K., & Morrison, A. (2004). The law and voluntary codes: Examining the "Tangled Web". In K. Webb (Ed.), *Voluntary codes: private governance, the public interest and innovation*. Ottawa: Carleton Research Unit for Innovation, Science and Environment.

Chapter 9
Online Service Providers: A New and Unique Species of the Firm?

Robert Wentrup and Patrik Ström

Abstract This chapter discusses Online Service Providers (OSPs) and gauges whether they can be defined as a new and unique species of the firm. It is concluded that OSPs have many typical features such as their intermediary role, the dependence on network effects, and a capacity to grow rapidly across borders via the online medium, but that there are few features that distinguish them as a unique species. Yet, it is argued that the intertwined social relation with end-customers, often in co-development milieu combined with the lack of a direct monetary relationship, could be one such unique feature. Therefore, an important challenge, particularly for countries in the global South, is to construct and implement policy frameworks in order to ensure that OSPs and the people using them are embedded in a dynamic but fair environment.

9.1 Introduction

A large part of information and services available on the Internet is produced by, or channelled through, profit-driven firms. In business and economic literature these firms are often called Online Service Providers (OSPs). But they are also known under names like *Internet firms* (Kim 2003; Chen 2006), *digital service firms* (Ekeledo and Sivakumar 2004), *e-commerce companies* (Luo et al. 2005), *e-commerce corporations* (Singh and Kundu 2002), *social media companies* (Okazaki and Taylor 2013), *siren servers* (Lanier 2013), *cybermediaries* (Caillaud and Jullien 2001), *Internet intermediaries* (OECD 2012), *social networks* (Boyd and Elison 2007) and more. This myriad of names and the lack of a distinct definition complicate the exercise of research and policymaking. For example, it is difficult to conduct quantitative studies without a coherent classification code. In

---

R. Wentrup (✉) • P. Ström
Centre for International Business Studies, School of Business, Economics and Law, University of Gothenburg, Gothenburg, Sweden
e-mail: robert.wentrup@handels.gu.se

M. Taddeo, L. Floridi (eds.), *The Responsibilities of Online Service Providers*, Law, Governance and Technology Series 31, DOI 10.1007/978-3-319-47852-4_9

national commerce registries there is no coherent classification code for OSPs. Webopedia (2015, no page) suggests a rather broad definition of OSPs:

> … any company, organization or group that provides an online service. These types of services may include Web sites, discussion forums, chat rooms, or Web mail. OSPs may also refer to a company that provides dial-up access to the Internet.

Some well-known OSPs are *Google*, *Facebook*, *Twitter* and *LinkedIn*. OSPs are young firms and have thrived since the end of the 1990s alongside the rapid growth of the Internet, which today accounts for more than 3.2 billion users worldwide. Facebook alone has almost half of them—1.5 billion users and annual revenues of 12,466 million USD in 2014, 55 % of which outside the US (Facebook 2015). Common for the successful OSPs are that they have large numbers of users; they are all dependent on the Internet for delivering their services; and they commoditize bits and bytes of information in some form, often personal information. Apart from these evident commonalities, are there any other unique characteristics for OSPs as firms?

In order for researchers and policymakers to understand how OSPs operate, and assess their impact and role in the economy, not least their influence over citizens, an important task is to define their distinct characteristics. Little has been written on this topic. The lack of knowledge in relation to ordinary firms has also been discussed in business media (Economist 2015). Through the development of technology and new forms of incentive structures, client relationships in many newly established firms seem to challenge the traditional perception of firms' development over time. The objective of the chapter is to fill the knowledge gap and attempt to distinguish OSPs from other types of firms. We discuss OSPs using a cross-disciplinary theoretical framework and look at some specific economic variables from a small set of US-based OSPs to see if they reveal any significant features. The chapter draws from literature of economics, international business, economic geography, and information technology. The contribution of the chapter is a number of distinct characteristics and a discussion around OSPs' newness and uniqueness as species of the firm.

9.2 The Birth and Infancy of OSPs

Social media or *social network sites* can be described as one type of OSP, and are often associated with Facebook, Instagram and Twitter. Boyd and Ellison (2007, p. 211) define social network sites as:

> web-based services that allow individuals to (1) construct a public or semi-public profile within a bounded system, (2) articulate a list of other users with whom they share a connection and (3) view and traverse their list of connections and those made by others within the system.

In a similar vein, Kaplan and Haenlein (2010) describe how social media stems from *Usenet* (a worldwide discussion forum) in the early 1980s, which allowed public posts of messages over the Internet. The whole idea of the Internet stems from the desire to communicate and connect people across borders, and the first recordings of social interactions over the network were a series of messages written by MIT scholars in 1962 (Leiner et al. n.d.). Sir Tim Berners-Lee and his colleagues at CERN in Switzerland are often recognized as the founders of the World Wide Web (WWW) in the late 1980s (ibid.). The main driver was to implement a system of automatic information sharing of research data between scientists and universities across borders. The protocol *HTTP* and its markup language *HTML* were born in the same period. Early attempts to transport digital communication over the Internet were influenced by a non-profit collaborative approach, where users could contribute with content and also code to online software. These collaborative features were also significant for what many understand as *Web 2.0*, where content and application code do not belong to a specific individual or firm, but instead are "open source", free for anyone to contribute to and access. One early example of this was the non-profit operating system (OS) *Linux* (created in 1991), originally created by the Finn *Linus Torvalds*, which became the antidote to *Microsoft's* OS (Windows).

The early days of the Internet were dominated by the universities and computer gurus. Unsurprisingly, it did not take long time before commercial spin-outs popped up. Search engines like *Netscape* and *Lynx*, and portals like *AOL* and *Yahoo!* were some of the first dominating OSPs. They were called *start-ups* or *dot.coms*, a term associated with a Silicon Valley garage and geeky enterprise style, often driven by passion for technology rather than financial motives. Silicon Valley became the birthplace for many pioneering OSPs (Zook 2002) and in terms of market capitalization and number of start-ups it is still the dominating geographical location (see Fig. 9.1). *Google* was founded here in the mid-1990s. *Facebook*, *Twitter*, *Airbnb*, *Uber* and *LinkedIn* were all born within a limited spatial realm in this area in the early 2000s. The fact that a single geographical space has such a leading position in one sector can be explained by cluster theory (Porter 1990) and the concepts of agglomeration (Dicken 2015, Leamer and Storper 2001, Rusten and Bryson 2010) and is not unique to the OSPs sector.

9.3 The Growth Phase of OSPs

When *Yahoo!* and *Netscape* were born there were merely 15 million Internet users in the world (Internet World Stats 2015), and most of them were located in the US. Some European, notably Scandinavian countries, as well as Japan were early adopters of broadband Internet infrastructure, but the masses of Internet usage and e-commerce did not really take off until the mid-2000s, when the total number of Internet users reached one billion. In the late 1990s, connection speed and devices were not adapted for smooth and rapid Internet "surfing". Neither did international e-commerce projects flourish (Forsgren and Hagström 2007). Yet, there were high hopes of the commercial potential of Internet-related firms, which drove financial markets to record heights and eventually contributed to the stock market crash in the

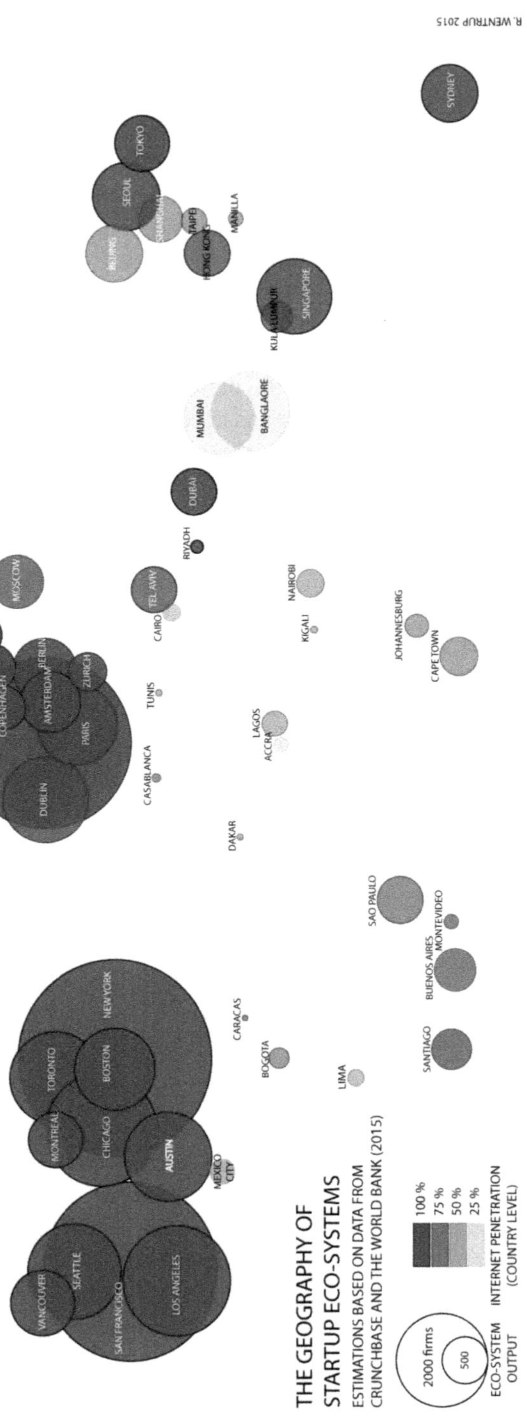

Fig. 9.1 A geography of worldwide start-up eco-systems based on data from Crunchbase (www.crunchbase.com) (Source: Authors' graphic)

year 2001, and consequently the first OSPs' apocalypse sometimes referred to as the dot.com collapse.

Today the OSPs landscape is more geographically distributed. On the supply side most of the largest OSPs are still located in California but there is an emerging stream of fast-growing OSPs from Asia, and particularly China, e.g. *Alibaba*, *Baidu*, and *Tencent*. Overall though, the global cities, e.g. New York, London, and Paris, give birth to most of the world's OSPs (Warf 2013). Smaller but rich countries like Israel and Sweden are other examples of locations that have been successful in producing OSPs, e.g. *Waze*, *Skype* and *Spotify*. Yet, it is incontestable that Silicon Valley has an edge in fostering OSPs, with global reach and high innovation capacity. Silicon Valley-based OSPs are recognized for their disruptive characteristics aligned with Schumpeter's (1934) theory of innovation. But the rest of the world is not lagging behind in being reactive in producing OSPs although they often have a less disruptive character and are more of a *Kirznerian* innovation type (Sundqvist et al. 2012), i.e. a local version of a Silicon Valley original. For example, there are local versions of Facebook in many countries, e.g. *VKontakte* in Russia. France has a local version of *Youtube* called *Daily Motion*, its own *LinkedIn* (*Viadeo*), its own *Spotify* (*Deezer*) and soon its own *Netflix* (Le Monde 2015). China has via its protectionist trade policy enabled the growth of local alternatives to the US originals (Anwar 2013). In Africa, OSPs working with specific local needs have emerged, e.g. *m-pesa* (Mbogo 2010). Although these local OSPs may not have the potential to scale on an international level, because they often address specific local needs not applicable on foreign markets, their existence demonstrates the emerging local competition in the OSPs market. Large international OSPs are forced to face local competition, and need to offer differentiated and localized online services, and adapt to local market prerequisites in order to overcome the liability of outsidership (Johanson and Vahlne 2009). In this respect, *born globals* theory (McKinsey 1993; Knight and Cavusgil 1996; Madsen and Servais 1997; Gabrielsson et al. 2008; Gabrielsson and Pelkonen 2008) could explain how OSPs reach a global audience swiftly, but with time the online service market is to an increasing extent influenced by local prerequisites, and requires local physical presence. An important feature in the OSPs market is geo-compatible services, which require sales staff on the ground to capture the local "buzz" and adapt the market offer accordingly. Hence, we can expect that the internationalization behaviour of OSPs will with time adhere to the global norm of conventional multinational enterprises, i.e. a gradual and increasing presence locally aligned, which is supported by the Uppsala school internationalization theory (Johanson and Vahlne 1977, 2009). Despite the geographical domination of Silicon Valley, there is an indication that there is room for local alternatives to grow, and that the dominance of Silicon Valley *digital monopolies* (The Economist 2014) might be diluted as the online markets become more local and also more regulated.

On a general level, the growth pattern indicates that the OSPs sector seems to follow the life cycle of how conventional industries evolve. If we denote the late 1990s as the pioneering phase, we have from the mid-2000s experienced a growth phase, which will continue for some time to come until Internet penetration rates reach saturation levels, and as e-commerce volumes reach maturity. Along this phase, there is likely to be increasing competition, also from various geographical regions. We will then most likely see a mature growth phase where competition will be significant and margins of OPS are likely to fall.

9.4 Geographical Distribution of OSPs Users

Above we concluded that both in nominal terms (number of start-ups) and in market capitalization the large OSPs have emerged from the global North, and particularly from the US. The geographical distribution of the users differs in nominal terms. On the user side we can conclude that the global North has the highest rates of Internet penetration (Internet users per capita), but due to demographic reasons most Internet users are (in nominal terms) located in the global South. China is the country with most Internet users (over 600 million). Although it can be debated whether the global South is catching up or not (Graham et al. 2015), and whether there will be a geographical distribution of power in the OSPs industry, a plausible scenario is that due to the user impact from the global South, this well certainly be reflected in the supply side and localization of the OSPs.

Global Internet traffic continues to grow at a high speed (20 % in 2012). However, comparing CAGR in 2015 with 2012 (39 %), it shows that the rate has slowed down (OECD 2015). This actually indicates that the digital divide gap still persists, which Table 9.1 also reveals. Measured in percent units, the Internet penetration increase has been the lowest in Africa in the last decade. Internet penetration is reaching saturation levels in the global North, where the Internet is affordable, but there are still immense digital shadows in many countries in Africa. The lack of Internet infrastructure and weak policy frameworks lead to low affordability and hinder mass-diffusion of Internet access. In Eritrea, for example, 1 year of Internet broadband costs approximately 50 years of an average salary (Graham et al. 2015). This obviously also has an impact on the OSPs side, i.e. OSPs' entrepreneurship. The time lag that passes before countries in the global South reach Internet levels allows OSPs in the global North to take a distant lead in the sector. In order to avoid a permanent techno-dependency of the global North (Sonaike 2004) it is, therefore, critical for countries in the global South to engage in putting Internet policies in place enabling local OSPs to grow and prosper.

Table 9.1 Global Internet penetration rates

	Europe	The Americas	Asia & Pacific	Africa	Arab states	CIS
Internet penetration rate 2005	46.3	35.9	9.4	2.5	8.3	10.3
Internet penetration rate 2015	77.6	66.0	36.9	20.7	37	59.9
Increase in percent units	31.3	30.1	27.5	18.2	28.7	49.6

Source: Data from ITU: http://www.itu.int/en/ITU-D/Statistics/Pages/default.aspx

9.4.1 Why Do OSPs Exist?

In order to gauge whether OSPs are a unique and new species of firm we must first define 'firm'. In economics, the theory of firm outlined by Coase (1937) states that firms exist to handle transaction costs since. It is expensive for individuals to organize these themselves by going to the marketplace and engaging in multiple simultaneous commercial relationships and contracts. Firms exist to coordinate these relationships and contracts, and can charge customers for that activity. The transaction costs or marketing costs are the cost of coordinating economic activity. People are willing to pay a price for a firm to organize different entrepreneurs to bring a service or a product to them. Deducing from Coase, we could thus state that OSPs exist since people are willing to pay a price for an agent to coordinate online information, e.g. social contacts, search algorithms, music or news, for them. However, for many OSPs the price mechanism is not monetized vis-à-vis the end-customer. In the case of OSPs, most people do not actually pay these firms directly. The bulk of revenues from *Google*, *Facebook*, and *Twitter* are generated via advertisement fees from other firms. Thus far there is often no monetized relationship between people and the successful OSPs although end-customers indirectly finance OSPs from purchases of other goods and services. Hence, OSPs are indirectly driven by end-customer consumption.

This business model is not new in the economy. For example, "free" newspapers and "free of charge" TV, which apply the same mode, have existed for some decades. The main difference between the free-of-charge media examples and OSPs, like *Facebook* and *Twitter*, is that the people actively co-create the content for these firms (Grabher et al. 2008, Perks et al. 2012), and de facto become agents within a co-development *user-OSP milieu*. The collaboration brings together both the supply side and demand side and creates what Jensen-Ørberg and Petersen (2014) call a value proposition developed by the service provider. This proposition can be different depending upon the specific line of business. Additionally, for social network OSPs, the business models are often built on real-life identities, and discourage pseudonyms or fake identities (Krombholz et al. 2012). The sustainability of the business model is dependent on functioning advertising generated by real humans, not machine-generated clicks. Although, this is only valid for a subset of OSPs, these particular features—co-creating and indirectly paying end-customers—is to our knowledge a unique feature of this type of OSPs (see Fig. 9.2).

What is then the implication of such a non-monetary relation between the OSPs and its users? One consequence could be an emotional fidelity from the end-customer to the OSP, potentially followed by an OSPs lock-in. For a Facebook or Twitter user with more than 500 "friends" or "followers", which has been proceeded by substantial personal investment, a potential divestment of the service would not be a random decision. Studies of users of legal online sharing music platforms have shown that interactivity, perceived ease of use, self-expression, social presence, and social identity are significant positive predictors of music sharing intention (Lee et al. 2011). For social network sites, self-representation, interaction and commu-

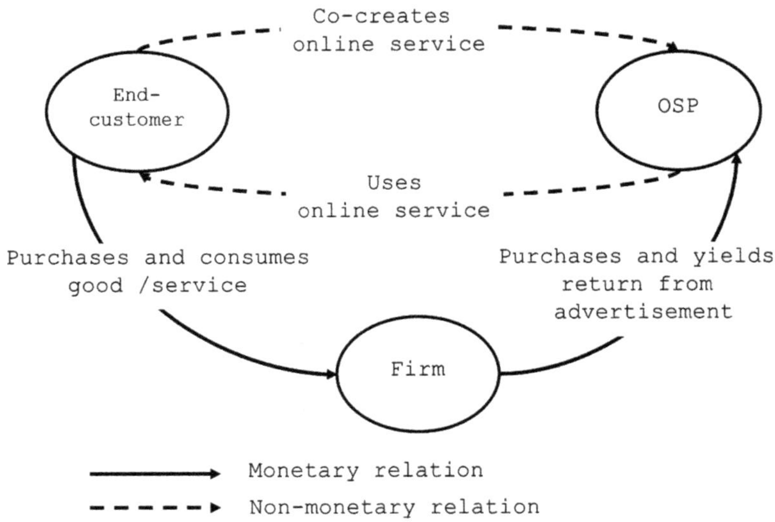

Fig. 9.2 The interflow between the end-customer OSPs and a conventional firm; authors' elaboration

nity identity are factors that stimulate online users to use these services (Krombholz et al. 2012). For OSPs, it is exactly this emotional and social "stickiness" that is the key in retaining the end-customer side of the market. Could we then imagine other types of services with this kind of relation? To our knowledge there is not any similar business sector where the relation between the firm and the user is applicable. Yet, if we look at social activities or public services we can see a similar pattern. People voluntarily contribute to the society, or within the boundary of an association. In some cases it is for purely altruistic purposes, but in most cases people expect a certain service in return. For example, we can ponder a referendum where people contribute with their vote, and in return expect a societal reaction based on this vote. The difference here though is that the agent is not a commercial firm, but a public body without any commercial interests. Thus, we conclude that the relation between the end-customer and social networking subsets of OSPs is based on a relation of a *social* rather than a *commercial* contract. The relation is based on a personal investment (co-creation), rather than pure consumption, and we argue that this is a unique feature for some OSPs in comparison to conventional firms.

9.5 Any Limits to Growth for OSPs?

Whereas Coase analysed why firms exist, Penrose (1959) accepted the existence of firms and was more interested in the growth of firms that could be explained in terms of resources and dynamic capabilities. She often used biological analogies when explaining how firms grew and expanded geographically. Human resources were central to Penrose, and she viewed the firm as a bundle of human resources, and found these critical to the firm's success and limits to growth. Buckley and Casson (2007 p. 153) interpreted Penrose growth theory as:

> the growth rate of the firm reflects a balance between the entrepreneurial dynamism which drives its diversification, and the difficulty of enlarging the firm's management team to exploit the resultant opportunities.

Penrose argued that there is no limit to the size to which a firm can grow. There might be a limit to size in a single market, but Penrose argued that firms can diversify from other markets and also diversify their product or service. This is apparent when we look at the internationalization behaviour of the examples of OSPs in Table 9.2 below. Penrose based her theory on the U-shaped average cost curve to the growth rate of the firm, not to the scale of production (see model A in Fig. 9.3). A key question, therefore, is to answer whether the average cost curves of OSPs are U-shaped or not.

One argument against the U-shaped average cost curve is that OSPs are not so labour-intensive dependent as, for example, other service firms like hotels, consulting firms, etc. We can also see from Table 9.2 below that the ratio of revenue per employee is high in OSPs. As firms grow due to changes in the rate of output, Penrose argues that firm have to dislocate resources; for example, change the job role of a manager. An increased demand for an online service on a specific market may lead an OSP to allocate a new data centre, and recruit (a few) staff on a new market. These changes increase the average cost of output. It is possible that given OSPs' relatively low labour-intensity, and the geographical concentration of core management functions, i.e. an ability to decouple production and online distribution at a low cost across borders (Sánchez-Peinado and Pla-Barber 2006), that the average cost curve is less steep on the right side, which would lead to a curve in between a U-shape and an L-shape, a boomerang-shape (see model B in Fig. 9.3). The implication for this hypothesis would be that OSPs' revenue growth pattern could be faster and geographically less limited than for conventional firms, which also has been the case for some successful OSPs. We argue that this feature of the OSPs' resources, i.e. being few and geographically concentrated, enables rapid scalability both in terms of revenue and geography. Our hypothesis is that the boomerang-shaped average cost curve would potentially be categorized as a *typical* feature of OSP. The complexity of networks and their connection to location has shown to be important in determining the development of knowledge-intensive industries, such as the video game industry (Balland et al. 2013; Storz et al. 2014). The functionality of the labour market and the preference for short distances to partners and cognitive similarity seem to be vital within the OSPs sector.

Table 9.2 Outlines some of the world's largest OSPs, all from the US. The data has been collected from the annual reports of the firms for the year 2014

OSPs	Founded year	Revenue (MUSD)	Revenue outside the US	Employees	Revenue / employee (MUSD)	Market offer	Business model	Competition	Geographical presence
Facebook	2004	12,466	55 %	9199	1.36	Social media platform, messaging service	B2B and B2C (mainly B2B advertising)	Monopoly status on most world markets except China, Russia, etc. Regional and local social networks	1.5 billion users around the world. 24 offices and 25 data centres in different countries
Google	1998	66,061	34 %	53,600	1.23	Search engine, email service, document service, etc.	B2B and B2C (mainly B2B advertising)	Monopoly status with 90 % market share. Vertical search engines and e-commerce websites, such as Kayak (travel queries), LinkedIn (job queries), WebMD (health queries), and Amazon and eBay (e-commerce).	World's largest search engine. Large international presence. 90 % of world market share of 3.2 billion users.
Twitter	2006	1400	40 %	4100	0.34	Ideas, content news and self-expression messaging service	B2B (advertising 90 %)	Facebook and Google, social and messaging services. Local competition, e.g. Kakao in South Korea.	78 % of users outside the US. More than 45 international offices.

LinkedIn	2002	2200	40 %	8700	0.25	Online professional network service	Diversified business model B2B and B2C with revenues coming from talent solutions, marketing solutions and premium subscription products	Members' professional networks. Local networks, e.g. http://www.sitepoint.com/social-networking-sites-for-business/	347 million members. 33 % of office space abroad. Offices in: Amsterdam, Bangalore, Barcelona, Beijing, Dubai, Dublin, Hong Kong, London, Madrid, Melbourne, Milan, Mumbai, Munich, New Delhi, Paris, Perth, São Paulo, Singapore, Stockholm, Sydney, Tokyo and Toronto.
Yahoo!	1995	4618	27 %	11,000	0.42	Search engine, email service, online advertising	Mainly B2B (advertising)	Online search engines, sites offering integrated Internet, Google AdSense, DoubleClick Ad Exchange, AOL's Ad.com products and services, social media and networking sites, e-commerce sites.	575 million users worldwide. Offices in 25 countries.

(continued)

Table 9.2 (continued)

OSPs	Founded year	Revenue (MUSD)	Revenue outside the US	Employees	Revenue / employee (MUSD)	Market offer	Business model	Competition	Geographical presence
eBay	1995	17,900	61 %	34,600	0.52	E-commerce marketplace and e-payments (PayPal)	B2C and B2B	Online and offline retailers. Distributors, liquidators, import and export companies, auctioneers, catalogue and mail-order companies, classifieds, directories, search engines, commerce participants (consumer-to-consumer, business-to-consumer and business-to-business),	65 % of staff in the US. Offices in 32 countries.
Netflix	2007	5504	27 %	2189	2.51	Internet delivery of TV shows and movies	B2C (mainly)	Multichannel video programming distributors (“MVPDs”), Internet-based movie and TV content providers (including those that provide pirated content) and DVD rental outlets.	Canada, Latin America and Europe—18,277 out of 53,391 members internationally (27 %)

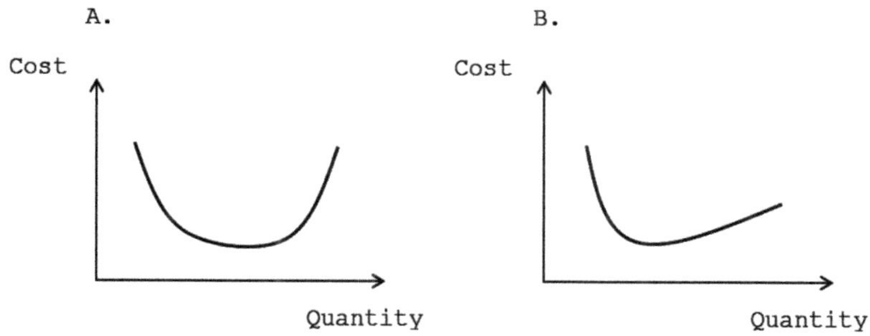

Fig. 9.3 The left-hand diagram (**a**) shows Penrose's original U-shaped average cost curve, and the right-hand diagram (**b**) shows a proposed boomerang-shaped average cost curve for OSPs

9.6 The Intermediary Role and Rapid Network Effects

The intermediary aspect is central for OSPs. Caillaud and Jullien (2001 p. 798) define cybermediaries as a "new type" of specialized intermediaries whose role is to:

> gather and process information on users that visit their website so as to help different classes of agents, in particular buyers and sellers of one specific good, to find each other. Although intermediaries may also provide services to facilitate transactions, a specificity of many of them is that they specialize on the pure informational aspects of intermediation, the physical part being left to sellers' distribution system.

An OSP providing search engine functionality has an intermediating role between users' search inquiries on the one hand, and websites containing the searched information on the other. An online music streaming OSP, like *Spotify*, intermediates between people interested in listening to music over the Internet, and music labels and musicians, which produce and market music digitally.

OSPs' markets are based on network externalities (effects) (Liebowitz and Margolis 1995; OECD 2015), and are two-sided (Armstrong 2006), or multi-sided (Hagiu 2014; OECD 2015). The OSPs must come up with business models to resolve the chicken and egg problem (Rochet and Tirole 2003). In order to prosper, and for the positive network effects to generate a surplus for the agents, the OSPs must bring the different sides of the markets on board. The worst-case scenario is the one of negative external externalities, when the surplus is destroyed. In the case of Spotify, the OSP needs to acquire a sufficiently large user base of music listeners and also needs to attract producers of music via record labels. Hence, they must reach, and then handle, what we argue is a *critical equilibrium* of keeping (at least) the two sides satisfied over time via the online medium (see Fig. 9.4). In order to reach that critical equilibrium the most common scenario for OSPs is to invest in building up a sufficient user base by investing in marketing and technology before it starts to monetize the service. The business model is either a direct charge for the online services to end-customers, or sales of online advertisement space to other firms using various models, e.g. pay-per-view, pay-per-click, etc. For those OSPs

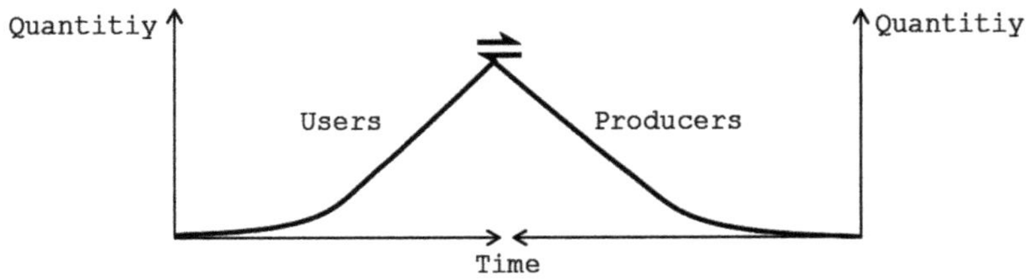

Fig. 9.4 An illustration of a critical equilibrium of a two-sided OSP; authors' elaboration

who charge their end-customers directly there is often some kind of freemium model in place, i.e. a limited range of the service is free of charge.

In the OSPs industry it is, therefore, generally accepted by investors to run an OSP with an un-monetized business model for a certain time period in order to reach the moment when the application generates traction and goes *viral* (Alvstam et al. 2016). The intermediary role and the network effects are typical, but not unique features for OSPs. Business models based on network externalities are nothing new in an economy, and there are countless examples of two-sided markets. In management literature, the concept of network effects has been used to describe standards and path dependence (Liebowitz and Margolis 1995). Nonetheless, the concept has attained increased interest due to the Internet's capacity to speed up the network effects. The particularities of two-sided Internet markets incline that successful OPSs are fast growing in terms of user base and OSPs can be highly valuated by investors before making any profit (Vogelsang 2010). A similar dynamic has been seen in the online computer game industry, where the possibility of bringing players in to the platforms is highly important, and where firms can then choose a subscription-based business model or a model that relies on items sold within the game (Ström and Ernkvist 2014). It is important to stress though that most OSPs never reach the stage of positive network effects and the critical equilibrium. We consequently claim that the intermediary role and network effects are typical, albeit not unique, characteristics for OSPs.

9.6.1 Examples of OSPs

9.6.1.1 Growth Pattern

Table 9.2 Some examples of US OSPs.

If we look at the year of inception of OSPs, we can see that they are all relatively young. All have been founded since 1995. The largest OSP, *Google*, has a life span of 17 years and already a turnover over 6.6 billion USD. *Google's* year-to-year growth rate in the last decade has been of exponential form, even though decreasing as its revenue increases (see Fig. 9.5). Of the top list of US firms there are few. Yet, such a fast growth path is not unique for OSPs. For example, on *Fortune's* list of the world's fastest-growing firms, there is a large spread of business sectors represented.

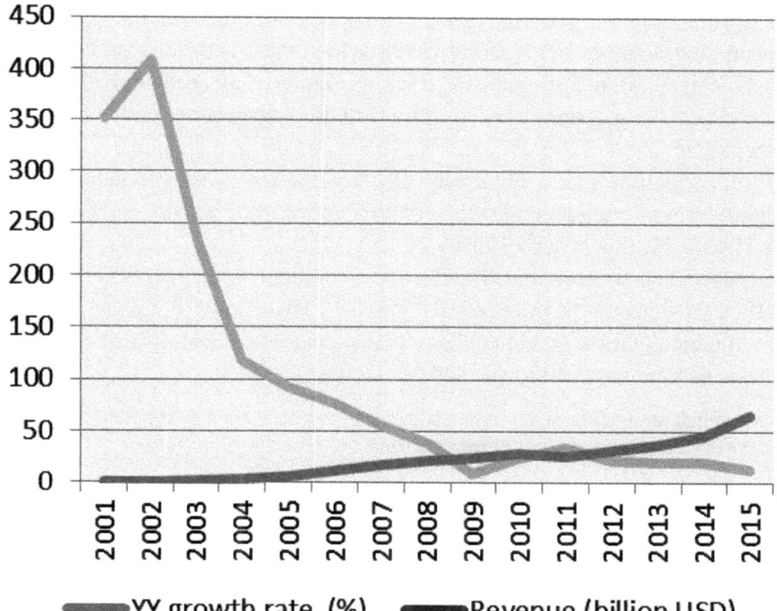

Fig. 9.5 Google's year-to-year growth rate and revenue 2000–2015

The top ranked in 2015 is a pharmaceutical firm (Fortune 2015). *Facebook* holds the 10th position of the ranking. In order to generalize, it is vital to analyse a longer time span and a larger dataset, but we are confident to discard the hypothesis that an exponential revenue growth pattern after inception is a unique feature for OSPs. However, as argued above, due to network effects and relatively weak labour-intensity (see column *Revenue per employee*), there is theoretically good potential for successful OSPs to experience a rapid growth shortly after inception.

If we look at the geographical expansion in Table 9.1, we find support for Penrose's theory of geographical diversity as a path to growth. The columns *Revenue outside the US* and *Geographical expansion* show that the OSPs have been successful in reaching a user base beyond the domestic spatial realm. Both *eBay* and *Facebook* have a majority of their turnover outside their home markets. Such internationalization behaviour is supported by the *born globals* literature and its argument that firms internationalize shortly after inception, and that they also can reach distant markets early in the internationalization phase (Luostarinen and Gabrielsson 2006). The online medium has been an accelerator for this behaviour (Yamin and Sinkovics 2006; Petersen et al. 2002; Chen 2006). The entry barrier for reaching an international user base has been substantially lowered with global online markets such as *AppStore* and *Android.*

Though there is ambiguity in literature over whether OSPs' internationalization pattern differs geographically in comparison to conventional firms, the examples in Table 9.1 show that the OSPs primarily have international offices in markets in the global North. Such internationalization behaviour is aligned with the Uppsala model (Johanson and Vahlne 1977, 2009; Luostarinen and Gabrielsson 2006), which emphasizes the gradual approach and the importance of markets with short 'psychic distance' first, i.e. countries with similar socio-economic and cultural characteristics.

We thus conclude that a feature of OSPs is the capacity to reach a large geographically spread audience shortly after inception, but we do not hold this as a unique feature for this group of firms.

It is interesting to note that despite being "online" firms, all OSPs in the table have office presence in the international markets. This shows that geography matters and that the localization aspect of the online service is central. It is aligned with the hypothesis of Leamer and Storper (2001, p. 641):

> We argue that the Internet will produce more of the same: certain forces for deagglomeration, but offsetting and possibly stronger tendencies toward agglomeration.

In a business where tacit knowledge is central for the innovation process, and where there is fierce competition to recruit the most skilful programmers and data engineers, the importance of 'being there' (Gertler 2003) and absorbing the 'buzz' (Storper and Venables 2004) is crucial.

9.6.1.2 Life Span

The average life span of a multinational corporation—*Fortune 500* or its equivalent—is between 40 and 50 years (Business Week 2015). Daepp et al. (2015) study of 25,000 companies across sectors from 1950 to 2009 showed that the typical half-life of a publicly traded company is about a decade, regardless of business sector, and that the most common reason a company disappears is due to a merger or acquisition. This regularity indicates that mortality rates are independent of a company's age. The lifespan is geographically dependent. For example, it is a well-known fact that several companies, especially in Japan and Europe, have existed for hundreds of years (Wikipedia 2015a). With current OSPs data we are not able to make such analysis. A closer look at the OSPs in the pioneering phase in the 1990s reveals that the life spans of the successful browsers were rather short, and that there has been a high turnover of market leaders (Wikipedia 2015b). The most successful OSPs seem to grow fast and have a short existence. Additionally, the acquisition pattern by leading OSPs has been aggressive. *Google* has transformed the informatics landscape with its growth and acquisition strategy (Robison 2008, see Google's acquisitions: Wikipedia 2015c), but has statistically also shortened the life span of approximately 185 acquired firms.

There is thus no indication that the life span should be longer than for other firms; on the contrary, we argue that there are indications pointing towards a shorter life expectancy for OSPs in comparison with other firms.

9.7 Conclusion and Further Research

In this chapter we outlined a few typical characteristics for OSPs: the intermediary role; the dependence on network effects; the concentration into a few geographical spaces in the global North as the production centre; but also the ability to rapidly reach out at a low cost, to a vast geographical user base beyond physical borders. We have not been able to determine many unique features, although we hold the social, rather than commercial relation with the end-customers within an entangled co-development milieu as a unique feature that will be of great importance for building the long-term customer base in order to develop sustainable business models. This new social feature in the relation between the OSPs and the consumers urges more research. There are many upsides with such an entwined user-producer relation, e.g. co-innovation, co-production, and cross-border collaboration. But herein lays a new dimension in the relation between the consumer and the producer. Issues like privacy, personal integrity, and intellectual property rights require more research, action and a subsequent international policy framework. Particularly in the global South, where digital policies and consumer rights are the least developed, people are exposed to potential exploitation. Research also shows that the institutional impact is of importance for the development of Internet-based industries in emerging economies. In the case of online games in China, the government tries to handle information flows, technical advancement and social control. This is difficult and can prove to be an institutional hurdle for further development of the industry (Ernkvist and Ström 2008). Tax is an additional challenge that needs to be addressed on a global level (OECD 2015). In sum, the policy framework has important catching-up work ahead in order to ensure that OSPs are operating in a stimulating but fair international environment.

We have also discussed the evolution of the OSPs industry, and depict it as rather traditional with distinctive growth phases. We are inclined to argue that OSPs, although they at first glance may seem to be a unique and new species, to a large extent resemble conventional firms. It should also be accentuated that most OSPs never reach the point of becoming *viral* or *digital monopolies* and remain modest both in terms of revenue growth and geographical scope. Most of them face the harsh reality of the profit-and-loss battle, which in many cases results in short life spans, probably shorter than for average firms. For the firms that establish themselves as major actors within one or more related Internet business areas, the challenge for competition authorities is to develop regulatory frameworks that can serve the purpose of both customers and firms and secure a healthy competitive situation in different parts of the global marketplace.

There are already signs that the DNAs of OSPs and conventional firms are converging. As OSPs grow internationally, local adaptability is needed, which requires staff on the ground, e.g. set-up of traditional management functions such as Human Resources. Without adapting to the offline space, OSPs risk to fall into the *virtuality trap* (Yamin and SInkovics 2006), which will harm their market position. The online space should thus not be disentangled from the offline space. Market offerings need to be adjusted to the local context. We foresee that local OSPs will continue to compete and collaborate with the large international OSPs. In parallel, conventional firms are to an increasing degree entering the international online space. E-commerce is blooming for manufacturers of goods, which are to an increasing degree blending into the service economy, which is referred to as the servitization of the traditional manufacturing industry (Bramklev and Ström 2011; Bao and Toivonen 2015). This also relates to the growth of the *Internet of Things*, where large amounts of data are generated through sensors in many industries that can at a later stage be used to develop new business opportunities. This technical advancement will also help to bring 'old' and 'new' firms closer together and it might be challenging to separate the service content from the hardware. Hence, we can expect OSPs and conventional firms to be even harder to distinguish from each other as the online space and the offline space progressively congregate. The evolutionary aspect of the industry in relation to its development and internationalization will be determined through the interaction with the surrounding business environment including the policymakers.

References

Alvstam, Claes-Göran, Patrik Ström, & Robert Wentrup. (2016). Heterogeneous economic space in a global archipelago: An economic geography perspective of emerging markets. Merchant, H. (Ed.) *Handbook of research on emerging markets*. Cheltenham: Edward Elgar Publishers.

Anwar, S. T. (2013). Icarus paradox or verto in prosperum: Chinese internet firms, business models and internationalization. *J World Investment Trade, 14*, 532–555.

Armstrong, M. (2006). Competition in two-sided markets. *RAND J Econ, 37*, 668–691.

Balland, P.-A., Vaan, M., & Boschma, R. (2013). The dynamics of interfirm networks along the industry cycle: The case of the global video game industry, 1987–2007. *J Econ Geogr, 13*, 741–765.

Bao, S., & Toivonen, M. (2015). Cultural differences in servitization: Nordic manufacturers in China. *J Sci Technol Policy Manage, 6*, 223–245.

Boyd, D. M., & Ellison, N. B. (2007). Social network sites: Definition, history, and scholarship. *J Comput-Mediat Commun, 13*, 210–230.

Bramklev, C., & Ström, P. (2011). A conceptualization of the product/service interface: Case of the packaging industry in Japan. *J Serv Sci Res, 3*, 21–48.

Buckley, P. J., & Casson, M. (2007). Edith Penrose's theory of the growth of the firm and the strategic management of multinational enterprises. *Manag Int Rev, 47*, 151–173.

Business Week. (2015). The lifespan of a company http://www.businessweek.com/chapter/degeus.htm. Accessed 25 Oct 2015.

Caillaud, B., & Jullien, B. (2001). Competing cybermediaries. *Eur Econ Rev, 45*, 797–808.

Chen, S. (2006). Are internet firms global? In*International marketing research*, *Advances in International Marketing* (Vol. 17, pp. 319–345).

Coase, R. H. (1937). The nature of the firm. *Economica, 4*, 386–405.

Daepp Madeleine I.G., Marcus J. Hamilton, Geoffrey B. West, & Luis M.A. Bettencourt. (2015). The mortality of companies. doi:10.1098/rsif.2015.0120.

Dicken, P. (2015). *Global shift – mapping the changing contours of the world economy*. London: Sage.

Ekeledo, I., & Sivakumar, K. (2004). The impact of e-commerce on entry-mode strategies of service firms: A conceptual framework and research propositions. *J Int Mark, 12*, 46–70.

Ernkvist, M., & Ström, P. (2008). Enmeshed in games with the government: Governmental policies and the development of the Chinese online games industry. *Game Cult, 3*, 98–126.

Facebook (2015). Facebook newsroom http://newsroom.fb.com/company-info/. Accessed 30 July 2015.

Forsgren, M., & Hagström, P. (2007). Ignorant and impatient internationalization? The Uppsala model and internationalization patterns for Internet-related firms. *Crit Perspect Int Bus, 3*, 291–305.

Fortune. (2015). The world's fastest growing companies http://fortune.com/100-fastest-growing-companies/. Accessed 15 Oct 2015.

Gabrielsson, M., & Pelkonen, T. (2008). Born internationals: Market expansion and business operation mode strategies in the digital media field. *J Int Entrep, 6*, 49–71.

Gabrielsson, M., Manek Kirpalania, V. H., Pavlos, D., Solberg, C. A., & Zucchella, A. (2008). Born globals: Propositions to help advance the theory. *Int Bus Rev, 17*, 385–401.

Gertler, M. (2003). Tacit knowledge and the economic geography of context or the undefinable tacitness of being (there). *J Econ Geogr, 3*, 75–99.

Grabher, G., Ibert, O., & Flohr, S. (2008). The neglected king: The customer in the new knowledge ecology of innovation. *J Econ Geogr, 84*, 253–280.

Graham, M., De Sabbata, S., & Zook Matthew, A. (2015). Towards a study of information geographies: (Im)mutable augmentations and a mapping of the geographies of information. *Geographical Environment, 2*, 88–105.

Hagiu, Andrei. 2014. Strategic decisions for multisided platforms *MIT Sloan Management Review*, 55: http://sloanreview.mit.edu/article/strategic-decisions-for-multisided-platforms/. Accessed 10 Oct 2015.

Internet World Stats. (2015). Internet Growth Statistics http://www.internetworldstats.com/emarketing.htm. Accessed 23 Oct 2015.

Jensen-Ørberg, P. D., & Petersen, B. (2014). Value creation logics and internationalization of service firms. *Int Mark Rev, 31*, 557–575.

Johanson, J., & Vahlne, J.-E. (1977). The internationalization process of the firm: A model of knowledge development and increasing foreign commitments. *J Int Bus Stud, 8*, 23–32.

Johanson, J., & Vahlne, J.-E. (2009). The Uppsala internationalization process model revisited: From liability of foreignness to liability of outsidership. *J Int Bus Stud, 40*, 1411–1431.

Kaplan, A., & Haenlein, M. (2010). Users of the world, unite! The challenges and opportunities of social media. *Bus Horiz, 53*, 59–68.

Kim, D. (2003). The internationalization of US Internet portals: Does it fit the process model of internationalization? *Mark Intell Plan, 21*, 23–36.

Knight, G. A., & Cavusgil, S. T. (1996). The born global firm: A challenge to traditional internationalization theory in S.T. Cavusgil and T.K. *Madsen's export internationalizing research—enrichment and challenges. Adv Int Mark, 8*, 11–26.

Krombholz, K., Merkl, D., & Weippl, E. (2012). Fake identities in social media: A case study on the sustainability of the Facebook business model. *J Serv Sci Res, 4*, 175–212.

Lanier, J. (2013). *Who owns the future?* London: Penguin.

Le Monde. 2015. Delphine Ernotte veut un Netflix de la création française http://www.lemonde.fr/economie/article/2015/10/29/delphine-ernotte-veut-un-netflix-de-la-creation-francaise_4799302_3234.html. Accessed 30 Oct 2015.

Leamer, E. E., & Storper, M. (2001). The economic geography of the Internet age. *J Int Bus Stud, 32*, 641–665.

Lee, D., Park, J. Y., Kim, J., Kim, J., & Moon, J. (2011). Understanding music sharing behaviour on social network services. *Online Inf Rev, 35*, 716–733.

Leiner, Barry M., Cerf Vinton G., Clark David D., Kahn Robert E., Kleinrock Leonard, Lynch Daniel C., Postel Jon, Roberts Larry G., & Wolff Stephen. (n.d). Brief history of the Internet.

http://www.internetsociety.org/internet/what-internet/history-internet/brief-history-internet Accessed 15 Sept 2015.
Liebowitz, S. J., & Margolis, S. E. (1995). Are network externalities a new source of market failure? *Res Law Econ, 17*, 1–22.
Luo, Y., Zhao, J. H., & Jianjun, D. (2005). The internationalization speed of e-commerce companies: An empirical analysis. *Int Mark Rev, 22*, 693–709.
Luostarinen, R., & Gabrielsson, M. (2006). Globalization and marketing strategies of born globals in SMOPECs. *Thunderbird Int Bus Rev, 48*(6), 773–801.
Madsen, T. K., & Per Servais, P. (1997). The Internationalization of Born Globals: An evolutionary process? *Int Bus Rev, 6*, 561–583.
Mbogo, M. (2010). The impact of mobile payments on the success and growth of micro-business: The case of M-Pesa in Kenya. *J Lang Technol Entrep, 2*(1), 182–203.
McKinsey & Co. (1993). Emerging Exporters Australia's High Value-Added Manufacturing Exporters McKinsey & Company and the Australian Manufacturing Council, Melbourne.
OECD. (2012). *Internet economy outlook*. Paris: OECD Publishing.
OECD. (2015). Addressing the tax challenges of the digital economy, Action 1–2015 Final Report, OECD/G20 Base Erosion and Profit Shifting Project. Paris: OECD Publishing.
Okazaki, S., & Taylor, C. R. (2013). Social media and international advertising: Theoretical challenges and future directions. *Int Mark Rev, 30*, 56–71.
Penrose, E. T. (1959). *Theory of the growth of the firm*. London: Basil Blackwell.
Perks, H., Gruber, T., & Edvardsson, B. (2012). Co-creation in radical service innovation: A systematic analysis of microlevel processes. *J Prod Innov Manag, 29*, 935–951.
Petersen, B., Welch, L. S., & Liesch, P. W. (2002). The Internet and foreign market expansion by firms: Theoretical questions and three predictions. *Manag Int Rev, 42*, 207–221.
Porter, M. E. (1990). *The competitive advantage of nations*. New York: Free Press.
Robison, R. (2008). Google: A chronology of innovations, acquisitions, and growth. *J Libr Adm, 46*, 5–29.
Rochet, J.-C., & Tirole, J. (2003). Platform competition in two-sided markets. *J Eur Econ Assoc, 1*, 990–1029.
Rusten, G., & Bryson, J. R. (2010). Placing and spacing services: Towards a balanced economic geography of firms clusters social networks contracts and the geographies of enterprise. *Tijdschr Econ Soc Geogr, 101*, 248–261.
Sánchez-Peinado, E., & Pla-Barber, J. (2006). Mode of entry in service firms: Strategic variables and characteristics of services influencing the internationalization process. *Int Mark Res (Adv Int Mark), 17*, 159–192.
Schumpeter, J. A. (1934). *The theory of economic development*. Cambridge, MA: Harvard University Press.
Singh, N., & Kundu, S. (2002). Explaining the growth of e-commerce corporations (ECCs): An extension and application of the eclectic paradigm. *J Int Bus Stud, 33*, 679–697.
Sonaike, A. S. (2004). The internet and the dilemma of Africa's development. *Int Commun Gaz, 66*, 41–61.
Storper, M., & Venables, A. J. (2004). Buzz: Face-to-face contact and the urban economy. *J Econ Geogr, 4*, 351–370.
Storz, Corneilia, Federico Riboldazzi, and Moritz John. (2014). Mobility and innovation: A cross-country comparison in the video games industry. *Research Policy*. http://dx.doi.org/10.1016/j.respol.2014.07.015.
Ström, Patrik, & Mirko Ernkvist. (2014). Product and service interaction in the Chinese online game Industry, Technol Innov Manage Rev May 2014.
Sundqvist, S., Kyläheiko, K., Kuivalainen, O., & Cadogan, J. W. (2012). Kirznerian and Schumpeterian entrepreneurial-oriented behavior in turbulent export markets. *Int Mark Rev, 29*, 203–219.

The Economist. (2014). http://www.economist.com/news/leaders/21635000-european-moves-against-google-are-about-protecting-companies-not-consumers-should-digital. Accessed 10 October 2015.
The Economist. (2015). http://www.economist.com/news/leaders/21676767-entrepreneurs-are-redesigning-basic-building-block-capitalism-reinventing-company. Accessed 27 Oct 2015.
Vogelsang, M. (2010). Dynamics of two-sided internet markets. *IEEP, 7*, 129–145.
Warf, Barney. (2013). Global geographies of the internet. e-book: Springer. doi: 10.1007/978–94–007-1245-4.
Webopedia. (2015). Online service provider http://www.webopedia.com/TERM/O/online_service_provider.html. Accessed 15 July 2015.
Wikipedia. (2015a). List of oldest companies http://en.wikipedia.org/wiki/List_of_oldest_companies. Accessed 26 Oct 2015.
Wikipedia. (2015b). Timeline of web browsers https://en.wikipedia.org/wiki/Timeline_of_web_browsers. Accessed 29 Oct 2015.
Wikipedia. (2015c). List of mergers and acquisitions by Google https://en.wikipedia.org/wiki/List_of_mergers_and_acquisitions_by_Google. Accessed 31 Oct 2015.
Yamin, M., & Sinkovics, R. R. (2006). Online internationalisation, psychic distance reduction and the virtuality trap. *Int Bus Rev, 15*, 339–360.
Zook, M. A. (2002). Grounded capital: Venture financing and the geography of the Internet industry 1994–2000. *J Econ Geogr, 2*, 151–177.

Chapter 10
Online Service Providers as Human Rights Arbiters

Rikke Frank Jørgensen and Anja Møller Pedersen

Abstract Within the EU, governments increasingly encourage online service providers to engage in co- and self-regulatory measures to prevent harmful and illegal content in the online sphere. As part of this tendency, governments shape schemes of liability for third-party content around the online service providers (intermediaries), giving them strong incentives to block, filter or take down content upon notification or request from public authorities in order to avoid liability. Concern has been raised, that the intermediaries are being used to implement public policy with limited oversight and accountability with severe implications on human rights. Drawing on case-studies of three EU directives, including the E-commerce directive, the chapter illustrates how measures of blocking, filtering, and take down of content in co- and self-regulatory frameworks interfere with the human rights standards related to freedom of expression and information. It further discusses current approaches towards human rights responsibilities of private actors, with a particular focus on the UN Guiding Principles on Business and Human Rights. The chapter concludes that self-regulatory frameworks are insufficient to counter the human rights challenges related to the involvement in online content regulation of intermediaries, who play a crucial role for online public participation.

10.1 Introduction

It is commonly argued that the internet has a profound value for freedom of opinion and expression, as it magnifies the voice and multiples the information within the reach of everyone who has access to it. As such it has become the central global

This chapter draws on the Case Study on ICT and Human Rights (Policies of EU), Work Package No. 2 – Deliverable No. 2.3, which was published in March 2016 as part of and funded by the EU FP7 FRAME project.

R.F. Jørgensen (✉)
Danish Institute for Human Right, Copenhagen, Denmark
e-mail: rfj@Humanrights.dk

A.M. Pedersen
Danish Institute for Human Rights, University of Copenhagen, Copenhagen, Denmark

M. Taddeo, L. Floridi (eds.), *The Responsibilities of Online Service Providers*,
Law, Governance and Technology Series 31, DOI 10.1007/978-3-319-47852-4_10

public forum (Kaye 2015). However, it is largely a commercial domain with infrastructure and services in the hands of private companies. To participate in public life online, individuals must engage with a variety of online service providers (intermediaries) such as access providers, search engines, social media platforms etc. and rely on them for exercising their fundamental rights, including freedom of expression and information. In this sense, online service providers set the limits and boundaries for public participation in the online domain.

Since the mid-nineties, EU regulators have enlisted these intermediaries in a variety of co- and self-regulatory schemes to disable or remove alleged illegal content from the internet. As part of this, the EU model of limited liability for third-party content has been developed, but also contested, not least because of its unclear legal provisions, which create incentives for the intermediaries to block more than necessary in order to avoid liability.

Each time content is blocked, filtered or taken down by an intermediary, that action has an impact on both the freedom of expression of the speaker and the freedom of information of the end-user that the information was intended to reach. Concern has been raised, that the intermediaries are being used to implement public policy with limited oversight and accountability with severe implications on human rights (Brown 2010; Callahan et al. 2009; Douwe Korff for the Council of Europe 2014; Kuczerawy 2015; Tambini et al. 2008).

Drawing on examples from EU regulation; Directive 2000/31/EC on electronic commerce (E-commerce directive), Directive 2011/93/EU on combating the sexual abuse and sexual exploitation of children and child pornography (child pornography directive), and Directive 2004/48/EC on the enforcement of intellectual property rights (IPR enforcement directive), the chapter examines EU content-regulation from the perspective of freedom of expression and rule of law. Any interference by the state with freedom of expression and information must comply with the rule of law and meet the strict criteria laid down in international human rights law; it must be prescribed by law, pursue a legitimate aim and be proportionate (Council of Europe 2014, para. 47). However, as argued by the authors, the current regulatory schemes are insufficient to provide the standards and compliance mechanisms required to meet these standards. Moreover, as human rights bind only states, not private actors, by allowing or even encouraging private actors to interfere with online expressions and information (content), without the necessary safeguards, states de facto neglect their human rights obligations and escape the strict requirements, which would have been otherwise incumbent upon them had they applied the restrictions themselves.

As an alternative (and widely promoted) approach, the authors discusses the UN Guiding Principles on Business and Human Rights that serve as the prevailing soft law standard for the human rights responsibility of private actors. In line with recent research on internet gatekeepers (Laidlaw 2012)[1] the chapter argues that self-regulatory approaches are insufficient to solve the above human rights challenges

[1]As of July 2015, Laidlaw's PhD thesis has been published under the title *Regulating Speech in Cyberspace*, by Cambridge University Press.

caused by intermediaries, not least when they have the capacity to impact democratic life in a way traditionally reserved for public institutions.

The remaining part of this chapter proceeds as follow. First, it introduces the challenges that online content regulation poses to the right to freedom of expression and information, focusing in particular on blocking, filtering and take-down of content. Next, these challenges are exemplified by EU-regulation related to co- and self-regulatory schemes of intermediary liability (the E-commerce directive) and the fight against alleged illegal content (child pornography and IPR infringements). Third, the UN Guiding Principles on Business and Human Rights are discussed, and finally, the authors provide specific conclusions on these challenges.

10.2 Online Freedom of Expression

The right to freedom of expression is codified in all major human rights treaties, at the international level, most notably in the Universal Declaration of Human Rights (1948) and the International Covenant on Civil and Political Rights (1966); and at the European level, in the European Convention on Human Rights and Fundamental Freedoms (ECHR) (1950) and the Charter of Fundamental Rights of the European Union (CFREU) (2000).

The ECHR defines a right to "to hold opinions and to receive and impart information and ideas without interference by public authority and regardless of frontiers" under article 10(1), and the possible restrictions to that right in Article 10(2).[2] These restrictions must be "prescribed by law", follow a "legitimate aim" and be "necessary in a democratic society", commonly referred to as the proportionality test (Council of Europe 2014, para. 47). Freedom of expression includes all forms of expression, without any distinction to content and through any medium (Jacobs 1961, p. 426). The European Court of Human Rights (ECtHR) has established that Article 10 applies fully to the internet (Perrin v. the United Kingdom 2005). This is also confirmed in the first UN resolution on human rights on the internet from 2012 (United Nations Human Rights Council, July 5, 2012).

Since international human rights law binds states, not private actors, only limitations to freedom of expression applied by public authorities constitutes 'interferences' in a legal sense. Accordingly, while states must abstain from interference with individuals' freedom of expression and information that does not meet the criteria laid down in Article 10(2) of the ECHR, the same obligations are not incumbent upon private actors. However, additionally to these *negative obligations*, the ECtHR has also developed a theorem on *positive obligation*s of Member States. The positive obligations concern a state's obligation to ensure that individuals may enjoy freedom of expression, even in conflicts with other private parties. In consequence,

---

[2] See (Council of Europe, 1950) Available at: http://www.echr.coe.int/Documents/Convention_ENG.pdf

the state responsibility for human rights violations may be invoked in cases where the state has failed to enact appropriate domestic legislation to ensure human rights protection in the realm of private actors (VgT Verein Gegen Tierfabriken v. Switzerland, 28 June 2001, para. 45). The ECtHR has recognised such positive obligations in relation to Article 10, for instance in cases where media plurality was at stake (Demuth v. Switzerland, 5 February 2003).[3]

The Court's doctrine on positive obligations does, however, draw limits to what can be expected of states, especially when state action premised on positive obligations touches upon other rights, such as property rights (Sluijs 2011, p. 20). A landmark case in this respect is *Appleby*, which concerned the alleged positive obligation of a state to protect protesters' freedom of expression on private property - a shopping centre - from which they had been removed (Appleby and Others v. the United Kingdom, 6 May 2003). The Court found that in a private sphere, a state's positive obligation only arises where the bar on access to the property effectively prevents effective exercise of the freedom, or if it destroys the essence of the right. Due to alternative ways for meaningful exercise of the right, the essence of the right was not destroyed in the present case, and the positive obligations were not triggered. In general, the Court has stressed, that the state must strike a fair balance between the concurring rights, and the burden on the state must not be disproportionate.[4]

In the online domain, interference with freedom of expression can take various forms, from inadequate guarantees of the right to privacy and protection of personal data, which may inhibit the dissemination of opinions and information, to measures that either remove the content (take-down), or disable end-users' ability to access it (filtering and blocking) (Deibert et al. 2010; La Rue 2011).[5]

As the digital infrastructure and online services are largely controlled by private companies, measures to remove or disable access to content require either cooperation from these service providers or coercion exercised upon them through 'new-school' techniques' (Balkin 2014, p. 3). These techniques are characterised by three features (1) collateral censorship; the state regulates one party (the intermediary) in order to control another, the speaker, (2) public/private cooperation and co-optation, and (3) new forms of digital prior restraints (Balkin 2014, p. 4).

[3] See, also, (Jean-François Akandji-Kombe, 2007)

[4] "In determining whether or not a positive obligation exists, regard must be had to the fair balance that has to be struck between the general interest of the community and the interests of the individual, the search for which is inherent throughout the Convention. The scope of this obligation will inevitably vary, having regard to the diversity of situations obtaining in Contracting States and the choices, which must be made in terms of priorities and resources. Nor must such an obligation be interpreted in such a way as to impose an impossible or disproportionate burden on the authorities" (Appleby and others v. UK 2003, para. 40)

[5] 'Blocking' refers to technical measures taken to prevent users from accessing specific websites, IP addresses, and domain name extensions. 'Filtering' refers to technical measures used to exclude pages containing certain keywords or other specific content from appearing when the end-user search for information (La Rue 2011, para. 29). 'Take-down' refers to situations where content is removed from webpages at the request of the owner of the content, a victim hereof, or public authorities on behalf of such (Delfi AS v. Estonia, 16 June 2015) para. 13.

Scholars have repeatedly warned against the many practical as well as principal problems related to current European practices of content regulation (Callahan et al. 2009; Kuczerawy 2015; McIntyre 2010; Tambini et al. 2008). As argued by Korff, measures such as filtering and blocking are inherently likely to produce false positives (blocking sites with no prohibited material) and false negatives (when sites with prohibited material slip through a filter). Moreover, the criteria for blocking specific websites, and the lists of blocked websites, are often secret and remedies little known or non-existent. As regards their effectiveness, most blocking is easy to bypass, even for not very technically skilled people (Douwe Korff for the Council of Europe 2014, p. 13).

Also, the UN Special Rapporteur on the promotion and protection of the right to freedom of opinion and expression has cautioned against the human rights implications of such measures and instructed that "requests submitted to intermediaries to prevent access to certain content, or to disclose private information for strictly limited purposes such as administration of criminal justice, should be done through an order issued by a court or a competent body which is independent of any political, commercial or other unwarranted influences" (La Rue 2011, para. 75).

There is an essential distinction between mandatory measures, which are based on law,[6] and voluntary measures (self-regulation). Law-based removal or disabling access to online content constitutes an interference with freedom of expression and information under Article 10 of the ECHR, and must adhere to the criteria laid down in Article 10(2). Arguably, it is legitimate to remove or block clearly identified illegal content, such as child pornography. However, the aim of the measure, and the means used to obtain it remain crucial to determining whether the measure is proportional and thereby lawful. In cases where the interference is law-based and/or clearly encouraged by the state (co-regulation), the potential violation of freedom of expression and information is attributable to the state, and the human rights conflict thus a vertical conflict between the state and the intermediary, the speaker or the end-users. At some point, however – in the zone between co- and self-regulation – potential human rights violations may only be attributed to the state, indirectly, through its positive obligation to protect against potential violations between private parties.

Voluntary measures of content regulation, in contrast, cause a horizontal conflict between the intermediary who imposes the measure and the speaker (who is being blocked) or the end-user (who is denied access). Horizontal conflicts must be resolved under the state's positive human rights obligations. Accordingly, what determines the outcome is whether the state has struck a fair balance between the freedom of the intermediary to conduct a business (provide internet services) and the right to freedom of expression and information of the speaker or end-user. These cases are not as clear-cut as the vertical ones.

[6] Including both content regulation, which is implemented and enforced directly by states, and content regulation, which is based on law, but implemented and/or enforced by private companies (co-regulation).

The state normally holds a wide margin of appreciation with regard to how it chooses to balance the rights of one individual against the rights of another. The protection of interests of the speaker against interference by the intermediary is normally considered to lie within the margin of appreciation of the state. Positive obligations to protect speakers from being blocked or filtered will only arise, when individuals are prevented from effectively exercising their right to freedom of expression and information or when pluralism of the information environment are clearly at stake. Accordingly, a clear example of when a strict positive obligation arises is where blocking or filtering by intermediaries deprives an online speaker from reaching an audience completely – or deprives an end-user completely from accessing certain content (J. v. Hoboken 2012, p. 148f).

In the following, the chapter will illustrate how EU policy dealing with illegal content encourages online service providers to apply measures that effectively interfere with freedom of expression and information, without the required human rights safeguards.

10.3 EU Regulation in Conflict with Freedom of Expression

"A growing amount of self-regulation, particularly in the European Union, is implemented as an alternative to traditional regulatory action. Some governments actively encourage or even place pressure on private business to self-regulate as an alternative to formal legislation or regulation which is inherently less flexible and usually more blunt than private arrangements" (MacKinnon et al. for UNESCO, 2014, p. 56).

Through a case-study of three EU directives; Directive 2000/31/EC on electronic commerce (E-commerce directive), Directive 2011/93/EU on combating the sexual abuse and sexual exploitation of children and child pornography (child pornography directive), and Directive 2004/48/EC on the enforcement of intellectual property rights (IPR enforcement directive), the authors explore the implications of EU content regulation on freedom of expression and information and the rule of law. While the E-commerce directive sets up the general limits for intermediary liability for user-generated content, irrespective of the nature of the liability (criminal or civil), the other directives deal with either criminal offences (directive 2011/93/EU) or civil liability (directive 2004/48/EC). All three contain obligations to either disable access to or take down alleged illegal content, and they all imply that intermediaries must play an active role. Yet, neither of them frame this as human rights issues.

Directive 2000/31/EC on E-commerce sets up an Internal Market framework for electronic commerce (European Parliament and Council of the European Union 2000). Pursuant to Article 2(a) and Recital 18, the directive is applicable to services provided by intermediaries, such as ISPs, search engines and social media platforms. Articles 12 – 14 set up 'safe harbours' under which an intermediary cannot be held liable for illegal content generated by its users, and protect private actors acting as intermediaries for three types of activities: 'mere conduit', 'caching' and 'hosting'.[7] Furthermore, Article 15 prevents Member States from imposing general monitoring obligations on the intermediaries. During a public consultation in 2010, it became apparent that the directive gave rise to various challenges, such as legal fragmentation and lack of clarity with respect to the liability provisions and their scope of application (European Commission 2010). The Commission has since launched several initiatives to improve the system, but without any result. The recent Digital Single Market Strategy promises an improvement of EU content regulation with due regard to its impact on freedom of expression and information (European Commission 2015), yet the initiative is still at an early stage, and results remain to be seen.

Blocking of alleged child pornography has been high on the EU policy agenda since the mid-nineties (European Commission 1996a, b). While earlier approaches prevailed towards self-regulation, the preferred path is now co-regulation (McIntyre 2010, p. 2010ff.). *Directive 2011/93/EU on combating the sexual abuse and sexual exploitation of children and child pornography* establishes minimum rules concerning the definition of criminal offences and sanctions for sexual abuse and exploitation of children (The European Parliament and the Council of the European Union 2011). Pursuant to Article 25 and Recital 47, Member States shall take all necessary measures to ensure the removal of, or disabling access to, websites containing or disseminating child pornography, and the directive is without prejudice to voluntary action (self-regulation) taken by the industry in that regard. Recital 25 and 47 require account to be taken to the rights of the end-users, and that the involved measures provide for adequate safeguards. However, the directive provides no guidance in the matter.

Turning to copyright, the music and film industries have - for more than a decade - searched for solutions to fight alleged infringements, for instance targeting file sharing in peer-to-peer networks. Earlier, lawsuits were the primary means, whereas regulation via intermediaries is now the preferred path (Brown 2010, p. 3). To prevent IPR infringements across Member States, the EU has adopted *Directive 2004/48/EC on IPR enforcement* (European Parliament and Council of the European Union 2004).[8] Without prejudice to the E-commerce directive, it concerns the measures, remedies and sanctions necessary to ensure IPR enforcement (Articles 9 – 13 and 16). Pursuant to Article 17 and Recital 29, Member States shall encourage the industry to take an active part in the fight against piracy and counterfeiting, for instance by developing codes of conduct dealing with the matter, thus encouraging self-regulation. To the extent intermediaries use of blocking, filtering and takedowns, freedom of expression limitations are likely to occur. Recital 32 of the directive recognises the respect for fundamental rights, but mainly IPR protection,

[7] Search engines are not formally covered by the E-commerce directive (Frydman 2004). However, most Member States grant search engines an exemption from liability, either by analogy with caching or hosting, or as derived from general principles of tort law, (Javier Martínez Bavière, p. 241).
[8] The EU has also adopted the copyright directive, (European Parliament and Council of the European Union 2001)

moreover, the directive offers no guidance in the matter. Besides lack of compliance with the rule of law and Article 10(2) of the ECHR, this may lead to massive blocking of legitimate file exchanges as the intermediaries are not the best placed to assess whether a specific use of IPR protected work is legitimate.

10.3.1 Vertical and Horizontal Human Rights Conflicts

The three directives are examples of the current EU 'co-regulatory' approach towards content regulation (Frydman et al. 2008, p. 5). 'Co-regulation' is a legal model in which the drafting, implementation and enforcement of norms is not under the sole authority of the state, but spread, voluntarily or not, between both private and public players (Frydman et al. 2008, pa. 1).[9] It may be restricted to cover regulation that contains a legally formalised role of public authorities (J. v. Hoboken 2012, p. 140), or include state participation in a broader sense (MacKinnon et al. for UNESCO, 2014, p. 56). It differs from a 'command and control' model in which drafting, implementation and enforcement is solely in the hands of public authorities, but also from 'self-regulation' where private actors make the rules and enforce them without any public intervention (Schulz and Held 2001, p. A-2), such as the UK model for countering child pornography (Brown 2010, p. 3).

Due to an increasing pressure on the intermediaries, the current co-regulatory practices may foster self-regulation, in particular as this is either presupposed (the E-commerce directive), accepted as a possible means (Article 25 and Recital 47 of the child pornography directive) or directly encouraged (Article 17 and Recital 29 of the IPR enforcement directive) by the three EU directives. Although self-regulation is praised as an effective tool to redress illegal or harmful speech (The United Nations (UN) Special Rapporteur on Freedom of Opinion and Expression, the Organization for Security and Co-operation in Europe (OSCE) Representative on Freedom of the Media, the Organization of American States (OAS) Special Rapporteur on Freedom of Expression, & the African Commission on Human and Peoples' Rights (ACHPR) Special Rapporteur on Freedom of Expression and Access to Information, June 1, 2011), it entails severe human rights and rule of law challenges (La Rue 2011, para. 38ff.).

In the following, these challenges are discussed, drawing on the analysis of the EU directives.

First, with respect to the human rights implications of co-regulation compared to self-regulation, the distinction between mandatory and voluntary content regulation is important. Mandatory content regulation is introduced and/or applied directly by the state and constitutes a *vertical human rights conflict* between the state and the intermediary, speaker or end-users. As such, the potential violation of freedom of expression is 'prescribed by law' and clearly attributable to the state. In contrast,

---

[9] Co-regulation may also be referred to as 'privatised law enforcement', (Douwe Korff for the Council of Europe 2014) p. 85.

voluntary content regulation in a self-regulatory scheme remains a *horizontal conflict* between the intermediary who imposes restrictions and the speaker/end-user who is subject to it. In practice, however, the distinction contains several grey zones. For instance, does active coercion of intermediaries to 'voluntarily' filter or block content, in the absence of a legal duty to do so, constitute an 'interference' with freedom of expression, and is it 'prescribed by law', including being sufficiently precise, accessible and foreseeable, in accordance with Article 10(2) of the ECHR? Moreover, at what stage may freedom of expression limitations be attributed to the state, when caused by 'voluntary' measures taken by intermediaries following state encouragement to do so?

Generally, the more informal the role of public authorities, the more difficult it is to argue that limitations to freedom of expression derive from public authorities and constitute 'interferences' with the right. Likewise, it is also difficult to argue that they fall within the positive obligations of the state

The positive obligations of the state are activated only when individuals are prevented from effectively exercising their right to freedom of expression, or when pluralism of the information environment are clearly at stake. States hold a wide margin of appreciation (J. v. Hoboken 2012, p. 148f.). However, states cannot simply disown measures (blocking, filtering etc.) by private entities that have such effects – especially not if the state de facto strongly encouraged those measures, and the states may then become responsible for *not* placing such a system on a legislative basis (Douwe Korff for the Council of Europe 2014, p. 73):

> There are serious doubts as to whether a blocking system that effectively imposes a restriction on most ordinary people's access to online information will ever be in accordance with the rule of law when it is chosen and operated by private parties, in the absence of public scrutiny, in the absence of a democratic debate, in the absence of a predictable legal framework, in the absence of clear goals or targets, in the absence of evidence of effectiveness, necessity and proportionality, and in the absence, either before or after the system is launched, of any assessment of possible counter-productive effects (Douwe Korff for the Council of Europe 2014, p. 72)

However, it is not – at least not on a general basis – that easy to establish, 'when' an individual is effectively prevented from exercising his/her freedom of expression and thus, when the positive human rights obligations are triggered. As such, the above statement appears quite optimistic.

Second, all of the above directives accept or encourage blocking and/or takedowns, and they all either accept or encourage these measures to be taken through self-regulation. Yet none of them frames such measures as limitations of / interferences with freedom of expression or acknowledges the human rights implications of such obligations / proposals. This might explain why self-regulation continues to be widely promoted by EU regulators, although it – from a rule of law perspective – is inherently imperfect due to lack of legal basis.

Third, the directives do not provide for common EU procedures for dealing with alleged illegal content, but only set up limits and suggestions for possible national means. Articles 12 – 14 of the E-commerce directive, for instance, do not stipulate that all Member States must implement a common EU liability scheme, but provide only for 'safe harbours' in which the Member States cannot impose liability on the intermediaries. Consequently, the liability schemes vary across Member States, resulting in great legislative fragmentation. This also became apparent during the 2010 public consultation on the directive (European Commission 2010) and has led to calls for further harmonisation (EDRI and others 2015). Likewise, with general reference to the human rights obligations in the ECHR and the CFREU, Article 25(2) and Recital 47 of the child pornography directive require that account is taken to the rights of the end-users, and that the measures provide for adequate safeguards. The directive, however, does not explain the potential human rights issues at stake and how to prevent human rights violations.

Fourth, the intermediaries to whom regulatory or judicial power is delegated are not necessarily the best placed to assess whether an allegation of illegal content is well founded, for instance, whether a specific use of an IPR protected work is illegal, or whether content related to child pornography is distasteful or illegal. They are therefore likely to rely on notifications or requests to block or take down content without challenge, in particular when facing liability for illegal third-party content. This might lead to 'over-compliance' resulting in blocking and takedowns of legal content, which thus no longer follow a legitimate aim, or meet the strict criteria of proportionality required by Article 10(2) of the ECHR.[10]

As argued by Korff, measures with an impact on fundamental rights, specifically blocking and filtering of websites, cannot ever be said to be 'necessary' and 'proportionate' to a 'legitimate aim' in a 'democratic society' if they are unsuited to achieve that aim, excessive in their effect, and lack procedural safeguards. For instance, blocking of child pornography related content: (i) does not stop sexual abuse of children or the sharing of images of such abuse; (ii) does stop access by the large majority of the population to sites that are not illegal; (iii) is based on secret criteria or lists that clearly do not have the quality of a 'law' in the ECHR sense; and (iv) is not subject to adequate remedy. This harms freedom of expression for those whose sites are wrongly blocked and for those who are effectively missing out on relevant information (Douwe Korff for the Council of Europe 2014, p. 74).

In sum, codification and encouragement of co- and self-regulation within the EU, focuses more on the (perceived) effectiveness that such schemes provide for in the fight against illegal content, than on the human rights conflicts and rule of law concerns raised by these practices.

[10] This presupposes, that such limitations happen within mandatory / co-regulatory frameworks in which the potential human rights violations derive from public authorities and constitute 'interferences'. In voluntary / self-regulatory frameworks the question is whether such limitations fall under the state's positive human rights obligations and thus, whether the state has ensured a proper balance between the rights of the individuals.

10.3.2 Intermediary Liability

Schemes of limited liability are one of the mechanism used to enlist intermediaries in online content regulation. The E-commerce directive codifies these expectations by stipulating the conditions under which intermediaries cannot be held liable for third party content. However, its lack of clarity put pressure on the intermediaries to act as 'gatekeepers' of the online sphere.

At a general level, 'gatekeepers' are entities that decide what shall or shall not pass through a gate (Laidlaw 2012, p. 44). Within regulatory studies, gatekeepers are non-state actors with the capacity to alter the behaviour of others in circumstances where the state has limited capacity to do the same. Regulation thus becomes 'decentralised' (Morgan and Yeung 2007, p. 280). As gatekeepers usually do not benefit from the misconduct they facilitate, although being in a position to prevent it, it can prove more effective to shape a liability scheme around gatekeepers as opposed to those breaking the rules (Laidlaw 2010, p. 264), which may also be referred to a 'collateral censorship'[11]. Hence, gatekeeping theory has been used to describe the tort doctrine of vicarious liability, for instance, in relation to the role of journalists and press institutions as gatekeepers of information (Laidlaw 2010, p. 264). Most recently, vicarious liability has been imposed on internet intermediaries to pursue peer-to-peer providers such as Napster and Pirate Bay for copyright infringements caused by illegal downloading by their users, and in relation to the notice and take-down provisions of the E-commerce directive.

However, the EU liability framework of the E-commerce directive entails several challenges, of which four will be pointed out below.

First, the liability distinctions of the directive remain rather unclear.[12] To benefit from a safe harbour, intermediaries must act 'expeditiously' to remove or to disable access to the information concerned, upon obtaining 'actual knowledge' of the illegality (as regards criminal liability) or 'awareness of facts or circumstances' from which the illegal activities or information is apparent (as regards civil liability) (Articles 13 and 14). It remains unclear, however, what constitutes 'actual knowledge' or 'awareness of facts or circumstances' or what it takes to act 'expeditiously'[13] and thus what it takes to benefit from a safe harbour.[14] The interpretation of the provisions thus differs across borders, leading to legal fragmentation (European Commission 2012, p. 32ff.) and creating incentives for over-compliance.

[11] Collateral censorship 'occurs when the state holds one private party A liable for the speech of another private party B, and A has the power to block, censor or otherwise control access to B's speech' (Balkin 2014)

[12] As discussed extensively by e.g. Hoboken, (J. v. Hoboken 2012)

[13] As regards the differences between the degree of knowledge in relation to civil and criminal liability, see also (Kuczerawy, 2015, p. 48)

[14] As stressed in the Digital Single Market Strategy (European Commission 2015) Para. 3.3.2.

Second, the directive regulates only exemptions from liability, but it does not protect the intermediaries from litigations aimed at injunctions, cf. Articles 12(3), 13(2) and 14(3). Consequently, Member States are not prevented from forcing the intermediaries to play an active role in law enforcement. Moreover, a court order to disconnect a user or block access to certain information is not necessarily accompanied by guidance to ensure compliance with the principle of proportionality in Article 10(2) of the ECHR and the risk of over-compliance is thus present.

Third, Articles 12–15 concern liability for illegal content, yet what counts as 'illegal' is determined by national law, and when assessed by an intermediary, who is not the best placed to do so, the risk of over-compliance is significant. Also, in contrast to the US approach, the directive does not distinguish between criminal and civil liability, but follows a horizontal approach defining one set of general rules applicable to any content; child pornography, IPR etc. (Frydman et al. 2008, p. 6). This may also lead to disproportionate handling of content (Kuczerawy 2015, p. 52).

Fourth, in relation to caching, it is required that the intermediary does not 'actively interfere' with the transmission (Article 13). However, if an intermediary, due to government pressure, applies filters to disable access to child pornography according to Article 25 of the child pornography directive, it runs the risk of invoking liability, as it then actively interferes with the transmission.

This paradox does not seem to be envisaged by the directive, but it has become even more present with the recent Delfi judgment (Delfi AS v. Estonia, 16 June 2015). This is the first case, in which the ECtHR examined a complaint regarding liability for user-generated content. It illustrates important aspects of the limited liability scheme of the E-commerce directive, but also liability schemes framed around intermediaries more generally, and seems to increase the responsibilities of the intermediaries.

The case concerned whether Delfi, a large news portal, could be held liable for offensive comments (of which the majority constituted hate speech) posted on its website by anonymous third parties. Delfi took down the comments upon notification, six weeks after publication, yet defamation proceedings were still launched against it. Delfi claimed to be an intermediary falling under the safe harbours of the national transposition of the E-commerce directive. The ECtHR found, however, that Delfi acted – not as an intermediary – but as a media publisher (due to its degree of editorial control) and so it could not benefit from the safe harbours.[15] Consequently, a key question became whether Delfi had been obliged to remove the comments before notification.

The Court considered that Delfi exercised a substantial degree of control over the comments, beyond that of a purely *passive* service provider, thus falling outside the safe harbour of the directive. The majority of the authors were anonymous (why liability could not be placed elsewhere), and the proceedings did not have any severe consequences for Delfi (Ibid, paras. 144–151). The Court, furthermore, interpreted

[15] The ECtHR, however, stressed that the findings could not be transferred to discussion forums or social media platforms (Ibid, para. 116).

national legislation in a way that in order to avoid liability, Delfi had not been obliged to *prevent* the uploading of comments. It would have sufficed to remove the comments without delay. Given its substantial degree of control over the comments, the Court considered that such interference with Delfi's freedom of expression was not disproportionate. Having regard to the ample possibilities for anyone to make his or her voice heard on the internet, the Court found that the obligation of a large news portal to take effective measures to limit the dissemination of hate speech and speech inciting violence could not be equated with private censorship. Particularly as the ability of a potential victim of hate speech to continuously monitor the internet is more limited than the ability of a large internet news portal to prevent or rapidly remove such comments (Ibid, paras. 157–158). With respect to clear hate speech and direct threats to the physical integrity of individuals, the rights and interests of others and of society as a whole, would thus entitle Member States to impose liability on news portals, failing to remove clearly unlawful comments without delay, even without notice from authorities or potential victims of the unlawful content, without contravening Article 10 of the ECHR (Ibid, paras. 159).

Several judges had dissenting opinions reflecting their concerns about the implications of the judgment. Some claimed that it implied a requirement of constructive knowledge on *active intermediaries*,[16] which would require, that all comments were monitored, and furthermore, that the fear of liability could lead to self-censorship (Delfi AS v. Estonia, 16 June 2015, Joint Dissenting Opinion of Judges Sajó and Tsotsoria, para. 1). Although not intending to censor expressions, by putting pressure and imposing liability on the intermediaries, governments are actually creating an environment of 'collateral' and private-party censorship (Balkin 2014). When liable for user-generated content, an intermediary has strong incentives to self-censor, limit and deny users access to communicate via the communications service it offers in order to avoid liability. It thereby imposes prior restraints on its users' expressions/information. Furthermore, the practices entail limited procedural safeguards, as action is taken not by a court, but private parties (Ibid, para. 3).

By confirming that Delfi should have *prevented* or rapidly removed the comments, the ECtHR seems to accept both pre- and post-publication liability. Without actual knowledge of their existence and nature, such duty in fact requires active intermediaries (news portals or blogs) to monitor all comments, similar to prior restraints (Ibid, paras. 34- 35) and prohibited by Article 15 of the E-commerce directive.[17]

To sum up, legal uncertainty around the provisions of the E-commerce directive and its national transpositions and the fear of liability for user-generated content may result in 'over-compliance' and thereby disproportionate limitations of freedom of expression and information. Moreover, it can lead to extra-legal pressure on

---

[16] Hosts who provide their own content and open their intermediary services for third parties to comment on that content.

[17] It could be argued, that the strict liability resulted from the clear illegal nature of the comments, (Delfi AS v. Estonia, 16 June 2015, Concurring opinion of Dissenting Judges Raimondo, Karakas, De Gaetano and Kjølbro) Concurring opinion by Dissenting Judges Raimondo, Karakas, De Gaetano and Kjølbro.

intermediaries to self-regulate by blocking and filtering or not challenge notifications and requests to block or take down certain content. Also, as the victims often do not have the resources to challenge the measures, content may be blocked or removed without any – administrative or judicial – review. Moreover, such self-regulatory frameworks may not even provide access to effective remedies, although required by Article 13 of the ECHR.

Without proper control or review mechanisms, nothing prevents human rights violations from occurring. Human rights obligations are briefly touched upon in the Preamble of the E-commerce directive,[18] but it does not deal with the nature potential human rights conflicts, and is thus not sufficient to protect freedom of expression, particularly not as the directive has to be transposed to national law in 28 Member States.

10.4 An Alternative Approach – Human Rights Responsibility of Private Actors

As illustrated, several challenges occur when law enforcement is delegated to intermediaries without sufficient rule of law safeguards. Moreover, and adding to the complexity, there is an increasing focus on the human rights responsibilities of online service providers, which put pressure on these private actors to both implement public policy (tackle illegal content) and respect human rights.

The key standard-setting document on human rights responsibilities of private actors is the United Nations 'Guiding Principles on Business and Human Rights'(United Nations Human Rights Council 2011), which is based on three pillars, ′Protect′, ′Respect′, and ′Remedy. The principles unpack the distinction between the state duty to protect human rights and the corporate responsibility to respect human rights. The principles maintain the primary (hard law) obligation of states to protect against human rights violations. At the same time, however, they give explicit recognition to the (soft law) responsibility of businesses to respect human rights (O'Brien and Dhanarajan 2015, p. 3).

The *first pillar* outlines the state duty to protect against human rights abuses by third parties, including private companies, which requires that states take appropriate steps to prevent, investigate, punish and redress human rights violations committed by private actors within their jurisdiction through effective policies, legislation and regulations and adjudications (United Nations Human Rights Council 2011, para. 1–10). The *second pillar* is the corporate responsibility to respect human rights, which implies that private companies should publish a policy commitment to respect human rights and act with due diligence in order to avoid infringing the

[18] Recital 9 of the E-commerce directive ties the free movement of information society services to Article 10 of the ECHR (freedom of expression and information), and recital 46 stipulates that with regards to 'the removal or disabling of access to the information concerned; the removal or disabling of access has to be undertaken in the observance of the principle of freedom of expression and of procedures established for this purpose at national level'.

human rights of others (United Nations Human Rights Council 2011, para. 11–24). Due diligence is envisaged to comprise of four steps, taking the form of a continuous improvement cycle (United Nations Human Rights Council 2011, para. 17–20). First, the company must assess the actual and potential impacts of its business activities on human rights (human rights impact assessment); second, remediate the findings of this assessment into company policies and practices; third, track how effective the company is in preventing adverse human rights impacts; and fourth, communicate publicly the due diligence process and its results. The *third pillar* addresses the need for greater access by victims of human rights infringements to effective – both judicial and non-judicial – remedy (United Nations Human Rights Council 2011, para. 25–31).

The Human Rights Council has stressed that a company's responsibility to respect human rights is a global standard that 'exists independently of states' abilities and/or willingness to fulfil their own human rights obligations, and does not diminish those obligations' (United Nations Human Rights Council 2011, para. 11). This was also reiterated by the former UN High Commissioner for Human Rights, Navi Pillay, in her report on 'The right to privacy in a digital age' to the UN General Assembly:

> The responsibility to respect human rights applies throughout a company's global operations regardless of where its users are located, and exists independently of whether the state meets its own human rights obligations (The Office of the United Nations High Commissioner for Human Rights 2014, para. 43).

In sum, states are obliged to prevent human rights violations by private actors, and private actors have a (moral – not legal) duty to respect human rights. States must ensure human rights compliant business practices via appropriate regulation, and each company has a responsibility to access their actual human rights impact. In the case of internet intermediaries, there is an extended ´sphere of influence´ compared to most companies. Not only does the intermediary have responsibilities in relation to its employees and community, its practices it may also affect, directly or indirectly, billions of internet users.

Although online service providers have a responsibility to assess the human rights impact of their activities in order to minimise their negative impact, states are not thereby 'lifted' from their human rights obligations. For instance, if an intermediary implements a filter with a view to block online content related to child pornography, as suggested in the directive on combating child pornography, the company has a responsibility to make a human rights impact assessment (HRIA) of that activity. If the HRIA shows that the filter results in the blocking of perfectly legitimate content, such activity has a negative human rights impact on freedom of expression of the speaker or freedom of information of the end-users and should be minimised accordingly.

The UN Guiding Principles have been widely appraised for their role in raising human rights awareness among businesses, yet their effectiveness has also been questioned not least due to their voluntary and non-binding character (O'Brien and Dhanarajan 2015, p. 3). Moreover, the UN Guiding Principles may move focus away from the state (legal) obligations to the (moral) responsibilities of businesses. From a human rights perspective, such a shift in attention from hard law obligations to soft law recommendations, is obviously a dangerous path, but close to the reality in EU Member States. As such, it is paramount, that states do not use or encourage self-regulation as a way to escape their own – hard law – human rights obligations, nor should they simply rely on the soft law responsibilities of businesses.

Laidlaw suggests that neither a corporate social responsibility (CSR) model, according to which businesses are responsible for human rights breaches within their sphere of influence, nor the UN Guiding Principles are sufficient to describe the human rights responsibilities of intermediaries in the online sphere. This is due to the intermediaries' role as gatekeepers for a number of practices essential to democratic participation (searching information, expressing opinions, participating in public debate, etc.) (Laidlaw 2012, p. 11). Online gatekeepers that have the capacity to impact democratic life are expected to serve the public interest, however, these companies are not imbued with the norms and requirements normally accompanying the exercise of public power. Moreover, they remain relatively isolated from legislative, executive and judicial oversight (Ibid, p. 46). To counter this challenge, it is suggested that the human rights obligations of these companies should increase according to the extent their activities facilitate or impact democratic culture[19] (Ibid, p. 241).

Within the UN system, attempts towards more legally binding human rights obligations for businesses have actually been taken. Although earlier attempts failed, the discussion on establishing such binding norms has continued among civil society and a group of states within the UN system (Lagoutte 2014, p. 8). In June 2014, the Human Rights Council adopted a resolution representing steps towards a legally binding instrument on human rights and business and established an intergovernmental working group to elaborate on the instrument (United Nations Human Rights Council 2014). The author of the UN Guiding Principles, however, has stated that the elaboration and adoption of a legally binding instrument will entail 'monumental challenges' in relation to institutions, enforcement etc. (Ruggie 2014, p. 3).

The increasing focus on the human rights responsibilities of private actors adds to the complexity of the online service providers as discussed above. Aside from the state pressure to implement public policy with a view to counter illegal online content, the intermediaries are also under an increasing pressure from states and the international community to respect human rights as elaborated in the UN Guiding Principles. An online service provider in good faith, who intends to comply with both public policy with a view to protect e.g. children against exploitation and copy-

[19] A distinction is made between micro-gatekeepers (certain content moderators), authority gatekeepers (Facebook, Wikipedia, portals), and macro-gatekeepers (ISPs, search engines). Macro-gatekeepers have the greatest democratic impact and thus the strongest human rights obligations (Laidlaw 2012, p. 60ff.). They are distinguished from the other levels, because users must inevitably pass through them to use the internet. As such, they engage all aspects of freedom of expression and information. Moreover, a shift from voluntary to more binding obligations is suggested (Laidlaw 2012) p. 241.

right holders against infringements, may easily interfere with freedom of expression and information of internet users. This is a complex landscape for the intermediaries to operate in, and with limited or no guidance from the EU regulator or Member States as illustrated above, this may result in a severe negative impact on freedom of expression and information.

Hence, it might prove difficult for service providers to meet these contradictory expectations, in particular when the implementation of the said directives happens in co- or self-regulatory frameworks that offer limited guidance on how to obtain human rights compliance.

10.5 Conclusions

The authors have pointed to a number of human rights challenges that occur at the junction of the EU policies dealing with illegal content on the internet, with a particular focus on self-regulation and intermediary liability.

As illustrated, internet intermediaries are increasingly being enlisted to impose – in a mix of mandatory and voluntary schemes – restrictions on freedom of expression and information; without the safeguards that would apply to state interference in similar situations. The UN Guiding Principles on Business and Human Rights have indicated the importance of addressing private actors' responsibility to respect international human rights law, yet, as soft law they do not solve the fundamental challenge raised in this chapter. Consequently, interference with EU citizens' freedom of expression and information largely occurs in a legal grey-zone with limited means of transparency and accountability.

Arguably, internet intermediaries have a significant impact on internet-users' ability to enjoy freedom of expression and information online. Yet, the intermediaries are subject to EU regulation, which does not maximise their adherence with international human rights standards, but creates ambiguity around liability for user-generated content and encourages self-regulation. Also and importantly, the measures deployed by the intermediaries do not comply with the rule of law test developed by the ECtHR in its case law on Article 10 of the ECHR.

The authors have highlighted how EU's policy related to intermediary liability, IPR enforcement, and combating child pornography etc., may influence negatively on users' freedom of expression and information online. As illustrated, the intermediaries operate in a legal grey zone with different and often conflicting expectations related to their role vis-à-vis content regulation. In practice, the intermediaries are expected to navigate between: (1) expectations of self-regulation (e.g. by blocking, filtering or take-down of content) as stipulated by the IPR enforcement directive and the child pornography directive; (2) liability schemes that expect them to expeditiously remove alleged illegal content in order to benefit from 'safe harbours' provisions; and (3) expectations of conducting human rights impact assessments to mitigate negative human rights impacts, as stipulated in the UN Guiding Principles on Business and Human Rights.

This zone of unclear expectations, norms and liability provisions is partly due to the character of the digital domain. With private actors in control of the digital infrastructure and services, it is no surprise that EU regulators and Member States have turned to these actors to regulate content, which is outside their direct sphere of control. Looking through the prism of the right to freedom of expression and information, however, this practice is problematic and calls for guidance and standards from EU regulators to ensure that the rule of law standards of Article 10(2) of the ECHR are protected when regulatory action is delegated to private actors. In the absence of such standards and guidance, the legal grey-zone presented by the directives are transposed to national law in 28 Member States, leading to legal fragmentation. In consequence, the intermediaries are left with self-devised codes of conduct while carrying out practices that affect users' fundamental rights.

EU policy tends to thrive towards a common and comprehensive EU approach when dealing with alleged illegal content, but fails to take into account some of the related concerns about freedom of expression and the rule of law. Arguably, there is a fundamental difference between the weight that is attributed to freedom of expression in the online environment from an economic free movement perspective as opposed to a traditional human rights perspective. Until recently, the underlying rationales for addressing human rights issues at the EU level have been economic in nature and human rights, which are by nature non-economic, have been addressed as auxiliaries to the establishment of an Internal Market. As a result, important policy concerns from the perspective of human rights end up being addressed indirectly or not addressed at all.[20]

Despite the goal set out by the EU Human Rights Guidelines on Freedom of Expression Online and Offline, according to which the EU is committed to respecting, protecting and promoting the freedom of opinion and expression within its borders (Council of the European Union, May 12, 2014, para. 7), this has yet not been implemented in secondary EU law (such as the three above mentioned directives). Up until now, the debate on co-regulation and self-regulation within the EU has shown limited attention to the freedom of expression issues evoked by such arrangements. In contrast, the Council of Europe has developed a number of standards pertaining to the use of internet filters, online freedom of expression, rule of law, etc. over the past 10 years. Also, a number of UN reports and resolutions adopted on freedom of expression on the internet since 2011 set standards in this field.

With the adoption of the Lisbon Treaty, fundamental rights play a more important role than ever within EU-law. Yet, EU standards and guidance for the protection and promotion of freedom of expression in the online domain is still lacking, particularly as it relates to the EUs internal policies. The recently launched Digital Single Market Strategy for Europe envisages among others an analysis of the need for new measures to tackle illegal content on the internet 'with due regard to the

[20] This point has also been raised by Hoboken (2014) in relation to privacy and data protection. See (J. Hoboken, August 15, 2014)

impact on the fundamental right to freedom of expression and information' (European Commission 2015, para. 3.3.2.). The ongoing debate on the strategy is a crucial opportunity to emphasise that fundamental rights are firmly situated as the baseline of the EU vision and practices in this field.

Bibliography

Akandji-Kombe, J. (2007). Positive obligations under the European Convention on Human Rights [online] Council of Europe Human Rights Handbook Series. Available at: http://www.echr.coe.int/LibraryDocs/DG2/HRHAND/DG2-EN-HRHAND-07(2007).pdf.

Appleby and others v. the United Kingdom. [2003, May 6] App. no. 44306/98 (European Court of Human Rights).

Balkin, J. M. (2014). Old-school/new-school speech regulation. *Harvard Law Review, 127*(8), 2296–2342.

Bavière, J. (2012). Injunctions against online search engine service providers [online]. International Bar Association. Available at: http://www.ibanet.org/Article/Detail.aspx?ArticleUid=7D602A2A-12FF-4E07-B525-32BE2993BE25.

Brown, I. (2010). Internet self-regulation and fundamental rights. *Index on Censorship, 1*, 98–106.

Callahan, C., Gercke, M., de Marco, E., & Dries-Ziekenheiner, H. (2009) Internet blocking. Balancing cybercrime responses in democratic societies.

Council of Europe. (2014). Recommendation CM/Rec(2014)6 of the Committeee of Ministers to member States on a guide to human rights for Internet users - Explanatory Memorandum. Strasbourg: Council of Europe.

Council of the European Union. (2014). EU Human rights guidelines on freedom of expression online and offline. Brussels: Council of the European Union.

Deibert, R., Palfrey, J., Rohozinski, R., Zittrain, J., & OpenNet, I. (2010). *Access controlled : The shaping of power, rights, and rule in cyberspace*. Cambridge, MA: MIT Press.

Delfi v. Estonia. [2015, June 16] App. No. 64569/09 (European Court of Human Rights).

Demuth v. Switzerland. [2003, February 5] App. no. 38743/97 (European Court of Human Rights).

EDRI and others. (2015). Open letter on intermediary liability protections in the digital single market to Vice-President of the European Commission Andrus Ansip [Online] Available at: https://edri.org/open-letter-on-intermediary-liability-protections-in-the-digital-single-market/ . Accessed 22 May 2016.

European Commission. (1996a). *Communication from the commission on illegal and harmful content on the internet (COM(1996) 487 final)*. Brussels: European Commission.

European Commission. (1996b). *Green paper on the protection of minors nd human dignity in audio-visual and information ervices (COM (1996) 483 final)*. Brussels: European Commission.

European Commission. (2010). Summary of the results of the Public Consultation on the future of electronic commerce in the Internal Market and the implementation of the Directive on electronic commerce (2000/31/EC). Brussels: European Commission.

European Commission. (2012). *Commisssion Communication to the European Parliament, the Council, The Economic and Social Committee and the Committee of the Regions. A coherent framework for building trust in the Digital Single Market for e-commerce and online services*. Brussels: European Commission.

European Commission. (2015). *Digital Single Market Strategy for Europe:Communication from the Commission to the European Parliament, the Council, the European Economic and Social Commitee and the Committee of Regions (COM(2015) 192 final)*. Brussels:European Commission

European Parliament and the Council. (2000). *Directive 2000/31/EC on certain legal aspects of information society services, in particular electronic commerce, in the Internal Market (Directive on electronic commerce). OJ L 178, 17.07.2000.*
European Parliament and the Council. (2001). *Directive 2001/29/EC on the harmonisation of certain aspects of copyright and related rights in the information society.* OJ 167/10, 22.06.2001.
European Parliament and the Council. (2004). Directive 2004/48/EC on the enforcement of intellectual property rights. OJ L 157, 02.06.2004.
European Parliament and the Council. (2011). Directive 2011/93/EU of 13 December 2011 on combating the sexual abuse and sexual exploitation of children and child pornography. OJ L 127, 29.04.2014.
Frydman, B. (2004). Coregulation: A possible model for global governance: In B. de Schutter & J. Pas (Eds.), *About globalization, view on the trajectory of mondialisation.* (pp. 227–242). Brussels: Brussels University Press.
Frydman, B., Hennebel, L., & Lewkowicz, G. (2008). 'Public Stategies for Internet Co-Regulation in the United States, Europe and China' *Working Papers du Centre Perelman de philosophie de droit*, No 2007/6.
Hoboken, J. (2012). *Search engine freedom : On the implications of the right to freedom of expression for the legal governance of web search engines.* Alphen aan den Rijn: Kluwer Law International.
Hoboken, J. (2014). The European Approach to Privacy [online] In: TPRC 42: The 42nd Research Conference on Communication, Information and Internet Policy. SSRN. Available at: http://papers.ssrn.com/sol3/papers.cfm?abstract_id=2418636## . Accessed 22 May 2016.
Jacobs, J. (1961). *The death and life of great American cities.* New York: Random House.
Kaye, D. (2015). *Report of the special rapporteur on the promotion and protection of the right to freedom of opinion and expression, David Kaye.* Geneva: United Nations.
Korff, D. for the Council of Europe. (2014). *The rule of law on the internet and in the wider digital world - Issue Paper.* Strasbourg: Council of Europe.
Kuczerawy, A. (2015). Intermediary liability & freedom of expression: Recent developments in the EU notice & action initiative. *Computer Law & Security Review: The International Journal of Technology, 31*(1), 46–56.
La Rue, F. (2011). *Report of the special rapporteur on the promotion and protection of the right to freedom of opinion and expression, Frank La Rue, A/HRC/17/27.* Geneva: United Nations.
Lagoutte, S. (2014). 'The State Duty to Respect Against Business-Related Human Rights Abuses. Unpacking Pillar 1 and 3 of the UN Guiding Principles on Human Rights and Business'. (*Matters of concern – Human rights' research papers series*, 2014/1).
Laidlaw, E. B. (2010). 'A framework for identifying internet information gatekeepers' *international review of law. Computers & Technology, 24*(3), 263–276.
Laidlaw, E. B. (2012). *Internet gatekeepers, human rights, and corporate social responsibility (PhD), London School of Economics and Political.* London: Science.
MacKinnon, R. et al. for UNESCO. (2014). *Fostering freedom online the role of internet intermediaries.* Paris: UNESCO.
McIntyre, T. J. (2010). Blocking child pornography on the internet: European union developments. *International Review of Law , Computers & Technology, 24*(3), 209–221.
Morgan, B., & Yeung, K. (2007). *An introduction to law and regulation: Text and materials.* New York: Cambridge University Press.
O'Brien, C. M., & Dhanarajan, S. (2015). 'The corporate responsibility to respect human rights: A status review' *NUS law working paper series*, 2015/005.
Office of the United Nations High Commissioner for Human Rights. (2014). *The right to privacy in the digital age. Report of the Office of the United Nations High Commissioner for Human Rights, A/HRC/27/37*.*
Perrin v. the United Kingdom. [2005, October, 18] App. no. 5446/03 (European Court of Human Rights).
Ruggie, J. G. (2014). A UN Business and Human Rights Treaty? *HARVARD Kennedy School.*

Schulz, W., & Held, T. (2001). *Regulated self-regulation as a form of modern government*. Hamburg: Verlag Hans-Bredow-Institut.
Sluijs, J. P. (2011). *From Competition to Freedom of Expression: Introducing Art. 10 ECHR in the European Network Neutrality Debate. TILEC Discussion Paper* 2011-040.
Tambini, D., Leonardi, D., & Marsden, C. T. (2008). *Codifying cyberspace : Communications self-regulation in the age of internet convergence*. London/New York: Routledge.
United Nations (UN) Special Rapporteur on Freedom of Opinion and Expression, the Organization for Security and Co-operation in Europe (OSCE) Representative on Freedom of the Media, the Organization of American States (OAS) Special Rapporteur on Freedom of Expression, & the African Commission on Human and Peoples' Rights (ACHPR) Special Rapporteur on Freedom of Expression and Access to Information. (2011, June 1). *Joint Declaration on Freedom of Expression on the Internet*.
United Nations Human Rights Council. (2011). *Report of the special representative John Ruggie. Guiding principles on business and human rights: Implementing the United Nations 'Protect, respect and remedy' Framework*. New York: United Nations. Geneva: United Nations.
United Nations Human Rights Council. (2012). *Resolution 20/8. The promotion, protection and enjoyment of human rights on the internet. A/HRC/20/L.13*. Geneva: United Nations.
United Nations Human Rights Council. (2014). *Resolution 26/... Elaboration of an international legally binding instrument on transnational corporations and other business enterprises with respect to human rights. A/HRC/26/L.22/Rev.1*. Geneva: United Nations.
VGT Verein Gegen Tierfabriken v. Switzerland. (2001, June, 28). App. No. 24699/94 (European Court of Human Rights).

Chapter 11
User-Generated Content: How Broad Licensing Terms Threaten the Web

Miloš Novović

Abstract Vast quantities of user content get uploaded online every single minute: pictures, videos, status updates, blog posts – they are the face of today's Internet. While a lot of this content is entitled to receive copyright protection under national laws, once uploaded, it is also subject to licensing clauses found in Terms of Service agreements. These clauses establish how platform providers can use the submitted content, and in many cases, language of these clauses is broad and ambiguous, to the point where users are left uncertain about who can use their content, in which ways, and for how long. This chapter argues that contractual ambiguity negatively affects users and platform providers alike. It examines the potential clashes between licensing clauses and privacy policies, deeming it necessary to ensure that in the case of conflict between the two, it is the privacy policy that should prevail. The chapter warns against the potential of invasive licenses to erode trust between users and platform providers, endangering content sharing which is the core of today's participative Web.

11.1 Introduction

Our online experience is shaped by the content that we share. What we type into search boxes, status update prompts, e-mail bodies; what we create and share with our friends and colleagues – be it a photograph, a video, a meme – is personal. But it is also endlessly creative and expressive. This content tells the story of who we are, who we were, and who we hope to become. As such, it is exceptionally valuable.

This brings us to a very important question: what are online providers actually allowed to do with the things that we post? Do I still own a photograph that I have uploaded to LinkedIn? If I do, what level of control do I retain? What kind of rights have I given them? Are they allowed to make a thumbnail out of that image? Can they put it in an e-mail inviting my colleagues to join their network? Print it out in

M. Novović (✉)
Faculty of Law, Institute for Private Law, University of Oslo, Oslo, Norway
e-mail: milos.novovic@jus.uio.no

M. Taddeo, L. Floridi (eds.), *The Responsibilities of Online Service Providers*, Law, Governance and Technology Series 31, DOI 10.1007/978-3-319-47852-4_11

a flyer, sell it on coffee mugs? Can my image be used to market goods to my friends, to promote political campaigns, to study aging? So far, the attempts to answer such questions have been focused on privacy law, and they almost completely neglect the *copyright law* dimension of the problem.

Copyright plays an important role in determining what platform providers may do with user-submitted content. Vast majority of the content that we upload online is entitled to receive copyright protection – and it does, without any active action on our part. This is because the Berne Convention (Berne Convention for the Protection of Literary and Artistic Works, 1886) an important international legal source currently ratified by 169 countries, (WIPO 2016) stipulates that copyright protection shall not be subject to any formalities: you are not required to, for example, bring a book that you wrote or a photograph that you took to a national copyright authority – the protection is granted automatically (Berne Convention Art. 5(2)). National legislation can, and very often does, impose some substantive requirements for copyright protection: but they are usually not too difficult to satisfy, usually boiling down to the requirement that the work is fixed in a certain medium – which user content always is – and that it passes a certain, often minimal, threshold of originality. And once that copyright protection is granted, you, as the author of the work, have a very broad control over how that work is used in the future. This is because copyright grants authors certain exclusive rights, ranging from economic ones – to make copies, to communicate the work to the public, to distribute it – to, in certain countries, moral ones – such as the right to be named as the author of the work. To ensure proper balance between the interests of the authors and the interests of the public, copyright law has developed an array of exceptions and limitations to these rights, which, despite the on-going harmonization attempts, vary between the countries (Rognstad 2008). But one thing is clear: once you have made a creative work, you get a substantial control over how it is used, and other people are not allowed to use it – exploit it – without getting your permission – a licence.

Copyright licence usually specifies which of the exclusive rights the other party gets, under which conditions, and for which period of time: if I tell you that you can make two copies of a photo I have sent you, I have given you a limited, non-exclusive licence to reproduce my work. Make three copies, and I have grounds to sue you. Nowadays, platform providers routinely insert licensing provisions in their Terms of Service agreements, requiring users to give them broad rights to exploit uploaded content. Text of the license can vary, but typically, it reads close to this:

> When you upload, submit, store, send or receive content to or through our Services, you give us (and those we work with) a worldwide license to use, host, store, reproduce, modify, create derivative works (such as those resulting from translations, adaptations or other changes we make so that your content works better with our Services), communicate, publish, publicly perform, publicly display and distribute such content. The rights you grant in this license are for the limited purpose of operating, promoting, and improving our Services, and to develop new ones.

This clause is modelled closely on the terms offered by Google; (Google ToS 2016) but it is by no means unique or more onerous than the provisions offered by the other platform providers. Consider, for example, a licensing clause found in Goodreads' (Goodreads ToS 2016) user agreement:

> By posting any User Content on the Service, you expressly grant, and you represent and warrant that you have a right to grant, to Goodreads a royalty-free, sublicensable, transferable, perpetual, irrevocable, non-exclusive, worldwide license to use, reproduce, modify, publish, list information regarding, edit, translate, distribute, publicly perform, publicly display, and make derivative works of all such User Content and your name, voice, and/or likeness as contained in your User Content, in whole or in part, and in any form, media or technology, whether now known or hereafter developed, and to grant and authorize sublicenses of the foregoing for any purpose at the sole discretion of Goodreads.

The practice is prevalent all across the industry – and it does not take an expert in copyright law to note just how broad these licenses are. There is not a single author's right that they do not strive to cover – be it through an endless enumeration of terms ("to copy, store, perform, display, reproduce, record, play, adapt, modify, distribute, prepare derivative works…") or through use of vague catch-all terms (such as "license to use" or "otherwise exploit"). Some of the agreements even stipulate that the authors waive all applicable moral rights, including the right to be named as the author of the work and to object to the work's derogatory treatment.

11.2 A License Unchecked: Harms of Overly Broad UGC Licenses

Broad licensing clauses may certainly seem intimidating to an average user. But are they actually *harmful?* Reaching the answer might not be as simple as it seems.

How am I harmed if Snapchat, an application which is used to distribute self-destroying images, gets a perpetual license to sub-license my content and make it available to their partners "for syndication, broadcast, distribution, or publication outside the Services"? (Snapchat ToS 2016). Better yet: suppose that there is a local printing company which starts advertising on Facebook and automatically receives my recently uploaded photos. Then, instead of sending me promotional e-mails with stock pictures, they automatically generate an e-mail offering me customized calendars and posters with several pictures of my family. I am not suffering a direct financial harm; arguably, I will be more tempted to see what that company has to offer – and isn't that what advertising is all about?

To answer these questions, we have to embrace a broader understanding of harm than the one routinely used today. The law is nowadays primarily preoccupied with direct financial consequences of the harm suffered by plaintiffs; in fact, many a lawsuit against platform providers, at least in the US, has been stopped precisely because the plaintiffs were not able to allege any financial harm (Cohen v. Facebook Inc. 2011; Perkins v. LinkedIn Corp. 2014). The very nature of user content comes into play here: individual pieces of content generally have very low economic value, rendering most of the remedies unavailable, either for legal or practical reasons. Yet, when we take a more holistic approach, looking beyond financial effects in individual cases, harmful effects of overly broad licenses become rather clear.

11.2.1 Legal Uncertainty

If they are known for anything, software contracts are famous for being very detailed, verbose, unfavourable to users and difficult to understand. This view has been widely acknowledged, and their drafting style has often been mocked in the popular culture. As far back as in 1990, in Good Omens, a book by Terry Pratchett and Neil Gaiman, a computer is purchased:

> …Along with the standard computer warranty agreement which said that if the machine (1) didn't work, (2) didn't do what the expensive advertisements said, (3) electrocuted the immediate neighbourhood, (4) and in fact failed entirely to be inside the expensive box when you opened it, this was expressly, absolutely, implicitly and in no event the fault or responsibility of the manufacturer, that the purchaser should consider himself lucky to be allowed to give his money to the manufacturer, and that any attempt to treat what had just been paid for as the purchaser's own property would result in the attentions of serious men with menacing briefcases and very thin watches, (Gaiman and Pratchett 1990).

More recently, a lot of semi-serious claims have been made that even the lawyers who write these agreements do not read them (MakeUseOf 2015). For instance, Apple's iTunes EULA has been famously and widely mocked for stating that users "also agree that you will not use these products for any purposes prohibited by United States law, including, without limitation, the development, design, manufacture or production of nuclear, missiles, or chemical or biological weapons" ("Apple - Legal - Licensed Application End User License Agreement" 2015).

While perceived as unintelligent by the general public, this drafting style is actually a rather common feature of the Common Law legal systems, and is by no means limited to software contracts. In those systems, detailed contracts that closely stipulate all aspects of the legal relationship between the parties have historically evolved due to the different factors, including the principle of *caveat emptor*, procedural features such as parole evidence rule, as well as general hesitance of Common Law judges to construe obligations which were not explicitly agreed upon by the parties (Cordero-Moss 2014, 11). Still, although lying down contractual obligations in great detail "heightens the impression that a well thought through formulation may solve all of the problems that may arise under the governing law" (Cordero-Moss 2014, 12), in reality, this is not always the case. Terms of Service agreements – like many other contracts – often contain terms which do not fully account for the effects they might have under the governing law, and end up creating considerable legal uncertainty.

11.2.1.1 Uncertainty Due to Interaction with Different Governing Laws

Part of the uncertainty brought on by the overly broad licensing clauses stems from the international, borderless nature of the Internet: in some jurisdictions, such clauses may be perceived as fundamentally unfair, and the broader they get, the easier it might be to successfully challenge them.

Platform providers seldom account for this, as they are often under the impression that their services are subject exclusively to American law. When drafting user agreements, they routinely insert choice-of-law and choice-of-forum clauses, stipulating that all potential disputes are to be resolved before the U.S. courts and under the U.S. law – and they believe in the absolute reach of these clauses. Such attitude is maybe most clearly reflected in Tumblr's user agreement, which starts with a bold proclamation: "*Tumblr is a U.S. company and subject only to U.S. laws and jurisdiction, [...]*" (Tumblr ToS 2016). Still, choice-of-law clauses, like all other contractual provisions, do not have an absolute reach – and platform providers can theoretically find the validity of their user terms assessed under as many laws as there are countries that their services are available in.

The most significant reason for this lies in the outer boundaries of party autonomy. Even though party autonomy in choice of law is widely recognized, its scope is not unlimited; and while parties are generally free to choose the laws governing *contractual* – and to an extent, tortious – elements of their relationship, there are other legal aspects, in which determination of the applicable law is based on the conflict principles, rather than party autonomy. (Cordero-Moss 2014) Classic examples of such scenarios are contracts for sale of immovable property, shareholder agreements, agency contracts, and other contracts with strong implications for third parties (Cordero-Moss 2014). Whereas parties to such contracts are free to select the law governing the contract *itself*, party autonomy cannot modify the underlying law and its third-party effects. Therefore, parties' choice of law governing the contract does not reduce the influence of the (otherwise applicable) intellectual property, company, consumer, employment, criminal, and other laws.

This is clearly quite important in the context of intellectual property licensing. If choice-of-law clause affects only the applicable *contract* law, that means that it has no effect on the *copyright* law applicable to content in the dispute – and this can have a vast bearing on the case, as some of the key questions might fall under the scope of copyright, rather than contract law.

Suppose that users want to bring a lawsuit against an online platform provider because their status updates were distributed in a new digital format, and claim that licensing clause was invalid, and their copyright infringed. Users' claims are partially based on the fact that copyright law of their home country has a prohibition against granting licenses for future modes of exploitation. The company, as a defence, raises a claim that statuses were not entitled to copyright protection, as they were too short and unoriginal.

Most aspects of such a case would be resolved under the governing copyright law, which is unaffected by the parties' choice of contract law. Whether the content in question has met protection requirements or not clearly falls under the scope of copyright law; so do the conditions under which certain economic rights may be licensed – the rules governing the validity of transfer. Now, if users challenged the validity of the license based on contract formation deficiencies – for example, by claiming that there was no adequate mutual assent – that would (generally) be considered to be a contractual issue, subject to choice of law.

But even in this limited field of "purely contractual" affairs, application of the American law may be challenged. Generally, wording of the choice-of-law clause itself may be ambiguous or restricted in scope, as to pertain only to certain categories of disputes – and the parties may argue that the current dispute does not fall under the scope of the choice-of-law clause. Even more importantly, laws of many countries subject the application of foreign laws to numerous restrictions. In the European Union, under the harmonized private international law framework, there are at least five mechanisms for applying law other than the one chosen in the agreement. First, under the article 10(2) of the Rome I Regulation, a party, in order to establish that it did not consent to a contract, may rely upon the law of the country in which it has habitual residence, if it appears from the circumstances that it would not be reasonable to determine the effect of their conduct in accordance with the law found in the choice-of-law clause. Secondly, in consumer contracts, formal validity of the agreement is governed *exclusively* by the contract law of consumer's country of residence (Rome I Regulation 9(4)). Thirdly, while *Rome I* does allow the parties to choose the law governing a consumer contract, it curtails the effects of such a choice. Namely, Regulation states that choice-of-law clause may not have the result of depriving a consumer of the protection afforded to him/her by the mandatory provisions of the law of consumer's residence. Therefore, an issue-by-issue comparison between the chosen law and the mandatory law of the consumer's habitual residence needs to be carried out. If the chosen law provides for more protection, it governs the contract; if it does not, the contract is governed by a mix of laws: the chosen law (insofar as it offers sufficient protections) and the mandatory provisions of the law of the consumer's residence (Ruhl 2011). Fourthly, legislative instruments on substantive consumer protection may limit the application of foreign law. In particular, article 6(2) of the Unfair Terms Directive states that "Member States shall take the necessary measures to ensure that the consumer does not lose the protection granted by this Directive by virtue of the choice of the law of a non-Member country as the law applicable to the contract if the latter has a close connection with the territory of the Member States". In other words, if user agreement contains certain terms which are considered to be unfair (and thereby void) under the Directive, but which would be valid under the chosen law, the choice-of-law clause is to be disregarded, and the specific term is to be invalidated. Fifthly, there are certain provisions of the law of the forum, which are deemed to be applicable to any situation falling within their scope, irrespective of the law otherwise applicable to the contract – thereby limiting the effects of a choice-of-law clause. These provisions – known as overriding mandatory norms – are meant to be applied only in cases where their application is necessary for safeguarding country's core public interests, such as its political, social or economic organization. Still, there are reasons to believe that courts might resort to using this mechanism in cases involving a dispute over user content licensing practices. Namely, website users are practically always a weaker contractual party – as authors, consumers, parties with significantly less bargaining power; and it is well-established in the EU that the norms protecting such parties can be – and indeed often are – deemed to be overriding.[1] This is primarily because protection of weaker parties can be seen as essential to safeguarding country's interests – and if such parties were to lose protection of their national laws due to their lack of bargaining power and a presence of a choice-of-law clause, those interests would be impaired.

This shows that there are many ways in which Terms of Service agreements may be subjected to the law of user's country of residence. But how does that impact platform providers and their user-content licenses?

As mentioned above, licensing rules differ around the globe. In some countries, the law requires that each mode of exploitation be listed in the licensing clause, alongside its duration and territory (*See e.g.* Art. 131-3, French Intellectual Property Code). Some laws state that statutory provisions may not be superseded by "shrink-wrap agreements" or other unilateral reservations (Rognstad in Reto Hilty and Sylvie Nérisson 2012). Several legal systems deem the grant of a license for an unknown type of use to be void (*See e.g.* Art. 31, German Copyright Act). A number of national laws require that the author remuneration be expressly included in the licensing agreement (*See e.g.* Art. 3, Belgian Copyright Act). Some of them even allow authors to claim an additional fee if the exploitation of work has led to substantial benefits to the other party (*See e.g.* Art. 131–5, French Intellectual Property Code). In some countries, moral rights cannot be waived (*Id.*), and in others, authors have the right to unilaterally rescind licences after some time (*See e.g.* US Copyright Act). On top of that, contract laws differ widely: some contract laws contain very broad fairness checks, aimed at protecting weaker contractual parties, or subject the conduct of the parties to an overarching principle of good faith – all of which might lead to invalidation of licences.

Unless platform providers account for this diversity of licensing requirements when drafting user agreements, they risk potential lawsuits; and the broader their licensing terms are, the easier it might be to find them unfair and thereby void. The consequences of this might differ – ranging from unenforceability of the term, to the obligation to pay back part of the profit made by exploiting user content. And there is only one way to avoid this risk: through drafting of jurisdiction-specific, narrowly-scoped licensing clauses.

This is easier said than done. When drafting user agreements, it is very tempting to write them as if they were solely governed by the U.S. law, and go for as broad clauses as possible; but in reality, this creates a major compliance risk. As any risk, it might never materialize, and the company may continue offering its features perpetually. But all it takes for this tower of cards to fall down, is one user who will

[1] Stone observes that "despite the reference in Article 9(1) to a country's public interests, and to its political, social or economic organisation, there is no reason to doubt that Article 9(2) can be applied to mandatory rules of the lex fori which are designed to protect weaker parties, such as small businesses when transacting with large or medium-sized businesses." *Id.*

want to test the (faulty) permission that in cyberspace, American law rules supreme. From that point on, the company may be engaged in a long and unpredictable legal battle; and ultimately, it will have to cover the full costs of non-compliance with local laws – but this time, on the foreign court's terms, and with volatile economic and reputational fallout.

11.2.1.2 Uncertainty Due to Wording of Licencing Clauses

Even if we assume that broad licencing clauses are subject to American law, there are still wide-ranging areas of uncertainty that stem from the ambiguous wording of licencing clauses. Let us take a look at a few licensing clauses. For instance, Google Terms of Service include the following clause:

> Some of our Services allow you to upload, submit, store, send or receive content. You retain ownership of any intellectual property rights that you hold in that content. In short, what belongs to you stays yours.
>
> When you upload, submit, store, send or receive content to or through our Services, you give Google (and those we work with) a worldwide license to use, host, store, reproduce, modify, create derivative works (such as those resulting from translations, adaptations or other changes we make so that your content works better with our Services), communicate, publish, publicly perform, publicly display and distribute such content. The rights you grant in this license are for the limited purpose of operating, promoting, and improving our Services, and to develop new ones. This license continues even if you stop using our Services (for example, for a business listing you have added to Google Maps). Some Services may offer you ways to access and remove content that has been provided to that Service. Also, in some of our Services, there are terms or settings that narrow the scope of our use of the content submitted in those Services. Make sure you have the necessary rights to grant us this license for any content that you submit to our Services. (Google ToS 2016)

First, we can first point out the great uncertainty as to which parties the license is actually given to. Who are those that Google "works with" – what does the term "work" entail? Advertising agencies, ISP, governments, other users? What happens when Google stops working with someone – does that mean license given to those third parties is automatically terminated? Does it automatically extend to new business partners?

This uncertainty as to the parties is found in a lot of Terms of Service agreements. Whether through ambiguous wording such as one found in Google ToS described above, or through stipulating that the license given by users is "transferrable" or "sublicensable" – as found in many other ToS agreements – users are effectively left uncertain as to whom they are giving rights to exploit their works.

This leads us to the second major area of uncertainty: the scope of the license. This is, once again, due to the unclear wording found in ToS agreements. What does it mean – Google can modify and create derivative works of your content so that it "works better with their services"? Better in which way? In terms of computational power efficiency, presentation to users, incorporation with other Google services? Once again, the ambiguity of the term "work" surfaces, albeit in different context. But the question persists: how can user content "work" at all?

Contractual text is vague in other relevant parts, too. The license given to Google is limited, but for the purpose of "operating, promoting, and improving our Services, and to develop new ones". Is there a single scenario that involves the use of user-generated content that does not fall within the scope of this clause?

Contractual language is not merely ambiguous and expansive, but it is also full of terms that do not make a lot of sense under many governing laws. What does the "right to use" user content mean? Copyright consists of certain exclusive rights – in the US, those contained in 17 USC §106 – and that copyright license is a contractual mechanism through which authors assign a particular right (or a set thereof) to someone else. In order for authors to be able to assign a right through license, therefore, they must have that right to begin with. But US copyright law does not recognize the "right to use" as an exclusive right – unlike patent law, which explicitly gives it to inventors – so what is the meaning of this clause? Should it be entirely taken out of consideration as invalid under both copyright and contract law, or should the courts attempt to find a way to construe its meaning? Wouldn't that constitute overstepping their constitutional authority, as *de facto* creating a new exclusive right?

The agreement asks users to "make sure you have the necessary rights to grant us this license for any content that you submit to our Services". Again, this strikes as a clause that makes little to no sense. Let us say a user e-mails a newspaper article to his colleague. How can s/he be expected to give Google (and those they "work with") a worldwide license to publish it? Should we only e-mail our own content?

Granted, certain level of ambiguity is expected and tolerated in contract law. Yet, if a contract is so vague as to give rise to the doubts about whether it is even properly formed, further difficulties inescapably emerge in the contract interpretation stage, when it becomes necessary for the courts to construe rights and obligation of each party. The clearest consequence of this is that platform providers will be left uncertain how they are allowed to use user content, and that users are left uncertain how their content may be used. In other words, parties will be bound by *something* – but it will be very difficult for them to say by what and for how long. This could result in a number of undesirable developments. Platform providers might, erring on the side of caution, decide not to implement new features, which would exploit content in ways that arguably fall outside the scope of the license. Users might decide against uploading content altogether, if they feel that license gives platform providers too many rights. The courts will struggle in interpreting the licensing clauses and their rulings will be as diverse as they will be unpredictable. At the end of the day, a broad and ambiguous license benefits no-one.

11.2.2 Removing Control from Users – Undermining the Trust

While legal uncertainty is clearly worrisome, more immediate concerns stem from the fact that licensing clauses can effectively deprive users of control over their own content. Giving a perpetual licence to use users' photos, e-mails, videos – in any way that a platform provider deems fit – takes away one of the primary benefits that copyright confers upon the authors: the right to decide how and when their content will be used. Broad licenses may therefore negatively affect the relationship between platform providers and their users: if users feel that they have relinquished control over their content, their trust in platform providers will be undermined.

We have already witnessed increased user concern for how the content they create and upload may be used by the platform providers. In 2012, Instagram users revolted against updated ToS offered by Instagram, as they feared Instagram would be allowed to sell their photos and display ads without identifying them as such. This prompted a retraction of the new set of terms by the Instagram, and one of its co-founders clarified in a blog post: "I want to be really clear: Instagram has no intention of selling your photos, and we never did. We don't own your photos – you do"(Instagram 2015).

Users are becoming increasingly more concerned that their content will be inappropriately exploited. Another testament to this is the "notice" which several times went "viral" on Facebook, as countless users were sharing a message, which typically read:

> [...] I declare that my rights are attached to all my personal data drawings, paintings, photos, video, texts etc. published on my profile and my page. For commercial use of the foregoing my written consent is required at all times.
>
> Those who read this text can do a copy/paste on their Facebook wall. This will allow them to place themselves under the protection of copyright. By this statement, I tell Facebook that it is strictly forbidden to disclose, copy, distribute, broadcast, or take any other action against me on the basis of this profile and or its content. [...]
>
> The content of my profile contains private information. The violation of my privacy is punishable by law (UCC 1-308 1-308 1-103 and the Rome Statute)" (Snopes.com 2015).

While the legal analysis offered by this notice leaves a lot to be desired – what does the Rome Statute have to do with user privacy? – we should focus on the fact that users seem quite afraid that their content will be commercialized without their further consent, and were more than willing to voice their concerns over this.

So far, platform providers' response to these user-voiced concerns has been largely unsatisfactory. When addressing these issues, platform providers usually present three arguments: that users are shielded by the privacy policies, that users still retain ownership over their content, and that platform providers would not risk their reputation through introduction of invasive features. None of these arguments truly hold up.

11.2.2.1 Privacy Policies Do Not Offer Sufficient Protections

Addressing users' protest over invasive licences, platform providers sometimes claim that they are bound by their privacy policies, and that this, on its own, reduces the scope of a licence significantly. However, it must be noted that user content can be aggressively exploited even without violation of privacy policies or data protection laws.

Suppose that you have a favourite brand of clothes – ACME – and that you have liked their page on Facebook. There are also quite a few pictures of you wearing ACME shirts, taken during your last vacation, and shared with your Facebook friends. One morning, you wake up, only to find out that your pictures appear in your friends' feeds – not only as something you have shared, but also as ACME commercials, with a caption that states that you like ACME. Given that this commercial is shared only with your friends – those who would be able to access the photo anyway – your privacy is not violated.

In addition, using copyrighted material for data-mining – for example, by processing user images to develop facial recognition algorithms – does not necessarily constitute a violation of privacy. Data acquired can be anonymised before sharing with advertisers – or stored in your user profile, so that it is accessible only to platform providers. Therefore, if you upload a picture in which you are drinking beer, the algorithm processes this picture and updates your advertising preferences "behind the scenes", proceeding to share them with advertisers in an anonymised manner – your privacy is, once again, technically not violated.

And finally, while copyright – in the EU – lasts for 70 years after the death of its author, privacy, as a deeply personal right, does not survive one's death. In theory, this leaves platform providers with treasure troves of licensed, copyrightable content over which no privacy law interests may be asserted.

Protections offered by privacy policies in such scenarios are therefore flimsy at best. But even in cases where privacy policies are directly applicable, they might not be the silver bullet that the users expect them to be. In some jurisdiction, their effects remain rather unclear under the law.

Practice of the US courts is illustrative of this: they have occasionally held that privacy policies do not constitute enforceable contracts, but rather, unilateral statements that are not binding. This defence was successfully invoked against consumers who have invoked their rights under the privacy policies that they have consented to. So, while the courts have repeatedly held that ToS agreements form enforceable contracts, privacy policies – and other data use policies that give certain rights to users – have a highly questionable legal status in the US.

For instance, in the case of *In re Northwest Airlines Privacy Litigation* plaintiffs were the customers of Northwest Airlines (In re Northwest Airlines Privacy Litigation 2004; *See also* In re Jetblue Airways Corp. Privacy Litigation 2004). After September 11, 2001, NASA requested that Northwest provide NASA with certain passenger information in order to assist NASA in studying ways to increase airline security. Northwest supplied NASA with records, which contained information such as a passenger's name, flight number, credit card data, hotel reservation, and any traveling companions. The basis for most of plaintiffs' claims was that Northwest's website contained a privacy policy that stated that Northwest would not share customers' information except as necessary to make customers' travel arrangements. Plaintiffs contended that Northwest's provision of records to NASA violated Northwest's privacy policy, giving rise to their legal claims.

The court held that the privacy policy did not constitute a binding contract and that the plaintiffs have failed to allege damages. As the court reasoned:

> The usual rule in contract cases is that 'general statements of policy are not contractual'. [...] The privacy statement on Northwest's website did not constitute a unilateral contract. The language used vests discretion in Northwest to determine when the information is 'relevant' and which 'third parties' might need that information. Moreover, absent an allegation that Plaintiffs actually read the privacy policy, not merely the general allegation that Plaintiffs 'relied on' the policy, Plaintiffs have failed to allege an essential element of a contract claim: that the alleged 'offer' was accepted by Plaintiffs. Plaintiffs' contract and warranty claims fail as a matter of law.

Even if the privacy policy was sufficiently definite and Plaintiffs had alleged that they read the policy before giving their information to Northwest, it is likely that Plaintiffs' contract and warranty claims would fail as a matter of law. Defendants point out that Plaintiffs have failed to allege any contractual damages arising out of the alleged breach. [...] Damages are an essential element of a breach of contract claim, and the failure to allege damages would be fatal to Plaintiffs' contract claims"(In re Northwest Airlines Privacy Litigation 2004).

In a similar case, a bank customer brought a class action against the bank, alleging that bank's disclosure of customers' e-mail addresses was in violation of its own consumer privacy policy and constituted a deceptive business practice under New York law, breach of fiduciary duty, breach of contract, and negligent misrepresentation. (Cherny v. Emigrant Bank 2009) The court dismissed all the claims, holding that the customer has not pled sufficient injury to state a claim for relief.

Even when a platform provider does something that is in clear violation of its policies, following the standards set in *Northwest*, users will have to prove that they have read the policies, that they relied on it, and that the breach caused them sufficient injury to bring a contractual claim. One might think that proving harm might be a rather easy task – yet it has proven to be extremely difficult. Privacy harm is generally very difficult to prove, (Solove 2015) and copyright harm might be difficult to prove because of the generally very low economic value of user generated content.

Finally, there are instances in which the terms will explicitly establish precedence of IP licenses over a privacy policy. For example, Match.com asks assent to the following licensing clause:

> By posting Content on the Website or as part of the Service, you automatically grant to the Company, its affiliates, licensees and successors, an irrevocable, perpetual, non-exclusive, fully paid-up, worldwide right and license to (i) use, copy, store, perform, display, reproduce, record, play, adapt, modify and distribute the Content, (ii) prepare derivative works of the Content or incorporate the Content into other works, and (iii) grant and authorize sublicenses of the foregoing in any media now known or hereafter created (Match.com ToS 2016).

This exact wording is also offered by Tinder, a mobile application owned by the same company that owns Match.com. What is of interest here is that privacy policies offered by Match.com and Tinder state:

> We may use information that we collect about you to: [...] enforce or exercise any rights in our terms and conditions.

This clearly results in a negative outcome for the privacy-conscious users. In a case where, for instance, Match.com/Tinder gives a sublicense to a third company to use user's profile pictures in its ads, users are left with very limited contractual recourse: Match.com has merely exercised its rights under the ToS. Of course, users can have other means of remedying this situation, but as discussed earlier, this will largely depend on the protections offered by the mandatory norms of the applicable law (s), which is anything but predictable.

Therefore, on one hand, there is a wide array of scenarios in which user content can be exploited contrary to their wishes, without violating the privacy policies: such as when using users' content to market goods to their friends, or using content after the death of its author. On the other, even in situations where direct breach of privacy policy can be proven, obtaining relief might be difficult in certain jurisdictions, either because of the court's reluctance to enforce privacy policies, or because of users' inability to demonstrate financial harm. Consequently, users' reliance on privacy policies is – misguided.

11.2.2.2 Restatements of Users' Ownership of Content Are Insignificant

Trying to reassure their users, several platform providers offer pleasant statements such as "you still own your data – what is yours stays yours" (See e.g. Google ToS 2016). The problem with this response is that it does nothing to address the core concerns posed by broad licensing: the problem is not that platform providers are *owners* of content, but that they might have the right to commercialize it without even notifying users. Platform provider's response can therefore be perceived as a populist line aimed at reassuring those who do not have much knowledge of intellectual property laws – as for a substantial part of users, it feels intuitive to conclude that platform providers cannot sell something that they do not own. Yet, if platform providers are free to sublicense user content and to use it for commercial purposes, who *owns* that content is largely insignificant.

Therefore, reiterations of content ownership are almost irrelevant from the legal standpoint. Yes, you are still the owner of the work you post online; you still get to sell your photos, give others the right to copy them, incorporate them in new works. But that does not – at all – address the question of *which* rights you have given to the platform providers via a license. In simple terms: if a friend asked you to sign an agreement allowing him to live in your apartment, redecorate it, lease it out, burn it to the ground – but the agreement came with a clause exclaiming "but you are still the true owner!", would you ever sign it?

11.2.2.3 Reputation Management Is Not an Effective Safeguard

Lastly, we might be tempted to think that platform providers' business model depends on having a large user base, and that they would therefore be hesitant to introduce overly invasive features, given the potential reputational fallout. So far, however, we find little evidence that reputational risk is a primary concern for platform providers: the fact that a large number of new profiles is created every day gives them the luxury of being able to alienate a certain number of users. When litigating cases against users, platform providers have shown a tendency to raise all the possible arguments in support of their position, regardless of the reputational risk.

One such case comes from France (Frederic v. Facebook 2016), where Frederic Durand-Baissas, an art teacher, had his Facebook account terminated after he posted an image of a famous nineteenth century painting, depicting a nude woman – which was arguably prohibited by a then-current version of Facebook's Community Standards. Frederic filed a lawsuit before the court in Paris, claiming that Facebook engaged in unlawful censorship, and asked for five thousand euros in damages. Facebook moved to dismiss the case, and claimed that the contract between them and Frederic was not a consumer contract, that there was no imbalance of power between them, and that Facebook never intended to target any users in France with their products.

On the other side of the Atlantic, Facebook was sued by Mustafa Fteja, (Fteja v. Facebook Inc. 2012) who claimed that his account had been terminated on discriminatory grounds. He claimed that, upon receiving notice of termination, he made every attempt to contact Facebook and find out what happened. He received no response. This led him to believe that his Muslim background had a role to play in the termination of his account. In response, Facebook asked for the case to be dismissed as meritless, as nothing in the *"Statement of Rights and Responsibilities"* obliged Facebook to provide any reason for shutting down user accounts, or to give users any advance warning.

What is evident both from *Frederic v Facebook* and *Fteja v Facebook* is that reputational risk was not a factor significant enough to affect company's behaviour towards users, or the arguments raised in the proceedings. There is no reason to suspect that their decision would be any different when it comes to licensing of user content – especially if billions of dollars are to be made from its exploitation.

11.3 Possible Solutions

What can be done to minimize the negative effects of the overly broad licences? To begin with, it is *essential* that platform providers start drafting more precise and narrow ToS agreements. While it might be very tempting to ask users for as many rights as possible, going too far is likely to be counter-productive. As we have seen, neither the users nor the platform providers stand to gain anything from broad, uncertain and likely unenforceable clauses.

Licensing terms should also be communicated more clearly, be precise and attempt to fully and clearly disclose all the ways in which platform providers will use the submitted content. This is not so difficult to achieve; in fact, it can be done through a series of illustrative steps in the sign-up process, during which users would be able to opt into certain features. For instance, some users might agree to "when you upload your photo, we can analyse your facial expressions to determine your mood" displayed alongside a sample photo; others might find this too invasive and decide not to opt-in. Granted, this would mean that platform providers would likely see a reduction in the available datasets – but still, users would actually be in control and their consent would be informed and meaningful.

However, relying on platform providers' initiative – or indeed, the ability – to reform the current user-content licensing framework on their own may be misguided. Contracts need to be interpreted under the governing law, and as we have seen, determining which law governs ToS and what the outcome of that interaction may be is a rather complex task.

That is why it is necessary for the legislators to step in. First and foremost, harmonization of rules of private international law seems to be necessary: we need increased predictability as to which copyright and contract law will apply to user content licensing. In the interest of predictability, and increased protection of users, it might be wise to curtail party autonomy in choice of law, and subject the licence solely to the laws of user's country of residence.

Secondly, legislators could impose heightened consent requirements, asking for greater levels of consent for broad licenses – for instance, by requiring the aforementioned series of illustrative steps during the sign-up process. While good in theory, the reach of this rule would be limited in practice: users would again be left with a "take it or leave it" scenario at one point, and when presented with that kind of deal, they would still likely consent to it. Still, such a solution would virtually impose mandatory transparency standards – which could go a long way towards preserving the trust between users and platform providers.

Legislators should also take steps to ensure that licensing clauses are subject to privacy policies. The potential problem here stems from the "data duality"; the fact that legislators are attempting to protect private data contained in a work of authorship, while still allowing that work to be licensed. This could have undesired effects when it comes to other instances of copyright licensing, as there are clearly cases in which authors' privacy rights should not allow them to invalidate a license: notably in the cases where author is a professional who agreed to license his/her work in the course of business. Therefore, creation of a legislative instrument that regulates solely user-generated, non-professional content might be the best option in this context.

Perhaps most critically, legislators could introduce mandatory author remuneration norms. They could require that the profits made by exploiting user content in *unexpected ways* be paid back to the content's author. However, legislators would have to ensure that the amount of remuneration paid is linked to the platform provider's profit, rather than the value of the work itself – as otherwise, typically low value of user-generated content would render such a remedy meaningless. By making a reference to users' reasonable expectations, such solution might offer long-term flexibility, and might ensure that loopholes in the current copyright, contract, privacy and private international law framework are not used to users' detriment.

11.4 Conclusion

This chapter strived to demonstrate that broad and ambiguous licensing clauses are a clear and present threat. On the surface, allowing platform providers to cast a wide net over user content can seem harmless – and can even, in certain condition, arguably foster innovation. Yet, broad licenses leave platform providers and users uncertain as to how content can be legally used; and this uncertainty benefits no one. In addition, if licensing clauses are not subject to privacy policy – and in many cases, this can be called into question – then platform providers have, in essence, a blank check to use UGC as they see fit. The ramifications of this are not only frightening in individual cases; they could lay ground to a broad erosion of trust between users and platform providers, endangering the current participative Web as such.

While no single step can offer a full solution to these problems, it is essential that the work on addressing these issues begins soon. Be it the more careful ToS drafting by the platform providers, a more transparent and streamlined sign-up process, or direct legislative action – the solution needs to be reached through wide coordination and active participation of all the stakeholders.

Acknowledgements Author wishes to thank professors Ole-Andreas Rognstad and Giuditta Cordero-Moss of the University of Oslo for their valuable guidance and support in the writing of his doctoral thesis. This chapter certainly benefited from their continuous feedback and kind encouragement. In addition, this chapter builds on the ideas developed while the author was a visiting scholar at The George Washington University School of Law. He therefore wishes to extend his warm thanks to its faculty, and particularly professors Susan Karamanian and Daniel J. Solove, whose feedback was deeply appreciated.

References

ACI. (2015). *The data explosion in 2014 minute by minute - infographic*. http://aci.info/2014/07/12/the-data-explosion-in-2014-minute-by-minute-infographic/. Accessed 16 May 2016.

Apple ToS. (2015). *Apple – Legal – Licensed Application End User License Agreement*. http://www.apple.com/legal/internet-services/itunes/appstore/dev/stdeula/. Accessed 16 May 2016.

The Berne Convention for the Protection of Literary and Artistic Works, 1886.

Cherny v. Emigrant Bank. (2009). 604 F.Supp.2d 605 (604 F.Supp.2d 605 2009).

Cohen v. Facebook, Inc. (2011). No. C 10-5282 RS, 2011 WL 5117164, (N.D. Cal. 2011).

Cordero-Moss, G. (2014). *International commercial contracts: Applicable sources and enforceability*. Cambridge: Cambridge University Press.

Frederic v. Facebook. (2016). TGI Paris, 4ème ch., 5 mars 2015, n° 12/12 401, http://www.legalis.net/spip.php?page=jurisprudence-decision&id_article=4515. Accessed 16 May 2016.

Fteja v. Facebook, Inc. (2012). 841 F.Supp.2d 829 (841 F.Supp.2d 829).

Gaiman, N., & Pratchett, T. (1990). *Good Omens: The nice and accurate prophecies of Agnes Nutter, Witch*. Reprint edition. HarperTorch.

Goodreads ToS. (2016). *Terms of use*. https://www.goodreads.com/about/terms. Accessed 16 May 2016.

Google ToS. (2016). *Google terms of service – Privacy & Terms – Google*. http://www.google.com/intl/en/policies/terms/. Accessed 16 May 2016.

Hilty, R., & Nérisson, S. (2012). *Balancing copyright – A survey of national approaches*. Zürich: Springer.
In re Jetblue Airways Corp. Privacy Litigation. (2004). 379 F.Supp.2d 299.
In re Northwest Airlines Privacy Litigation. (2004). Not Reported in F.Supp.2d, 2004 WL 1278459.
Instagram. (2015). *Updated terms of service based on your feedback*. http://blog.instagram.com/post/38421250999/updated-terms-of-service-based-on-your-feedback. Accessed 16 May 2016.
MakeUseOf. (2015). *10 ridiculous EULA clauses that you may have already agreed to*. http://www.makeuseof.com/tag/10-ridiculous-eula-clauses-agreed/. Accessed 16 May 2016.
Match.com ToS. (2016). *Match.com - Find singles with Match.com's Online Dating Personals Service*. http://www.match.com/registration/membagr.aspx. Accessed 16 May 2016.
Perkins v. LinkedIn Corp. (2014). 53 F. Supp. 3d 1190, 2014 WL 2751053, (N.D. Cal.)
Rognstad, O-A. (2008). *Opphavsrettens balanse. Avtale eller lovregler?*, Festskrift till Marianne Levin (EST-2008-ml-521).
Ruhl, G. (2011). *Consumer protection in choice of law*. SSRN scholarly paper. Rochester: Social Science Research Network. http://papers.ssrn.com/abstract=1816641. Accessed 16 May 2016.
Samuelson, P. (1999). Privacy as intellectual property. *Stanford Law Review, 52*, 1125.
Snapchat ToS. (2016). *Terms of service – Snapchat*. https://www.snapchat.com/terms. Accessed 16 May 2016.
Snopes.com. (2015). *FALSE: Facebook Privacy Notice: Snopes.com*. http://www.snopes.com/computer/facebook/privacy.asp. Accessed 16 May 2016.
Solove, D. J. (2013). Privacy self-management and the consent dilemma, 126 Harvard Law Review 1880 (2013)
Solove, D. J. (2015). *Why the law often doesn't recognize privacy and data security harms*. https://www.linkedin.com/pulse/article/20140702054230-2259773-why-the-law-often-doesn-t-recognize-privacy-and-data-security-harms. Accessed 16 May 2016.
Tumblr ToS. (2016). https://www.tumblr.com/policy/en/terms-of-service. Accessed 16 May 2016.
WIPO. (2016). *Contracting Parties to Berne Convention*. http://www.wipo.int/treaties/en/ShowResults.jsp?treaty_id=15. Accessed 16 May 2016.

Part III
Users' Rights & International Regulations

Chapter 12
Online Service Providers' Liability, Copyright Infringement, and Freedom of Expression: Could Europe Learn from Canada?

Federica Giovanella

Abstract Imposing a high level of liability to Online Service Providers may result in over-enforcement against allegedly infringing conducts, with a "chilling-effect" on freedom of speech.

The regime for OSPs' liability under Dir. 2000/31 was conceived as a well balanced approach. However, the interpretation of the European Court of Justice and of some national courts has raised the bar on providers' duties. The absence in Dir. 2000/31 of a notification system of infringing activities is surely one of the reasons why this scenario has come into existence. As a result, the Commission has been considering the adoption of a new regime of "notice & action".

An interesting approach is the Canadian one, based on the so-called "notice and notice" system. Upon receiving notification from the copyright owner, the OSP is not obliged to remove the content; it must only forward that notification to the alleged infringer. Furthermore, non-compliance leads only to the payment of statutory damages.

"Notice and notice" has proved an efficient and effective tool that correctly balances freedom of expression and providers' needs. In this paper, I compare the European and Canadian regulatory frameworks for OSPs' liability, and investigate whether the Canadian approach could offer a solution to some of the persisting problems of the European one.

All the indicated websites were last visited on October 15th, 2015.

F. Giovanella (✉)
Post-doc Fellow in Comparative Private Law, University of Trento, Trento, Italy
e-mail: federica.giovanella@unitn.it

M. Taddeo, L. Floridi (eds.), *The Responsibilities of Online Service Providers*,
Law, Governance and Technology Series 31, DOI 10.1007/978-3-319-47852-4_12

12.1 Copyright Infringement and Free Speech in the Online Environment: The Key Role of OSPs

Online Service Providers (OSPs) functions are indissolubly intertwined with the development and prosperity of the Internet. The way the Internet has developed and is currently organized makes it impossible to imagine a World Wide Web without intermediaries.

OSPs' actions are vital not only as far as the basic functioning of the Internet is concerned, but also in order to allow anyone and everyone to connect to the Internet web and engage in any of the activities it offers. The latter aspect is clearly the most crucial in terms of the possibility to exercise what in the USA would be the champion of the First Amendment Rights, namely: freedom of speech, currently protected under art. 10 of the European Convention of Human Rights and art. 11 of the Charter of Fundamental Rights of the European Union.

The Internet represents an unprecedented opportunity for anyone to express themselves in a worldwide arena, with probably the greatest resonance ever imagined. The same resonance also implies that when a copyrighted work is used without the copyright owner's permission the damage caused can be many times greater than it would have been had the same infringement occurred "off-line". In addition, violating copyright has become easier than it was in the past: the easiness with which files can be uploaded and shared, or distributed in whatever way, is again something that could never have been previously imagined and copyright holders are facing a new and different infringement wave. In order to put an end to this onslaught, OSPs collaboration has become crucial: very often only OSPs can access the infringing material posted online and then stop the infringing activity or at least they are in the best position to implement measures aiming at such a result. In addition, in some cases OSPs are the only subject identifiable and accessible to which an infringement claim can be addressed.

The way OSPs' liability legislation is structured, therefore, becomes crucial: imposing a high level of liability on OSPs may result in over-enforcement against allegedly infringing conducts, with a so called "chilling-effect" on freedom of speech. On the contrary, a light approach might induce an increase in infringing activities, with negative effects on the protection of copyrighted works.[1]

Finding the correct balance between the two ends of the spectrum has proven itself to be a very difficult task. In the early days of the Internet, given the lack of specific provisions, judges applied the classical rules of liability, for the actions of both users and intermediaries.[2] However, as soon as the Web became a mass phe-

[1] For an economic analysis of providers' liability: Schruers 2002, pp. 231 ff.

[2] Sometimes OSPs were considered as publishers; see, for instance, in UK: Godfrey v Demon Internet Service [2001] Queen's Bench (QB) 201; in Italy: Trib. Naples, 9.8.1997, in Riv. Dir. Industriale 1999, II, 38; in USA: Stratton Oakmont, Inc. v. Prodigy Services Co., 1995 WL 323710 (N.Y. Sup. Ct. 1995).

nomenon, policymakers found it necessary to create an *ad hoc* liability regime for intermediaries. The first specific regulations were introduced as early as the end of the Nineties.[3]

Providers' liability is a well-known hypothesis of "liability for others", dictated by the uniqueness of the environment within which wrongful actions occur. The imposition of a third-party liability on OSPs was the easiest way to obtain a clear and precise framework for liability: it is often the case that this type of liability is imposed after encountering difficulties in tracking down the actual wrongdoer. In the meantime, this approach also complies with the so called "deep-pockets theory".

OSPs' liability includes a wide range of circumstances: from copyright infringement to defamation, from trademark violation, to privacy and personal data protection breach. The present chapter concentrates only on copyright cases for two main reasons. First, the Canadian regulation is written as part of copyright law and it therefore refers only to cases involving this right. Secondly, cases reporting copyright violation are those where freedom of expression assumes the most critical role. Although European Court of Justice's copyright cases involving OSPs' liability are increasing in number, I shall nonetheless refer to some cases concerning other intellectual property rights in order to give a better account of the current interpretation of the European framework for OSPs' liability.

Despite the fact that a large number of countries worldwide adopted similar regulations, divergences are apparent.[4] This chapter deals with the European framework and compares its key characteristics with the rules adopted in the Canadian context. The aim is to illustrate the Canadian approach to the intermediaries' liability and to propose the adoption of a similar framework in the European context, in light of the recent proposal by the European Commission to adopt a new regime of "notice & action". To this end, the chapter will first illustrate the current European framework and its interpretation by the European Court of Justice (§ 2), demonstrating the need for a notification system (§ 3). It will then describe the peculiarities of the Canadian framework regarding OSPs' liability for third party copyright infringement (§ 4), concluding with the previously mentioned suggestion to adopt similar rules at a European level (§ 5).

12.2 The European Way: Directive 2000/31

The European framework for OSPs' liability stems from Directive 31/2000 on Electronic Commerce.[5] The entire regulatory structure is centred on four articles, which exempt OSPs from liability whenever they meet specific requirements. If a

---

[3] The Agreed Statements in Art. 8 of the WIPO Copyright Treaty could be considered as a first step towards consensus for OSP protection (Margoni, Perry 2013, pp. 1202–1203).

[4] De Beer, Clemmer 2009.

[5] Directive 2000/31/EC of the European Parliament and of the Council of 8 June 2000 on certain legal aspects of information society services, in particular electronic commerce, in the Internal Market [2000] Official Journal (OJ) L 178, 17.7.2000, 1–16. For comments: Julià-Barceló, Koelman 2000; Baistrocchi 2003; Verbiest et al. 2007.

OSP complies with the specific conduct required by the law, it will not be held liable for a third party's conduct. As it is well-known, the Directive subdivides OSPs' actions into three different categories: mere conduit, caching, and hosting (arts. 12–14). These three activities entail an increasing level of involvement by the intermediary; this implies, for example, that it is harder for a hosting provider than for a mere conduit provider to be exempt from liability in relation to a user's wrongful behaviour. The Directive also specifies that Member States cannot impose on intermediaries a general obligation to monitor (art. 15).

It is often highlighted how the Directive contents were greatly inspired by the USA regulation introduced through the "Digital Millennium Copyright Act" (DMCA)[6] and its safe-harbours provisions introduced by the "Online Copyright Liability Limitation Act" (OCILLA).[7] A meaningful difference between the European and the American approach is that the latter applies only to copyright infringement, while the former applies to all kinds of civil wrongdoing. In 1996 the USA had indeed already enacted another statute, namely the "Communications Decency Act" (CDA).[8] Among other provisions, section 230 of the CDA specifically addresses the issue of civil liability for providers, stating that "[n]o provider or user of an interactive computer service shall be treated as the publisher or speaker of any information provided by another information content provider". The OCILLA and CDA are surely the most renowned pieces of legislation covering the issue of provider liability in the USA, however, other regulations are also in place, regarding other areas– for instance – trademarks.[9]

In the same vein as the US DMCA, the European Directive classifies providers into the three previously mentioned separate categories. Under art. 12 of the Directive, providers that offer only a connection to the Web are to be considered as "mere-conduit" providers. The provision defines "mere-conduit" providers as those whose service consists in "transmission in a communication network of information provided by a recipient of the service, or the provision of access to a communication network". Articles 13 and 14 of the Directive provide definitions of "caching" and "hosting" providers. The former implies "the automatic, intermediate and temporary storage of that information, performed for the sole purpose of making more efficient the information's onward transmission to other recipients of the service upon their request". The latter involves "storage of information provided by a recipient of the service". Each type of intermediary is required to comply with specific behaviour, in order to be exempt from liability.

A closing provision, which has proved to be a lifeline for intermediaries in some European Court of Justice's decisions, is contemplated in art. 15. The norm provides that "Member States shall not impose a general obligation on providers [...] to monitor the information which they transmit or store, nor a general obligation

[6] Pub. L. 105–304, 112 Stat. 2860 (Title 17 U.S.C. § 512).
[7] For an overview: Patten 2007; Brown 2008; Reese 2009. For a comparison between the EU and USA approaches: Julià-Barceló 2000; Peguera 2009.
[8] Pub. Law 104–104; 110 Stat. 56. For an overview: Cannon, 1996. For a comparison between CDA and DMCA: Band, Schruers 2002; Gasser, Schulz 2015.
[9] Lemley 2007, pp. 105–107.

actively to seek facts or circumstances indicating illegal activity". The same article, however, includes the possibility that Member States establish an obligation for providers to inform the public authorities of alleged illegal activities undertaken by their customers. From the point of view of the balance between copyright enforcement and freedom of speech, art. 15 is of absolute importance: the absence of a general obligation to monitor aims at avoiding the infamous "chilling-effect" above mentioned.

The text of the Directive has been criticized for many reasons, including, but not limited to, the lack of a clear system of notification and the absence of "search engines" and similar activities among the contemplated OSPs categories,[10] which creates problems in determining the specific functions carried out by providers.[11] The framework introduced by Dir. 2000/31 was probably a good starting point, crafted, as it was, for the Internet of the Nineties. However, the speed with which technology and more specifically in this case the Internet, evolved, meant the Directive was quickly made obsolete.

Provider functions nowadays go beyond what their role was considered to be at the time of the Directive enactment. Currently, one of the biggest matters of contention relates to hosting providers and the specific requirements they have to meet in order to enjoy the liability limitations included in art. 14. Specifically, art. 14 states that a hosting provider is exempted on condition that it: (a) "does not have actual knowledge of illegal activity or information and, as regards claims for damages, is not aware of facts or circumstances from which the illegal activity or information is apparent" or (b) "upon obtaining such knowledge or awareness, acts expeditiously to remove or to disable access to the information".

More and more often, Member States' courts request preliminary ruling by the Court of Justice of the European Union for the interpretation of the Directive. The initial structure and overall conception of the Directive have been questioned and rearranged through a number of CJEU's decisions. CJEU's interventions have been crucial not only in harmonizing the interpretation of the Directive,[12] but also in helping to update the Directive to the evolving Internet environment, by filling the existing gaps. The CJEU's judgments have also pointed out some of the limits of the Directive, demonstrating that the time has come for reconsidering the current OSPs' liability framework. However, the evolving interpretation of the CJEU and of some national courts has raised the bar on providers' duties, with a negative impact on their business and, most importantly, on freedom of expression. The current legal framework coupled with the courts' interpretation result in a high level of uncertainty that providers face with regards to what may or may not cause their liability.

---

[10] Julià-Barceló 2000, pp. 111–118.

[11] Van Eecke 2011, pp. 1468–1472; Elkin-Koren 2014, p. 29; Gasser, Schulz, 2015, *passim*.

[12] Scholars notice the tendency of national courts to apply national laws instead of the European ones: Margoni, Perry 2013, p. 1204; Van Eecke 2011, p. 1460.

12.2.1 *Interpretative Developments of the Court of Justice of the European Union*

The most important decisions made by the CJUE on the matter of OSPs' liability have been delivered in the last four years, even though the Court had already dealt with cases somehow involving the role of providers in the enforcement of copyright in previous years.[13] Actually, the most relevant CJEU cases dealing with copyright enforcement and OSPs do not rely specifically or solely on Dir. 2000/31; they often rely on Dir. 2004/48 relating the enforcement of intellectual property rights, which includes some provisions that involve providers in the enforcement of copyright in the digital environment.[14] Also for this reason, I shall later refer briefly to some seminal cases relating to OSPs' liability and other IP rights enforcement.

Among the most relevant issues faced by the Court of Justice in the realm of copyright are the questions of traffic monitoring and hyperlinking. Regarding the former, two fundamental decisions are of note, namely two cases known as the "Scarlet" and "Netlog" cases.[15] The latter issue is illustrated by the more recent judgement of "UPC Telekabel Wien".[16]

The Scarlet case involved the Belgian provider Scarlet Extended SA and the Belgian society for authors and publishers Société Belge des Auteurs, Compositeurs et Editeurs SCRL (SABAM). SABAM had asked Scarlet to implement a filtering system on peer-to-peer communications with the aim of impeding the sharing of copyrighted files. The lawsuit arose from the refusal of Scarlet to comply with the requests of SABAM. The Brussels Court of Appeal referred for a preliminary ruling asking whether the EU law allowed the imposition on a provider "to install, for all its customers, in abstracto and as a preventive measure, exclusively at the cost of that ISP and for an unlimited period, a system for filtering all electronic communications [...] in order to identify on its network the movement of electronic files containing a musical, cinematographic or audio-visual work [...] and subsequently to block the transfer of such files, either at the point at which they are requested or at which they are sent".[17]

---

[13] See Productores de Música de España (Promusicae) v. Telefónica de España SAU, C-275/06, January 29, 2008; LSG-Gesellschaft zur Wahrnehmung von Leistungsschutzrechten GmbH v. Tele2 Telecommunication GmbH, C-557/07, February 19, 2009. More recently, on the same issue: Bonnier Audio AB et al. v. Perfect Communication Sweden AB, C-461/10, April 19, 2012.

[14] Directive 2004/48/EC Of The European Parliament Aand of The Council of 29 April 2004 on the Enforcement of Intellectual Property Rights [2004] Official Journal (OJ) L 195, 2.6.2004, 16–25. See spec. art. 8.

[15] Scarlet Extended SA v. Société belge des auteurs, compositeurs et éditeurs SCRL (SABAM), C-70/10, November 24, 2011; Belgische Vereniging van Auteurs, Componisten en Uitgevers CVBA (SABAM) v. Netlog NV, C-260/10, February 6, 2012.

[16] UPC Telekabel Wien GmbH v. Constantin Film Verleih GmbH e Wega Filmproduktionsgesellschaft mbH, C-314/12, March 27, 2014.

[17] Scarlet v. SABAM, par. 28.

Even though the decision involved more than one European Directive, what is of relevance for the purposes of this work is the reasoning linked to art. 15 of Dir. 2000/31, on the absence of an obligation for OSPs to monitor the traffic. A filtering system such as the one requested by SABAM would have imposed on Scarlet a duty to survey the traffic of each single user in order to prevent further infringements. Letting aside other rights involved, such as users' privacy and freedom of expression, the CJEU considered the need to balance copyright holders' rights against OSPs' right to conduct a business. Given that a filtering system like the one requested by SABAM would impose very high costs on Scarlet, the Court ruled that it would be in contrast with European law.

In 2012, the CJEU decided on a very similar case, again involving SABAM. The Belgian society had sued Netlog – the owner of an online social networking platform – requesting that a filtering system for the information stored on its platform should be established in order to prevent files being made available which infringe copyright. The question was almost the same as the one decided in the previously described Scarlet case. The divergence lay mainly in the different function of the provider: whereas Scarlet was a mere conduit provider, Netlog was a hosting provider. The filtering system requested by SABAM would need the hosting provider to identify, among all the files stored on its servers by its users, the ones which infringed copyright; and then prevent those files from being made available through its services. Following the reasoning given in the *Scarlet* case, the CJEU found this system to be in contrast with art. 15 Dir. 2000/31.

In 2014 the CJEU had to cope with the intricate issue of hyperlinking to copyrighted materials. In the *UPC Telekabel Wien* case, two film producing companies requested the Austrian court to order the mere conduit provider UPC Telekabel to block its customers' access to a website that illicitly made available movies created by the two producing companies to the public. The Austrian *Oberster Gerichtshof* requested a preliminary ruling in order to question the compatibility with EU law of an order prohibiting an access provider to allow its customers to access certain websites "if the access provider can avoid incurring coercive penalties for breach of the prohibition by showing that it had nevertheless taken all reasonable measures".[18] The Austrian court also asked about the compatibility with EU law of a measure that would make it more difficult for providers' customers to access the previously mentioned website if the measure would require not inconsiderable costs. The Austrian court did not specify the measure that should be implemented, because the injunction at issue would have left its addressee to determine the specific measure to adopt.

The CJEU interpreted EU law as not precluding a court to issue such an injunction, provided that this does not unnecessarily deprive Internet users of access to information and that the measure prevents unauthorized access to protected material. The fact that the measure could be chosen by the same provider certainly weighed in favour of the lawfulness of such an injunction.

---

[18] UPC Telekabel Wien v. Constantin, par. 17.

In relation to filtering systems, *UPC Telekabel* goes, at least partially, in the opposite direction of the previous cases *Scarlet* and *Netlog*. It clearly proves how OSPs' role is under pressure and shows how the entire structure of Dir. 2000/31 can be (and is) undermined by the application of other directives.[19] Since these judgements are the most relevant regarding intermediaries liability for copyright infringement, a concise illustration of certain cases pertaining providers' liability for other IP rights infringement is fundamental. These cases help to shed light on the role of intermediaries and on the key issue of "knowledge".

In joined cases *Google France v. Louis Vuitton Malletier*, *Google France v. Viaticum Luteciel* and *Google France v. CNRRH*,[20] the CJEU decided on the so-called "AdWords" Google service. The service enables economic operators to obtain the placing by Google of "sponsored links", displayed in different positions and in alternative ways than those appearing as the "natural" results of search. These links are displayed in correspondence to one or more words that the economic operator "reserves" in an agreement with Google. The sponsored links also include a short commercial message ("ad").

The Louis Vuitton company realized that, while searching for its product through Google research, among the resulting sponsored links some actually readdressed to websites selling counterfeited products. It was ascertained that Google allowed advertisers to choose keywords such as "imitation" or "copy". Louis Vuitton sued Google to obtain a declaration that Google had infringed its trademarks.[21] The *Cour de Cassation* asked the CJEU if Google behaviour constituted trademark infringement, and, if so, whether the provider of the paid referencing service could qualify under art. 14 of Dir. 2000/31 and therefore enjoy liability limitations.

Once ascertained that Google activities as a search engine could be included in those described by art. 14,[22] the Court of Justice recalled recital 42 of Dir. 2000/31, which states that the exemptions only apply to those providers whose activity is *merely technical, automatic and of passive nature*, implying that the "provider has neither knowledge of nor control over the information which is transmitted or stored". In this actual case, Google had determined the display order of the results, based also on the remuneration paid by the advertisers. The fact that the service was subject to payment was not in itself sufficient to deprive Google from the exemptions provided by Dir. 2000/31. Meanwhile, the correspondence between the keywords selected and the search terms entered by users could not determine the knowledge of or control over the data entered into Google's system by advertisers

---

[19] Consider also the seminal and much criticized decision of the CJEU in "Google Spain": Google Spain SL v. Agencia Española de Protección de Datos (AEPD), C-131/12, May 13, 2014.

[20] Joined cases C-236/08 (Louis Vuitton Mallettier SA v. Google France SARL), C-237/08 (Viaticum SA e Luteciel SARL v. Google France SARL), C-238/08 (Google France SARL v. Centre national de recherche en relations humaines (CNRRH) SARL, Pierre-Alexis Thonet, Bruno Raboin, Tiger SARL), March 23, 2010.

[21] The same issue was at the base of the other joined cases.

[22] Louis Vuitton v. Google, cit., par. 110; Opinion of the Advocate General, Joined Cases C-236/08, C-237/08 and C-238/08, September 22, 2009, pars. 134–135.

and stored on the provider's servers. However, the role of Google in drafting the commercial message to be combined with the sponsored links should be taken into consideration.

In this case, the CJEU did not clarify whether this specific function of Google could be covered or not by the liability exemptions of art. 14. The Court only provided guidance for the national court to consider this aspect. In this sense, the intervention of the CJEU was far than useful in explaining when a hosting provider ceases to be neutral and starts being aware of the infringing activity. On the contrary and questionably, Advocate General Poiares Maduro was of the opinion that art. 14 could not be applied to the features of "AdWords" because Google is not neutral when transmitting the information.[23]

The same question was central in the 2011 case *L'Oréal SA v. eBay International*.[24] The lawsuit centred around the sale of L'Oréal products on the eBay website without L'Oréal's consent. The reference for preliminary ruling included ten different questions, the most relevant of which in this context are those concerning the interpretation of art. 14 of Dir. 2000/31 and the application of injunctions on intermediaries for the protection of trademarks infringed by users.

As for the former, the High Court of Justice of England and Wales asked whether eBay could be considered as a hosting provider according to the meaning of art. 14 and if so, in what circumstances would the provider have "awareness" or "actual knowledge" of the infringing activity under the same article.

The CJEU noticed that even if eBay undoubtedly stored information transmitted by its customers, this was not sufficient to conclude that its activity fell within the scope of art. 14. When a provider does not act neutrally but plays an active role that gives it knowledge of or control over the data transmitted, then it can no longer qualify for art. 14, as the case of *Luis Vuitton v. Google* had already clarified. eBay clearly provides assistance to its customers, and it processes the data entered by the user-seller. Meanwhile, the mere fact that an intermediary only stores data, sets the terms of services, provides general information to its customers and is remunerated for these services does not constitute a certain index of the inapplicability of the exemption from liability provided by Dir. 2000/31. However, when an operator has given specific assistance and optimised the sellers' bids, it should be considered as having an active role that provides itself with knowledge of or control over stored data. In the latter case the provider cannot enjoy the liability limitations introduced by art. 14.

Advocate General Jääskinen expressed doubts about the interpretation given by the CJEU in *Luis Vuitton v. Google* and the concept of "neutral" provider. The Advocate was clear in stating that eBay should enjoy the exemptions under art. 14 for paid Internet referencing service, even though it was involved in drafting the contents of listings and guided this listing in its system with technical means.[25]

---

[23] Opinion of the Advocate General, Joined Cases C-236/08, C-237/08 and C-238/08, cit., pars. 137 – 146.

[24] L'Oréal SA v. eBay International AG, C-324/09, July 12, 2011.

[25] Opinion of the Advocate General, C-236/08, C-324/09, December 9, 2010, pars. 138 – 151.

As for the possession of "awareness" by the operator of the online marketplace, it should be noted that art. 14 states that the intermediary is exempted from liability when they had no "actual knowledge of illegal activity or information" and were not "aware of facts or circumstances from which the illegal activity or information is apparent" or, if in possession of such knowledge or awareness, acted expeditiously to remove the information or disable access to it.

For the provider to lose its exemption from liability, the Court considered it sufficient for it to "have been aware of facts or circumstances on the basis of which a diligent economic operator should have identified the illegality in question and acted in accordance with Article 14(1)(b) of Directive 2000/31".[26] The Court considered the provider as becoming "aware" in the event that the intermediary uncovers an illegal activity or information either due to its own investigation or due to the notification by a third party. However, the CJEU advised that such a notification cannot automatically preclude the exemption from liability, since "notifications of allegedly illegal activities or information may turn out to be insufficiently precise or inadequately substantiated".[27] National courts should nevertheless take notifications into account when determining the awareness "of facts or circumstances on the basis of which a *diligent economic operator* should have identified the illegality".[28]

A partially different view was taken by Advocate General Jääskinen, who explained what should in his view be considered as "actual knowledge".[29] In his words "legally 'knowledge' may refer only to past and/or present but not to the future [...] the requirement of actual knowledge seems to exclude construed knowledge. It is *not* enough that the service provider ought to have known or has good reasons to suspect illegal activity".[30] This would ultimately mean that constructive knowledge would not be enough. To have an actual knowledge an OSP must have been notified of infringing uses. As to which kind of notification would be appropriate, the Advocate General did not give further explanation.

Regarding the question of injunctions, the preliminary ruling analysed whether Member States should grant trademark owners injunctions against intermediaries in order to prevent further infringement by third parties, and if so what scope should such an injunction have. The Court of Justice held that such injunctions would be lawful under EU law, unless they are in contrast with art. 15 of Dir. 2000/31 regarding the absence of an obligation to monitor and on the condition that there should be a fair balance among the different rights and interests at stake.

Even though this decision relates to trademarks, it is particularly relevant to the present analysis. The judgment clarifies the importance of third party notices in relation to OSPs' awareness of illegal activities. Both judgements on trademarks

[26] L'Oréal v. eBay, cit., par. 120.
[27] L'Oréal v. eBay, cit., par. 122.
[28] L'Oréal v. eBay, cit., par. 122. Emphasis added.
[29] Opinion of the Advocate General, C-236/08, C-324/09, pars. 154 – 168.
[30] Opinion of the Advocate General, C-236/08, C-324/09, pars. 162 – 163. Emphasis added.

summarized here, refer to the concept of providers' "neutrality". Such a conception, that proved to be misleading, has been read as a misinterpretation of recital 42, which should be applied only to mere conduit and caching providers, while it has been mainly applied to hosting intermediaries.[31]

12.2.2 The Need for a Notification System

It follows on from what has been illustrated so far that it is currently impossible to categorically state when OSPs have awareness or knowledge of infringing activities. This issue is mainly related to hosting providers, in light of art. 14 and of its requirements. As previously explained, art. 14 states that a hosting provider can be held liable if it has either "actual knowledge" or "awareness" of the existence of infringing content. In case of "awareness", the intermediary could be held liable only for damages. However, the provision does not describe in which situation there can be "awareness" or "actual knowledge".[32] The absence of specific parameters has lead courts – including the CJEU – to fluctuating decisions: should providers be held liable for general knowledge or only when they are aware of specific infringing activities? Some Member State courts ask for specific and detailed information about the infringing content;[33] some others accept a generic indication as well.[34] It is clear that each provider has a "general knowledge" that there might be some infringing contents among all the contents hosted. However, such interpretation would jeopardize the entire exemption from liability for hosting providers.[35]

Some Italian courts have introduced the category of "active hosting provider", meant as a provider that not only serves as an intermediary but has an active role in the management and delivery of the contents, acting also with the aim of obtaining a profit.[36] "Active providers" exist neither in Dir. 2000/31 nor in its Italian implementation (d.lgs. 9.4.2003, n. 70). They might be a further development of the above reported *eBay* and *Louis Vuitton* cases. The CJEU specified that providers cannot enjoy the limitations under art. 14 when they cease to be *neutral*, which might amount to being *latu sensu* actively involved in the infringing activity.

---

[31] Van Eecke 2011, pp. 1481–1482.

[32] Some national implementations do not even ask for "actual knowledge" but introduced different requirements. See: European Commission, Commission Staff Working Document, 2012, pp. 32 ff. Knowledge is an even more critical factor with reference to juridical persons: Van Eecke 2011, p. 1475; Burk 2011.

[33] For instance: Juzgado de lo Mercantil n. 7 of Madrid, 23.7.2008, n. 320; Trib. Turin, 5.5.2014 (unpublished).

[34] See e.g. Trib. Rome, 16.12.2009, Dir. Informatica, 2010, 268. See also Commission Staff Working Document on Online services, cit., 32 ff.

[35] Mantelero 2014, 796.

[36] Trib. Rome, 16.12.2009, cit.; Trib. Milan, 7.6.2011, in Dir. Informatica, 2011, 660; Trib. Milan, 9.9.2011, in Riv. Dir. industriale, 2011, II, 364; Trib. Milan, 25.5.2013, in Resp. Civ. Prev., 2013, 1996. *Contra*: Court of Appeal Milan, 7.1.2015, in Resp. Civ. Prev., 2015, 1245.

The paradox of such an interpretation is that if a provider implements some specific tools to improve its capabilities of detecting infringing contents, it might be qualified as "active" and therefore not enjoy the limitations under art. 14. This so-called "Good Samaritan" paradox implies that the more a provider tries to prevent infringing activities, the less limitations it obtains with regards to its liability, with the consequence that providers would be discouraged from implementing mechanisms to prevent infringement.[37]

The ambiguity in the interpretation and application of the Directive leads to a high level of uncertainty for providers with regard to their liability regime. In addition, providers are not in the best position to know whether a given content is covered by IP rights or not. Such an imposition would be a burden for OSPs, especially in terms of costs.

It is not by chance that the DMCA includes a provision introducing a federal cause of action for misrepresenting infringing posted material: the person abusing the notification system asking for the removal of a non-infringing content is the individual who would be liable. On the contrary, the provider would be shielded from liability for this specific conduct.[38]

In absence of a similar provision under the European framework, a provider, which does not know precisely if and how it would be held liable for the contents hosted on its servers, might act in a precautionary manner and remove the suspect material. Such a conduct would definitely harm freedom of expression in the online environment.

This vagueness calls for the introduction of a notification system that was not contemplated in Dir. 2000/31. Art. 21 required re-examination of the Directive 3 years after it came into force,[39] with reference also to the need for "notice and take down" procedures,[40] as well as the liability following the taking down of the allegedly infringing content. Art. 16, likewise Recital 40, encouraged the adoption of self-regulation to enhance the potentiality of the Directive and, if possible, to make it work more smoothly.

As a result, some Member States have introduced notification regimes on their own initiative. For instance, Finland introduced a specific notification regime, but it is limited to copyrighted infringement cases;[41] a similar approach was also implemented in Lithuania.[42] Belgium, however, relied on a cooperation with

[37] Van Eecke 2011, pp. 1483–1484. See also: Regional Court of Hamburg, 5.3.2010, ref. no. 324 O 565/08. In the *eBay* case, the fact that filters had been installed to detect information that might contravene the terms of use of the site and implemented a notice and take-down system (Cf. L'Oréal v. eBay, cit., par. 46), might have weighed in favor of the applicability of exemptions to the provider.

[38] § 512(f).

[39] Art. 21, par. 2, Dir. 2000/31.

[40] "Notice and take down" is the name given to the notification system under the DMCA, § 512(c)(3).

[41] Commission of the European Communities 2003, pp. 14 ff.

[42] Verbiest et al. 2007, p. 15.

providers[43], and in Italy a special regulation was introduced in 2014 that includes a particular notification system involving an administrative authority.[44] Some of these procedures apply to any kind of providers, others only to specified kinds; procedures also have different time-frames and requirements.[45] Providers have also adopted best practises and codes of conduct that can be regarded as a privately-implemented notification system.[46]

Surveys have highlighted the lack of uniformity among Member States, as well as the difficulties faced by providers in dealing with this uncertain framework.[47] The same public critiques have also emphasized the need to introduce a notification regime, which takes freedom of expression sufficiently into account.[48]

More recently, the EU has launched other public polls [49] in order to assess the issue of OSPs' liability more effectively, including the possible introduction of "notice-and-action", i.e. mechanisms aiming at eliminating infringing or otherwise illegal content from the Internet.[50] The Commission itself held that it will "adopt a horizontal initiative on notice and action procedures",[51] without modifying Dir. 2000/31.[52]

The Commission's declarations simply made explicit what the case law had already implicitly highlighted: the European framework for OSPs' liability needs a notification system. In my view, to be effective it needs to address – at least – the following issues:

1. Contents and requirements of notices;
2. Time-frame for OSPs' reaction to notices;
3. Attribution of responsibilities in the event of notices' misuse;
4. Clarity on what constitutes "knowledge" for OSPs.

The following paragraph will illustrate the Canadian experience as an alternative at which EU should look if and when implementing a new regulatory framework for OSPs' liability.

---

43 Belgian Internet Service Providers Association Code of Conduct at: http://www.ispa.be/code-conduct-fr/

44 Marini Balestra, Tremolada 2014; Tosi 2015, pp. 125 ff.

45 European Commission, Commission Staff Working Document, 2012a, b, pp. 40–44. See also Annex II of the same document.

46 Memorandum of Understanding on the Sale of Counterfeit Goods on the Internet, May 4, 2011, at: http://ec.europa.eu/internal_market/iprenforcement/docs/memorandum_04052011_en.pdf

47 European Commission, Communication, 2012a, b, p. 13.

48 Summary of the results of the Public Consultation on the future of electronic commerce in the Internal Market and the implementation of the Directive on electronic commerce (2000/31/EC), 2010, pp. 10 ff., at: http://ec.europa.eu/internal_market/consultations/docs/2010/e-commerce/summary_report_en.pdf

49 Public consultation on the future of electronic commerce in the internal market and the implementation of the Directive on electronic commerce (2000/31/EC), at: http://ec.europa.eu/internal_market/consultations/2010/e-commerce_en.htm; A clean and open Internet: Public consultation on procedures for notifying and acting on illegal content hosted by online intermediaries, at: http://ec.europa.eu/internal_market/consultations/2012/clean-and-open-internet_en.htm

50 European Commission, Communication, 2012a, b, p.13 fn. 49.

51 European Commission, Communication, 2012a, b, p. 15. See Kuczerawy 2015.

52 European Commission, Commission Staff Working Document 2013, pp. 17 ff.

12.3 The Canadian Approach to OSPs' Liability for Copyright Infringement

While the European framework significantly draws on the US DMCA, Canada has developed its own approach to the issue of OSP's liability. Before the amendments to the Copyright Act in 2012, Canada did not have any specific statutory provision to regulate OSPs' liability for third party infringement, that was until then considered a matter of case law.

Secondary liability in the Canadian context emerges from Sections 3 and 27 of the Copyright Act. The former lists copyright owners' rights to produce, reproduce, perform and so on, including the right to "authorize any such act". S. 27 states that it is an infringement of copyright to do anything that the Act itself reserves only for the owner, without the owner's consent. In its famous case "CCH",[53] the Canadian Supreme Court interpreted S. 27 as including the "right to authorize". CCH – a publishing company, sued the Law Society of Upper Canada claiming a violation of CCH's copyright on its publications. The Law Society maintains and runs a library in Toronto, where it provides a "request-based photocopy" service for Law Society members, the judiciary and other authorized people. CCH contended that the Law Society had breached copyright either by providing the custom photocopy service with which the publisher's works were reproduced and sent to patrons upon request, or by maintaining the self-service photocopiers and copies of the works in the Library.

The main point was to understand what was meant by "authorize". The Supreme Court held that this word has to be interpreted in a strict way, in its "strongest dictionary meaning, namely, '[g]ive approval to; sanction, permit; favour, encourage'".[54] Judges should presume that in authorizing an activity, a person does so only in so far as it is in respect of the law.[55] This presumption could be overreached by showing that there is a certain relationship or degree of control between the authorizer and the infringer. The Supreme Court stated that the Law Society lacked sufficient control over the users to conclude that it approved the alleged infringement. Given these premises, it concluded that there had been no authorization by the Society,[56] and therefore no secondary liability.

The Supreme Court tested its reasoning relating to Internet intermediaries in that same year within the *Society of Composers, Authors & Music Publishers of Canada (SOCAN) v. Canadian Association of Internet Providers (CAIP)* case.[57] SOCAN wanted to impose royalties on providers because, in the plaintiffs' view, their activi-

[53] CCH Canadian Ltd v Law Society of Upper Canada, [2004] 1 SCR 339.
[54] CCH Canadian Ltd. v. Law Society of Upper Canada, cit., par. 38.
[55] Citing *Muzak Corp. v. Composers, Authors and Publishers Association of Canada, Ltd.*, [1953] 2 S.C.R. 182, 193.
[56] CCH Canadian Ltd. v. Law Society of Upper Canada, cit., parr. 39–45.
[57] 2 S.C.R. 427 (2004).

ties constituted infringement of the copyright owners' right to communicate and to authorize such communication.

The Supreme Court refuted SOCAN claims, on the base of *CCH* interpretation of "authorization". The OSPs' awareness that someone can use its technology to infringe copyright does not necessarily constitute authorization. On the contrary, approval or encouragement is needed to be able to define the OSP's role as giving authorization. If authorization was considered broadly, this would put providers in the difficult position of understanding whether a specific content is infringing or not.[58] The activity of providers in communicating (even infringing) copyright works was therefore considered "neutral", including caching activity.[59] Neutrality would cease in the event that the provider notices the existence of infringing material on its system and fails to take remedies, such as asking the customer through a "take down notice" to remove the infringing content. Only if, upon receiving a request the provider does not comply with, can it be considered as authorizing the infringing activity.[60]

The Court applied also a statutory defence known as the "common carrier exception":[61] If a person only provides the means through which a communication is made, then that person is not herself communicating, and therefore is not violating, the copyrighted work.

This provision could be applied also to providers, when they operate only as a conduit for the transmission of the information and do not modify the content of the communication. On the contrary, when a provider only offers ancillary services, it is covered by this exception. The possible provider's knowledge that someone could use its facilities to infringe copyright does not entail liability.[62] This provision applies only to liability for communication to the public by telecommunication, but not to other copyright owner's rights.[63]

In the same decision, the Supreme Court also looked forward to the introduction of a notification regime such as those adopted by the EU (*sic*) and the USA.[64] Indeed, as previously mentioned, before the Copyright Modernization Act of 2012 amending the Copyright Act, Canada had no statutes regulating this intricate issue. Furthermore, even though today Canada has a specific legislation, this relates only to copyright matters; for other cases, such as defamation ones, *SOCAN* and *CCH* rules still apply.

[58] SOCAN v. CAIP, cit., par. 127.
[59] SOCAN v. CAIP, cit., *passim*.
[60] On this matter, the Supreme Court referred to the Canadian Association of ISPs' Code of Conduct. Cf. SOCAN v. CAIP, cit., par. 110.
[61] Section 2.4(1)(b) of the Copyright Act.
[62] SOCAN v. CAIP, cit., parr. 32–33. The Court (par. 96) compared ISPs to the owners of telephone wires citing the old case *Electric Despatch Co. v. Bell Telephone Co.*, 20 S.C.R. 83 (1891).
[63] Scassa, Deturbide 2012, p. 428.
[64] SOCAN v. CAIP, cit., par. 127.

The Canadian statute considers three kinds of providers, which can be traced back to the same categories listed by European legislation. The regulation opens with the provision of S. 31.1(1), which focuses on access providers and states: "A person who, in providing services related to the operation of the Internet or another digital network, provides any means for the telecommunication or the reproduction of a work or other subject-matter through the Internet or that other network does not, solely by reason of providing those means, infringe copyright in that work or other subject-matter". The wording is more open to interpretation than its European counterpart: it applies to "services related to the operation of the Internet or another digital network", therefore including services that cannot strictly be considered as OSPs, such as peer-to-peer intermediaries. This provision has been read as an "explicitly broader common carrier exemption" than the S. 2.4(1)(b).[65]

The definition is a starting point to shape the other types of providers as well. Under S. 31.1(2) a caching provider is not liable for the act of caching or doing "any similar act in relation to it", if it does not modify the copyrighted content, other than for technical reasons; it ensures that its activities are compliant with industry practice; it does not interfere with the use of technology that is lawful and consistent with the same practices in order to obtain data on the use of the work.[66]

By virtue of S. 31.1(4) a hosting provider is meant as "a person who, for the purpose of allowing the telecommunication of a work or other subject-matter through the Internet or another digital network, provides digital memory in which another person stores the work or other subject-matter". The host provider cannot be considered liable unless it knows of a *court decision* or a *decision from other competent jurisdiction* ruling that the host provider's storage of a work or other subject-matter amounts to an infringement of copyright.

A closing provision, applicable to all kinds of providers, states that liability limitation does not apply to an act that constitutes copyright infringement under subsection 27(2.3), known as the "enabling provision".

This subsection states that it is an infringement of copyright to provide a service primarily for the purpose of enabling acts of copyright infringement, if an actual infringement occurs by means of the use of that service. A typical case would be peer-to-peer technology.[67]

Unlike the European context, Canadian provisions do not specifically proclaim that OSPs are exempted from liability. Rather, they state that certain providers' activities do not constitute copyright infringement. No extrajudicial remedy such as the DMCA notice and take-down exists: the provider is liable only when it knows

[65] Hagen 2010, p. 377.

[66] This definition is consistent with the interpretation given by the Copyright Board in SOCAN Statement of Royalties, Public Performance of Musical Works 1996, 1997, 1998, (Re) 1 C.P.R. (4th) 417, par. 32.

[67] The provision does not apply to "off-line" devices (cf. Hagen, 2010, p. 380). S. 27(2.4) provides a non-exhaustive list of factors that courts can take into account to determine whether there was copyright infringement under S. 27(2.3).

of an order coming from a court or other judicial authority. However, the Copyright Modernization Act introduced a so-called "notice and notice" regime[68].

The notice and notice system applies to all types of providers and requires them to forward notices of copyright infringement from copyright owners to the user that uploaded the allegedly infringing content. The process starts with a notice of claimed infringement sent by the copyright owner to the provider. S. 41.25(2) describes in detail the contents and the form of the notification, that shall additionally identify the work allegedly infringed and specify the infringement claimed. Upon receipt, the provider shall "as soon as is feasible forward the notice electronically to the person" identified in the notice, and inform the claimant. The provider shall also retain records that allow the identification of the alleged infringer.[69] The Act also states that the Minister in charge may fix a maximum fee that providers can charge for performing this duty.[70]

In the event that providers do not comply with this regime, they are liable only for statutory damages, which are limited in amount by S. 41.26(3). This is a very relevant difference with the European legislation, where assuming that the copyright owner's notice is sufficient to generate the provider's "actual knowledge", a non-compliant-OSP loses the liability exemption.

The Canadian framework also includes a specific regime for search engines,[71] by which these services are regarded mainly as other kinds of provider. However, S. 41.27(3) limits the applicable immunity within a specific time period of 30 days, which has been criticized for introducing *de facto* a notice and take-down regime.[72]

Before the implementation of this framework, Canadian OSPs had applied a self-imposed code of conduct,[73] as well as a notice and notice systems agreed with the Canadian Cable Television Association and the Canadian Recording Industry Association (CRIA). The regime introduced by the Modernization Act is largely based on this procedure, but it represents an improvement because it encompasses all kinds of copyrighted material, while the self-adopted regime applied only to music.[74]

---

68 SS. 41.25, 41.26, 41.27(3).

69 S. 41.26(1)(b).

70 S. 41.26(c).

71 S. 41.27.

72 Hamilton 2005, p. 393.

73 CAIP "Code of Conduct", at: http://www.cata.ca/Communities/caip/codeofconduct/CodeConduct.html. See Bernstein, Ramchandani 2002; Nesbitt 2003; Chwelos 2006.

74 Hamilton 2005, p. 296.

12.4 Conclusions, Could Europe Learn from Canada? Suggestions for a Much Needed Notification Regime

As it has been seen, Canada approaches the question of OSPs' liability in a very different way from the European Union. Even though the Canadian framework clearly has some flaws, it nevertheless promises better performances that those experienced under Dir. 2000/31.

As illustrated, the Canadian system provides specific details about the form and content of the notification, which constitutes the source and the proof of the provider's actual knowledge of the infringing activity. The statute does not state what happens when the notice lacks some of the requirements. However, it might be assumed that the provider is not obliged to contact the user and cannot be held liable, even for statutory damages. An incomplete notice might not constitute proof of the intermediary's actual knowledge, exactly as the CJEU clarified in the *eBay* case. The introduction of standardized notices in European legislation would benefit both providers and right holders: currently the former bear the risks linked to imprecise or insufficient notices, while the latter bears the burden of proving the notice to be sufficient and precise.[75]

The Canadian statute does not clearly state the time-frame for providers' reaction to notices. The words adopted ("as soon as feasible") are rather vague, in the same vein as the European text ("expeditiously"). However, Canadian regulation leaves the door open for providers to ask for a fee for performing their obligation.[76] Such a possibility allows a better balancing of the different rights, additionally taking into account providers' freedom to conduct a business; meanwhile, the fact that providers could charge a fee might be an incentive for them to supply a better, faster service.

Canada does not explicitly rule on the case of notices' misuse either. Since the current framework largely draws on pre-existing standard procedures, it might be the case that misuse had not been registered before. Indeed, this different approach probably affects this issue. When a system requires the provider to "take-down" the material once a notice has been received, a copyright holder might be motivated to send as many notices as possible since a dispute with the content provider seems very unlikely. Using the Canadian approach, the notice is sent to the OSP to be forwarded to the content provider, who will then decide whether to take the content down or not. It might not be as easy to obtain a "take-down", as it would be in the first example: this may be a deterrent for an abuse or misuse of notices by copyright holders.

It follows on from the conducted analysis that the Canadian system does not entirely address all the concerns that the current European framework raises. A better approach would include specific norms related to notices' abuse, as well as a specific time-frame for providers' compliance with the notice. Nonetheless, the

[75] Van Eecke 2011, p. 1485.
[76] S. 41.26 (2).

Canadian method has the virtue of allowing for a good balance between copyright holder's claims, users' freedom of speech, and providers' freedom of entrepreneurship.

This chapter focused mainly on copyright infringement, due to the fact that the Canadian approach applies only to copyright issues. Nonetheless, a clear notification system is needed in all the cases of providers' liability, also outside the realm of IP rights. I consider the Canadian approach not quite a template, but at least a good starting point for new European legislation. A complete revision of the rules governing civil liability in the online environment would be advisable to improve the European approach and the adoption of a more effective OSPs' liability framework would be a first, serious step towards better regulation that takes into account all the rights at stake, first and foremost being freedom of speech.

It remains to be seen whether international treaties currently under discussion, such as the Comprehensive Economic and Trade Agreement or the Trans Pacific Partnership Agreement, will affect the issue analysed here and to what extent.[77]

References

Baistrocchi, P. (2003). Liability of intermediary service providers in the EU directive on electronic commerce. *Santa Clara Computer & High Technology Law Journal, 19*, 111–130.

Band, J., & Schruers, M. (2002). Safe harbors against the liability hurricane: The communications decency act and the digital millennium copyright act. *Cardozo Arts & Entertainment Law Journal, 20*, 295–320.

Bernstein, A., & Ramchandani, R. (2002). Don't shoot the messenger! A discussion of ISP liability. *Canadian Journal of Law & Technology, 1*, 77–85.

Brown, B. (2008). Fortifying the safe harbors: Reevaluating the DMCA in a Web 2.0 world. *Berkeley Technology Law Journal, 23*, 437–467.

Burk, D. L. (2011). Toward an epistemology of ISP secondary liability. *Philosophy & Technology, 24*, 437–454.

Cannon, R. (1996). The legislative history of Senator Exon's communications decency act: Regulating barbarians on the information superhighway. *Federal Communications Law Journal, 49*, 51–94.

Chwelos, P. (2006). Internet service providers report. https://www.ic.gc.ca/eic/site/ippd-dppi.nsf/vwapj/ISP_Report_2006_01_20.pdf/$file/ISP_Report_2006_01_20.pdf

Commission of the European Communities. (2003). First report on the application of Directive 2000/31/EC of the European Parliament and of the Council of 8 June 2000 on the directive on electronic commerce. http://www.europarl.europa.eu/registre/docs_autres_institutions/commission_europeenne/com/2003/0702/COM_COM(2003)0702_EN.pdf

De Beer, J., & Clemmer, C. D. (2009). Global trends in online copyright enforcement: A non-neutral role for network intermediaries? *Jurimetrics, 49*, 375–409.

Elkin-Koren, N. (2014). After twenty years: Revisiting copyright liability of online intermediaries. In S. Frankel & J. G. Daniel (Eds.), *The evolution and equilibrium if copyright in the digital age* (pp. 29–51). Cambridge: Cambridge University Press.

[77] See Art. 5.5 CETA, version published on September 26, 2014; TTP, Section I, Leaked Version of October 5, 2015, at: https://wikileaks.org/tpp-ip3/WikiLeaks-TPP-IP-Chapter/WikiLeaks-TPP-IP-Chapter-051015.pdf

European Commission. (2012a). A coherent framework for building trust in the digital single market for e-commerce and online services. http://eur-lex.europa.eu/resource.html?uri=cellar:87375c7c-1bd0-445d-b251-60599af8c73b.0009.03/DOC_1&format=PDF

European Commission. (2012b). Commission staff working document: Online services, including e-Commerce in single market. Accompanying the document: A coherent framework for building trust in the digital single market for e-commerce and online services. http://ec.europa.eu/internal_market/e-commerce/docs/communication2012/SEC2011_1641_en.pdf

European Commission. (2013). E-commerce action plan 2012–2015. State of play 2013. http://ec.europa. eu/ internal_market/e-commerce/docs/communications/130423_report-ecommerce-action-plan_en.pdf

Gasser, U., & Schulz, W. (2015). Governance of online intermediaries: Observations from a series of national case studies. The Berkman Center for Internet & Society Research Publication Series. http://cyber.law.harvard.edu/publications/2015/online_intermediaries

Hagen, G. R. (2010). "Modernizing" ISP copyright liability. In M. Geist (Ed.), *From "Radical Extremism" to "Balanced Copyright". Canadian copyright and the digital agenda* (pp. 361–394). Toronto: Irwin Law.

Hamilton, S. N. (2005). Made in Canada: A unique approach to internet service provider liability and copyright infringement. In M. Geist (Ed.), *In the public interest: The future of canadian copyright law* (pp. 285–308). Toronto: Irwin Law.

Julià-Barceló, R. (2000). On-line intermediary liability issues: Comparing E.U. and U.S. Legal Frameworks. *E.I.P.R.* 22: 105–119.

Julià-Barceló, R., & Koelman, J. K. (2000). Intermediary liability: Intermediary liability in the E-Commerce directive: So far so good, but it's not enough. *Computer Law & Security Review, 16*, 231–239.

Kuczerawy, A. (2015). Intermediary liability & freedom of expression: Recent developments in the EU notice & action initiative. *Computer Law & Security Reviews, 31*, 46–56.

Mantelero, A. (2014). Responsabilità aquiliana per uso della Rete e responsabilità del provider. In F. Delfini & F. Giusella (Eds.), *Diritto privato dell'informatica* (pp. 785–828). Torino: UTET.

Margoni, T., & Perry, M. (2013). Deep pockets, packets, and harbors. *Ohio State Law Journal, 74*, 1195–1216.

Marini Balestra, F., & Tremolada, R. (2014). Enforcement of online copyright in Italy: The new regulation adopted by the Italian communications authority. *Intellectual Property Quarterly*, 143–158.

Nesbitt, S. (2003). Rescuing the balance? An assessment of Canada's proposal to limit ISP liability for online copyright infringement. *Canadian Journal of Law & Technology, 2*, 115–133.

Patten, L. B. (2007). From safe harbor to choppy waters: YouTube, the digital millennium copyright act, and a much needed change of course. *Vanderbilt Journal of Entertainment & Technology Law, 10*, 179–214.

Peguera, M. (2009). The DMCA safe harbors and their European counterparts: A comparative analysis of some common problems. *Columbia JL & Arts, 32*, 481–512.

Reese, A. R. (2009). The relationship between the ISP safe harbors and ordinary rules of copyright liability. *Columbia Journal of Law & the Arts, 32*, 427–443.

Scassa, T., & Deturbide, M. (2012). *Electronic commerce and internet law in Canada*. Toronto: CCH.

Schruers, M. (2002). The history and economics of ISP liability for third party content. *Virginia Law Review, 88*, 205–264.

Tosi, E. (2015). *High tech law: The digital legal frame in Italy*. Milano: Giuffrè.

Van Eecke, P. (2011). Online service providers and liability: A plea for a balanced approach. *Common Market Law Review, 48*, 1455–1402.

Verbiest, T., et al. (2007). Study on the liability of internet intermediaries. http://ec.europa.eu/internal_market/e-commerce/docs/study/liability/final_report_en.pdf

Chapter 13
Non-financial Disclosures in the Tech Sector: Furthering the Trend

Peter Micek and Deniz Duru Aydin

Abstract Facing pressure from stakeholders, companies in the information and communications technology (ICT) sector have begun to release reports tabulating and detailing government requests they receive to restrict access to content and hand over user data. While these reports initially aimed to reassure customers, privacy advocates, and other observers that the companies deserve trust and defend user interests, their rapid spread and growth following revelations of mass surveillance in 2013 has led to transparency reporting becoming an expected, normal channel of communication to the public, and an indicator of ethical considerations by tech and telecom firms. This chapter charts the rise of transparency reporting, noting its foundations in financial regulation, human rights norms, and corporate social responsibility discourses. We investigate the diverse array of academic, civil society, and regulatory stakeholders promoting and analysing the reports, as well as legal and discretionary limitations to the practice. We posit transparency as an enabler for an ethical framework for online service providers (OSPs), as well as a functional way to assuage user bases regarding privacy and freedom of expression concerns. Expanded and standardized reporting on more categories of corporate activity could further illuminate the opaque decisions that OSPs make about content and user data, and lead to more ethical outcomes.

P. Micek (✉)
Global Policy & Legal Counsel, Access Now, New York, NY, USA

School of International and Public Affairs, Columbia University, New York, NY, USA
e-mail: peter@accessnow.org

D.D. Aydin
Oxford Internet Institute, Oxford, England

M. Taddeo, L. Floridi (eds.), *The Responsibilities of Online Service Providers*,
Law, Governance and Technology Series 31, DOI 10.1007/978-3-319-47852-4_13

13.1 Introduction

At the beginning of 2015, there were more than three billion Internet users around the world, with global Internet penetration rates exceeding 40 % ("Measuring the Information" 2015). Every day, massive amounts of user data are generated and stored in the servers of major technology companies that offer services globally, often free of charge. This phenomenon is usually referred as the age of "Big Data," which – as argued by boyd and Crawford – "is less about data that is big than it is about a capacity to search, aggregate, and cross-reference large data sets" (boyd and Crawford 2012, p.663). From online advertising to search algorithms, user information is utilized and leveraged by various private actors on a daily basis as a fundamental part of their business models (Mendel et al. 2012, p.12–19). However, not only private actors but also governments try to make use of and control the data generated daily by the billions of users of the major Online Service Providers (OSPs) and telecom companies. Given the public's growing tendency to use electronic communications for nearly every aspect of social, professional, political, and cultural life – including for organizing criminal activities, exercising human rights, and sharing cute penguin videos – law enforcement and intelligence agencies along with many other government officials are increasingly demanding OSPs handover user data and restrict access to or remove content (Mendel et al. 2012, p. 96). Governments can request access to various types of data including but not limited to subscriber data, stored or real time traffic data (metadata), or content data (Malby et al. 2013). Governments also submit requests asking or demanding companies to restrict access to information, delete user accounts, and remove or deprioritize content. For example, Turkey and Russia, countries that both have a troublesome record for press freedoms and free expression, accounted for the majority of requests Twitter received to take down content ("Twitter Transparency Report" 2015). At times, these requests are lawful or motivated by a genuine concern over the privacy or security of a country's citizens, though such requests to remove content online could also represent unlawful or overbroad censorship. Coined as "networked authoritarianism," by Rebecca MacKinnon, governments increasingly use such mechanisms of intermediary liability – usually in addition to filtering practices deployed by Internet Service Providers (ISPs) – to control online content (MacKinnon 2010).[1]

In addition to legal measures to restrict content on the Internet, various government ministries, agencies, and executives also submit ad hoc requests to companies. At times, governments shut down or block the company's services in retaliation for non-compliance with these extra-legal requests. The most notorious example happened during the early 2011 uprising in Egypt, as the Mubarak government successfully demanded telecommunications companies shut down mobile phone networks and disseminate pro-regime propaganda via SMS (Howard and Hussain 2013; Ghonim 2012, p. 212; Oates 2011), reportedly upon threat of retaliation against corporate staff.

[1] Deibert et al. (2008) offers an overview of filtering practices in 26 countres around the world.

The current state of affairs, characterised by out-dated law and opaque corporate policy, creates problems for all stakeholders. Legal, reputational, and financial liabilities result when companies are perceived to actively participate in, or passively enable, violations of user privacy or free expression. On a basic level, these disruptions and blocking prevent customers from reaching the content and services they want, and, increasingly, require in order to exercise their rights and function in society. Unstable and insecure services do not deliver strong and sustainable financial returns for companies, nor do they satisfy users' needs. Media and policy circles' attention on the sector's alleged or proven contributions to dragnet government spying programs are leading to legal, financial, and reputational harms to internet and telecom firms like IBM and Cisco (Stempel 2013; Bort 2011), the cloud computing industry (Castro 2013; Castro and McQuinn 2015), and U.S. telecoms hoping to expand to Europe (Severin et al. 2014; Gryta et al. 2013). Indeed, "American companies have reported declining sales overseas and lost business opportunities, especially as foreign companies turn claims of products that can protect users from NSA spying into a competitive advantage" (Greene et al. 2014).

On the content regulation side, OSPs have suffered economic and reputational harms in the face of overbroad censorship imposed by governments. In addition, with regard to ensuring freedom of expression online, the U.S. government – as the home country of major OSPs – became one of the actors pushing for companies to take up a stronger stance in favor of the free flow of information online. In early 2006, four U.S. companies were scrutinized by Congress after being accused of succumbing to the requests of the Chinese government to censor specific websites, offer an entirely new version of their search engine or facilitate government surveillance and intimidation of journalists ("Net firms criticized" 2006; Taddeo and Floridi 2015).[2]

Recognizing these threats to the integrity and success of their business model, Internet and telecom firms started to look for ways to regain the trust of the public (Goldberg 2016). How does the sector as a whole respond to user concerns and maintain their business models, while also complying with legal orders to hand over data or restrict content?

Transparency has been one of the pillars companies lean on in this challenge. More specifically, companies provide non-financial disclosures in the form of a transparency report, which we define as an official, periodical statement issued by an ICT company (including OSPs, telecom companies, Internet Service Providers (ISPs), and mobile network operators), which discloses data on government and other third party requests for user information, network disruptions, and content restriction, and information on other policies and practices impacting privacy and freedom of expression online. As early as April 2010 (more of this background can be found under the section History of Transparency Reporting), Google published a transparency report where it reported on government requests to access user information or restrict content. In 2013, soon after the revelations of mass surveillance exposed by former National Security Agency contractor Edward Snowden,

[2] There is an ongoing case against Cisco, see ("Doe I v. Cisco," 2016).

the largest Internet firms filed a lawsuit demanding the U.S. government to allow them to disclose more about the extent of data requested by intelligence agencies. They also formed cross-industry coalitions such as Reform Government Surveillance, which asks for governments to endorse principles and reform laws to achieve a variety of ends that include extended transparency over surveillance powers (Chavez and Salgado 2013; Taddeo and Floridi 2015, p. 20). The firms settled the lawsuit and followed through with greater disclosures on security and surveillance topics, propelling the wind in favour of openness into a gale, as today more than 60 ICT companies worldwide release so-called "transparency reports" on the scope and scale of their contribution to government surveillance and content restriction.[3] Transparency reporting represents one of the main avenues tech and telecom companies choose to repair their business models and regain public trust.

In the rest of this chapter, we investigate the notion of the transparency reporting as a best practice for online service providers to adopt in order to protect their users' rights and defend their business practices. We start by analysing its driving forces, followed by a topography of the current state of transparency reporting in technology and telecom sectors, including its limitations and gaps. Furthermore, we argue that transparency reporting contributes to creating an evidence-based internet policy and governance discourse, and serves as one of the first steps to create an industry-wide corporate social responsibility norm grounded in an understanding of companies' human rights responsibilities, which could in turn help define and clarify the ethical responsibilities of OSPs.

13.2 History of Transparency Reporting: Actors and Drivers

As seen in Table 13.1, currently, more than sixty companies release transparency reports ("Access Now Transparency Reporting Index" 2016). In this section, we provide an overview of the different actors and drivers behind this trend, which started in 2010 when Google published the first-ever transparency report (Miller 2010).

13.2.1 The Ideological Roots

Corporate social responsibility (CSR) "encompasses the economic, legal, ethical and discretionary (philanthropic) expectations that society has of organizations at a given point in time" (Caroll 2004, p.116). Therefore, transparency reporting can be easily investigated within a CSR framework.

The push for transparency from the private sector is well-rooted, and even precedes today's CSR mechanisms. In the United States, corporate transparency and financial disclosures form the basis of the nation's foundational financial laws, the

[3] A regularly updated list of companies that publish transparency reports can be found at Access Now's Transparency Reporting Index: https://www.accessnow.org/transparency-reporting-index

Table 13.1 List of companies that publish transparency reports, current as of May 2016 ("Transparency Reporting Index" 2016)

Company name	Company platform	First release	Frequency	Country of headquarters	Region of headquarters
AOL	Internet	2/13/14	6 months	USA	North America
Adobe	Internet	3/12/15	Annually	USA	North America
Amazon	Internet	6/12/15	6 months	USA	North America
Apple	Internet	6/16/13	6 Months	USA	North America
AT&T	Telco	4/17/14	6 months	USA	North America
Cheezburger	Internet	2/5/15	Annually	USA	North America
Cisco	Hardware	4/21/15	6 months	USA	North America
Cloudflare	Internet	2/27/14	6 months	USA	North America
Comcast	Telco	3/20/14	6 months	USA	North America
Credo	Telco	1/8/14	6 months	USA	North America
CyberGhost	Internet	5/13/14	Monthly	Romania	Europe
Daum Kakao	Internet	1/23/15	N/A	South Korea	Asia-Pacific
Deutsche Telekom	Telco	5/5/14	Annually	Germany	Europe
Dreamhost	Internet	3/11/15	Annually	USA	North America
Dropbox	Internet	12-May	6 months	USA	North America
Etsy	Internet	7/14/15	Annually	USA	North America
Evernote	Internet	14-Mar	Annually	USA	North America
Facebook	Internet	8/29/13	6 months	USA	North America
GitHub	Internet	4/16/15	Annually	USA	North America
Google	Internet	9/24/10	6 months	USA	North America
Internet Archive	Internet	2014	Annually	USA	North America
Kickstarter	Internet	4/8/15	Annually	USA	North America
Leaseweb	Internet	4/11/13	6 months	Netherlands	Europe

(continued)

Table 13.1 (continued)

Company name	Company platform	First release	Frequency	Country of headquarters	Region of headquarters
LinkedIn	Internet	1/27/14	6 months	USA	North America
Lookout	Internet	2014	Annually	USA	North America
Lyft	Internet	7/9/09	Annually	USA	North America
Mapbox	Internet	6/26/13	3 months (quarterly)	USA	North America
Medium	Internet	1/5/15	Annually	USA	North America
Microsoft	Internet	3/21/13	6 months	USA	North America
Naver	Internet	1/1/15	6 months	South Korea	Asia-Pacific
Nest	Internet of Things	6/17/15	N/A	USA	North America
Orange	Telco	5/27/15	Annually	France	Europe
Pinterest	Internet	3/10/14	3 months (quarterly)	USA	North America
Posteo	Internet	5/5/14	Annually	Germany	Europe
Proxy.Sh	Internet	N/A	Real-time	N/A	N/A
Reddit	Internet	1/29/15	Annually	USA	North America
Rogers Communications	Telco	6/5/14	Annually	Canada	North America
Sask Tel	Telco	9/14/14	Annually	Canada	North America
Silent Circle	Internet	11/15/13	3 months (quarterly)	Switzerland	Europe
Slack	Internet	5/1/15	Annually	Canada	North America
Snapchat	Internet	4/2/15	6 months	USA	North America
Sonic.net	Telco	4/13/12	Annually	USA	North America
SpiderOak	Internet	5/17/12	Annually	USA	North America
Sprint	Telco	1/1/15	Annually	USA	North America
T-Mobile US	Telco	7/1/15	Annually	USA	North America
TekSavvy	Telco	6/4/14	N/A	Canada	North America
Telenor	Telco	5/12/15	Annually	Norway	Europe
TeliaSonera	Telco	6/22/14	6 months	Sweden	Europe

(continued)

Table 13.1 (continued)

Company name	Company platform	First release	Frequency	Country of headquarters	Region of headquarters
Telstra	Telco	3/7/14	Annually	Australia	Asia-Pacific
Telus	Telco	9/18/14	Annually	Canada	North America
Time Warner Cable	Telco	3/21/14	6 months	USA	North America
Trade Me	Internet	7/16/13	Annually	New Zealand	Asia-Pacific
Tumblr	Internet	2/3/14	6 months	USA	North America
Twitter	Internet	7/2/12	6 months	USA	North America
Twilio	Internet	7/1/15	6 months	USA	North America
Uber	Internet	4/15/16	6 months	USA	North America
Verizon	Telco	1/22/14	Annually	USA	North America
Vodafone	Telco	6/6/14	Annually	United Kingdom	Europe
Wickr	Internet	2/25/13	6 months	USA	North America
Wikimedia	Internet	8/6/14	6 months	USA	North America
WIND Mobile	Telco	Fall 2014	Annually	Canada	North America
Word Press	Internet	4/17/14	6 months	USA	North America
Yahoo	Internet	6/17/13	6 months	USA	North America

Securities Act of 1933 and the Securities Exchange Act of 1934. Responding to the conditions that led to the Great Depression, the Acts continue to enforce standards for corporate reporting on financial data. The U.S. Congress has also amended the Acts to require corporate reporting on a growing range of non-financial matters, such as the use of conflict minerals ("Fact Sheet" 2014; "SEC Proposes Rules" 2015).

In 1999, just before the start of the decade when corporate social responsibility "flourished," Cynthia A. Williams argued that the "Securities and Exchange Commission (SEC) can and should require expanded social disclosure by public companies to promote corporate social responsibility" (Caroll 2004, p. 116; Williams 1999, 1199). In the years since, regulatory regimes in the U.S. and Europe have embraced, in fits and starts, various forms of non-financial reporting. Investors, not to mention advocacy organizations, have pushed for standardization of the metrics and reporting mechanisms.[4] Companies in several sectors, including extractives, mining, and food and beverage have fought regulatory burdens, but also competed to excel in transparency initiatives. According to a 2015 report by the consulting company KPMG, ninety percent of the global top 250 companies (G250) engage in corporate responsibility reporting ("Currents of Change" 2015).

The idea of "expanded social disclosure" as a fundamental concept has matured and become a norm in corporate social responsibility through regulatory and advocacy oriented action at the national, regional, and international level, which in turn had effects on the telecom and technology sectors as well as on transparency reporting.

13.2.2 *Initiatives for Increased Corporate Transparency*

In terms of concrete initiatives in the broader business ethics concept, one of the earliest examples is the OECD's Guidelines for Multinational Enterprises in 1976. (Ruggie 2007, p. 819). Two decades later, the UN Global Compact was launched in 2000, as a "voluntary initiative engaging companies and civil society in promoting UN principles in the areas of human rights, labor standards, environmental protection, and, since 2004, anti-corruption," with more than 8000 companies and 4000 non-business actors currently participating (Ruggie 2007, p. 820).

More recently, the European Union amended the Accounting Directive in 2014 to require non-financial disclosures by all public-interest businesses of 500 or more employees. Member states must transpose the Directive into national law by the Dec. 6, 2016, and the measures take effect the following year ("EU Directive" 2014). By requiring companies to disclose relevant and material environmental and social information in their annual reports, the Directive sets a floor, not a ceiling. More stringent national laws, like France's far-reaching and robust Grenelle 2, for example, passed in 2010 and implemented in 2012, remain in place and enforceable (Baddache 2012).

13.2.2.1 UN Guiding Principles on Business & Human Rights

Relatedly, the UN Guiding Principles on Business and Human Rights (UNGPs), which were endorsed unanimously by the UN Human Rights Council in 2011, counsel corporations to "account for how they address their human rights impacts"

[4] Examples include Oxfam's Behind the Brands Campaign (for the food industry), The Extractives Industry Transparency Initiative, and the anti-sweatshop movement of the 1990s, which led to the White House Apparel Industry Partnership in 1995 ("Extractive Industries Transparency Initiative" 2003; Ratute, 2010). Non-sector-specific examples include the Dow Jones Sustainability Group Index and the FTSE4Good Index.

through reporting and external communications.[5] The UNGPs elaborate the Ruggie "Protect, Respect, and Remedy" Framework established by John Ruggie in 2008, in his capacity as UN Secretary-General's Special Representative for business and human rights. The Guidin Principles are not sector-specific, but rather aim to apply broadly to all industries. According to the UNGPs, businesses should, "both know and show that they respect human rights in practice. Showing involves communication, providing a measure of transparency and accountability to individuals or groups who may be impacted," including investors.[6] The UNGPs note that corporate communications take various forms, including "non-financial reports," and "sector-specific indicators," which we would argue characterize the criteria emerging across ICT sector transparency reports. As for corporate attention to the UNGPs, which by design carry no legal weight, sector and geographical divides remain: multinational telecommunications companies are much more likely to reference the UNGPs than are U.S. Internet platforms.[7] One way through which the ICT sector adopted this outlook was to create industry coalitions, including the Reform Government Surveillance coalition, the Telecommunications Industry Dialogue and the Global Network Initiative.

13.2.2.2 Global Network Initiative

In September 2005, Yahoo! was shown to have handed over journalist Shi Tao's email account data to Chinese authorities, leading to Tao's conviction and sentencing to 10 years in prison (Schonfeld 2006). U.S. Congress summoned the company's executives for public shaming (Rampell 2007). In response, Yahoo! created an internal body to address the privacy and free expression impacts of it operations, launched as its Business and Human Rights Program in 2008. The company released ethical guidelines, stating its intent to "work with industry, government, academia and NGOs to explore policies to guide industry practices in countries where content is treated more restrictively than in the United States and to promote the principles of freedom of speech and expression" (MacKinnon 2006). In keeping with this statement, Yahoo! met with a diverse external set of corporate, academic, civil society, and investor stakeholders who together created the Global Network Initiative (GNI).

Responding to this situation, where a government pressured a company to contribute to a human rights violation, the GNI Principles declared:

> The collaboration between the ICT industry, investors, civil society organizations, academics and other stakeholders can strengthen efforts to work with governments to advance

---

[5] See principle 21 of the UNGPs.

[6] See principle 23 of the UNGPs.

[7] See for example AT&T, http://about.att.com/content/csr/home/frequently-requested-info/governance/transparencyreport.html, MTN https://www.mtn.com/Sustainability/Documents/Digital_Human_Rights.pdf and Vodafone https://www.vodafone.com/content/sustainabilityreport/2015/index/operating-responsibly/human-rights.html, while searches of Google and Facebook corporate communications turn up no references.

> freedom of expression and privacy globally. For these reasons, these Principles and their accompanying Implementation Guidelines establish a framework to provide direction and guidance to the ICT industry and its stakeholders in protecting and advancing the enjoyment of human rights globally. ("Global Network Initiative Principles" 2008, para. 6)

On transparency, one of the GNI Principles requires a bifurcated set of disclosures for participating company members: "Participants will be held accountable through a system of (a) transparency with the public and (b) independent assessment and evaluation of the implementation of these Principles."[8]

The idea behind the entity appears to be that together, stakeholders can push back on harmful government requests, while providing assessors and the public with firmer grounds to trust and verify that the companies are behaving ethically, within their constraints. The GNI continues today with the platforms Yahoo!, Microsoft, Google, LinkedIn and Facebook as its major company members. GNI is currently collaborating with the Telecommunications Industry Dialogue, a group of eight telecom companies that jointly address the challenges to freedom of expression and privacy in the telecommunications sector.

While the GNI focuses on the ICT sector, it is only one of many multi-stakeholder bodies intended to facilitate dialogue, regulate behaviour, and set norms for various industries.[9] The initiatives vary in their standards, membership requirements, and processes – and their effectiveness. Sceptics abound, especially within civil society, who feel the mechanisms do not sufficiently bind members to uphold their codes of conduct, enforce penalties for non-compliance, or provide adversely impacted stakeholders with adequate access to remedy. They also can simply waste precious time for accountability advocates (Green and Miller-Davis 2014). With their self-regulatory approaches, many of the bodies do not require greater transparency than the companies must provide under law. Nonetheless, related efforts continue to strengthen, sharpen, and integrate indicators for reporting on environmental, social, and governance matters, with increased targeting by business sector, including ICTs.

While transparency reports remain voluntary in the sense that they are not required by regulators or the law, a variety of U.S. and Canada-based stakeholders apply public and private pressure toward institutionalization of these commitments. For example, international NGO Access Now[10] maintains a Transparency Reporting Index that "features a record of transparency reports from today's leading internet companies and telcos" ("Transparency Reporting Index" 2016). The advocacy group accompanies its Index with language extolling the reports as, "one of the strongest ways for technology companies to disclose threats to user privacy and free expression," intentionally applying an ethical dimension to its informational listing.

---

[8] The GNI membership includes academics, investors, human rights organizations, companies and others, but not governments.

[9] Other examples include United Nations Global Compact, Global e-Sustainability Initiative (GeSI), Telecommunications Industry Dialogue.

[10] One of the authors, Peter Micek, is employed by Access Now.

The U.S.-based New America Foundation's Open Technology Institute created an Transparency Reporting Toolkit, in collaboration with Harvard University's Berkman Center for Internet & Society (Budsh et al. 2016). The project identifies best practices and looks to set normative standards through a "template" or model transparency report and disclosure guidelines. Also housed within the New America Foundation is Ranking Digital Rights (RDR), a research initiative which ranked 16 major internet and telecommunications companies on "their public commitments and disclosed policies affecting users' freedom of expression and privacy," in its first installment ("Indicators- Ranking Digital Rights" 2015). The project's first Corporate Accountability Index was based on an original set of criteria across three categories: Commitment, Privacy, and Freedom of Expression. Several of the RDR criteria refer to the types of information commonly found in transparency reports. By scoring companies higher for disclosing such information regularly, RDR asserts transparency reporting as a best practice in corporate respect for human rights.[11] A similar initiative is Electronic Frontier Foundation's annual "Who Has Your Back?" report, in which one of the categories under consideration for evaluating companies is transparency reporting ("Who has your back?" 2015).[12]

In Canada, academics and officials moved quickly following the Snowden disclosures to pressure local telecom and tech companies to report, and have led the way toward regulation-backed transparency reporting standards. Academic experts run the Telecom Transparency Project, which "investigates how telecommunications data is monitored, collected, and analyzed for commercial, state security, and intelligence purposes" ("The Telecom Transparency Project" 2015). The Canadian Office of the Privacy Commissioner, on the other hand, compiles data from companies, but notes that coherence is lacking between the different company reports, and points at government-established reporting guidelines (which civil society and academia found to be significantly problematic) as a solution. ("Transparency Reporting by Private Sector" 2015; "Transparency Reporting Guidelines" 2015; Parsons 2015).

13.2.3 Effects of Snowden Revelations

The revelations of mass surveillance by former NSA contractor Edward Snowden, beginning in June 2013, heralded the beginning of a new era of transparency in the ICT sector. Companies understood that their business models were threatened by the widespread lack of confidence in their services. Following the summer of 2013, there was a sharp rise in the number of companies publishing transparency reports. In 2012, only six companies published transparency reports. In 2014, this number increased to 39.

---

[11] See for example indicators F7 "Data about government requests," F8 "Data about private requests," and P11 "Data about third-party requests for user information."

[12] The project started in 2010.

But the quality of those reports differs, and the companies also face restrictions on what they may legally disclose. For example, the Reform Government Surveillance coalition, which promotes "Transparency about government demands" as one of its requests to "the world's governments" states that:

> Transparency is essential to a debate over governments' surveillance powers and the scope of programs that are administered under those powers. Governments should allow companies to publish the number and nature of government demands for user information. In addition, governments should also promptly disclose this data publicly. ("Reform Government Surveillance" 2015)

As a result of ongoing dialogues and legal battles, in 2014, the U.S. government reached an agreement with many of the coalition's members as to the specific granularity, timing, and scope of the statistics on national security requests for customer information that the companies may legally publish. A letter from the U.S. Deputy Attorney General James Cole (known as the DAG Letter) iterated a rather complicated, two-option scheme with different requirements based on the type of legal process issued by the U.S. government to access the data (James M. Cole, personal communication, January 27, 2014). Companies can choose one of the two options, the first of which lets them report the number of requests in bands of 999, while the second allows bands of 250.

Interestingly, the DAG Letter also put new products, services, and businesses at a transparency disadvantage: the agreement set forth a 2-year delay on reporting about the first order received for a platform, product, or service that had not previously received an order. Also prejudicing smaller firms, the lowest number of requests for customer data that the DAG Letter allows companies to disclose is 0-249, rather than, say, zero.

A creative way to circumvent this restriction has arisen in the form of a short statement, periodically issued, that the entity has *not received* certain types of requests. Dubbed a "warrant canary," this statement partially circumvents the "gag" or non-disclosure orders that accompany certain forms of legal process. The term refers to the proverbial "canary in a coal mine," a bird used to detect poisonous gases whose death would warn miners to evacuate; in this case, if the entity no longer makes the statement, astute readers can assume the firms have received a request. While warrant canaries are largely employed by small tech firms, larger platforms like Reddit and Apple noticeably stopped issuing warrant canaries, suggesting they had received secret national security or intelligence agency legal requests for the first time (Volz 2016; Roberts 2014).[13] As of mid-2016, the legality of warrant canaries has not been tested in U.S. courts. It should also be noted that Twitter did not agree to the DAG Letter, preferring to continue in court to press its case for the ability to reveal the exact number of requests it receives (Gullo 2015).

---

[13] A no-longer-maintained list of warrant canaries was compiled by the Canary Watch initiative at canarywatch.org

13.3 Furthering the Transparency Reporting Phenomenon: Current Outlook

Previously, we have defined the transparency report as an official, periodical statement issued by an ICT company (OSPs or telecom companies including Internet Service Providers (ISPs) or mobile network operators), which discloses data on government and other third party requests for user information, network disruptions, content restriction, and information on other policies and practices impacting privacy and freedom of expression online.

The format and depth of these reports vary widely, for several reasons. Without a single legal, compliance, or regulatory standard to adhere to, the content of the reports is left to companies to decide. In addition, borrowing from CSR literature, "organizational transparency consists of two underlying sub-dimensions – capability and intent – both of which must be present for highest levels of transparency regarding a particular information object," and can change from company to company (Watts 2015, p. 26). Responding to a variety of external and internal stakeholders, the firms weigh several factors that shape the nature and extent of their "intent," including what disclosures are germane to their audience of investors, regulators, users, competitors, the press, and others; which indicators reflect positively on their efforts, or best illuminate their challenges; and their values as a company. Their intent is counter-weighted by their "capability:" which statistics and data they are able aggregate internally, and what they are legally allowed to disclose. As nearly all OSPs as well as major telecom companies operate globally, they face opportunities as well as limitations that shape the content of their transparency reports (DeNardis and Hackl 2015). First, each company responds to different types of requests according to the jurisdictions they operate in. In addition, they face different limitations on the scope of data the company may publish legally in these jurisdictions. The legacy of the sector also plays a role: there are stark differences in the transparency reports published by technology companies and telecommunications operators, whose corporate cultures and regulatory regimes evolved separately and differ greatly, though many argue are converging (Bohlin et al. 2004; Huang et al. 2012).[14]

However, all of the reports include data relevant to users' privacy concerns, namely, the number of requests for user data coming from governments, police, or other law enforcement agencies. In some cases, the reported numbers would also include requests coming from intelligence agencies, whose reporting is often subject to limitations. For example in the United States, as demonstrated previously, distinction is made in transparency reports between criminal requests (including subpoenas, search warrants, emergency disclosure requests, pen register orders, wiretap orders) and requests made by intelligence agencies for national security investigations (including National Security Letters and Foreign Intelligence

[14] For example, telecommunications sector has been subject to standardization in CSR reporting through the Global Reporting Initiative and its sector-specific guidelines ("Global Reporting Initiative" 2016); Tschopp and Nastanski 2014, p. 154).

Surveillance Court (FISC) requests). Reporting on the latter category, by law, is subject to delays and limitations on granularity (such as the requirement to report in bands of 250, 500, or 1000 rather than exact numbers), although we observe reform in these non-disclosure requirements with the passage of the USA Freedom Act.[15] In addition, the complex nature of cross-border data requests, namely the Mutual Legal Assistance mechanism, leads to inaccuracies in reporting, as "many American companies regularly count foreign MLA requests as U.S. Government requests for data" (Woods 2015, p. 11).

For the case of telecommunications operators and ISPs, similar limitations exist and change drastically from country to country. First of all, ISPs are not transparent on "the extent to which they carry out filtering, their policies on filtering, or explanations about the legal requirement of filtering" (MacKinnon et al. 2015, p.88). Multinational telecom providers appear more sensitive to the legal requirements and limitations on transparency in different jurisdictions, a sensitivity they justify by citing the higher risks they face than internet platforms for non-compliance. Telecoms usually own physical infrastructure and place personnel on the ground in the countries where they operate, representing greater capital investments and costly assets subject to forfeiture. In addition, they operate under legal licensing agreements wherein they agree to comply with government requests under penalty of law. Often, internet platforms carry neither the asset nor legal exposure that telecom companies commonly do, a difference that telecoms tend to announce at every opportunity.[16] In fact, the first Vodafone transparency report in 2014 came with a Legal Annex, where the company provided country-by-country information on the laws regulating government access to user data as well as local laws restricting the transparency and accountability of surveillance. Thus, the reports can present telecoms with an elevated stage to air their grievances with governments and push back against laws they see as unfair and overly restrictive.

For OSPs and specifically for online social networks such as Twitter and Facebook whose business models are based on user-generated content, the reports also include the numbers of requests for removing user-generated content. When it comes to removing content on a case by case basis, as required by a government entity, the level of granularity reported varies greatly from company to company. As an example, while Facebook only provides a single category of content restrictions, Google distinguishes between different reasons behind content removal requests: Defamation, Privacy and Security, Drug Abuse and Other. In addition, Google reports on Right to be Forgotten Requests on a per-country basis, and includes compliance rates per URL ("European privacy requests" 2016).

---

[15] Microsoft states in the U.S. National Security Orders section of its latest transparency report that it is providing "recent data using the new, [narrower] range" ("U.S. National Security Orders Report" 2016). Yahoo went a step further by publicly disclosing the three National Security Letters it received from the FBI (Masden, 2016).

[16] Although this seems to change, especially in the EU context as more and more OSPs become subject to government and EU-level legal scrutiny in various areas including antitrust, tax or data protection.

Companies appear to relish the opportunity that transparency reporting provides them as a platform to establish and build brand identity and allegiance, and express corporate values. Kakao, a popular South Korean social media and messaging company, for example, publishes a "Personal Information Lifecyle" infographic under the proclamation that:

> Privacy protection' is at the top of the list in all processes that handle personal information. We at Kakao will do our best so that users can use Kakao's services without having to worry about the invasion of their privacy. ("Personal Information Lifecycle" n.d.)

Twitter includes the slogan "Communicate fearlessly to build trust" in its report ("Twitter Transparency Report" 2015). On a more lighthearted note, social humor platform Cheezburger brings a unique approach to transparency reporting by blogging about how reports can be used creatively in order to establish brand identity via interesting content ("Cheezburger Transparency Report" 2015). The company included karaoke songs performed by staff at its holiday party in its first-ever transparency report (Huh 2015; "Cheezburger Transparency Report" 2015).

13.4 Transparency as an Enabler for an Ethical Framework for OSPs

We have posited the transparency reporting phenomenon within a multidisciplinary understanding at the intersection of business and human rights and corporate responsibility. As argued by Wettstein (2012), the intersection of these two fields had not received significant academic attention until the Ruggie framework (Wettstein 2012, p. 4741). As recognized in the Ruggie "Protect, Respect, Remedy" Framework and elaborated in the UN Guiding Principles on Business and Human Rights, businesses impact human rights, and have responsibilities to respect human rights. These responsibilities can be reflected in national or international legal instruments, but also extend beyond legal or compliance requirements. In the debate and literature surrounding the business and human rights framework, one of the proposed ways through which non-state actors and specifically corporations could be held accountable for human rights commitments is to design a "'new international normative regime,' instead of reinterpreting existing domestic and international legislation" (Monshipouri et al. 2003, as cited in Wettstein 2012).

Similarly, we use the term "norms" to encapsulate this broader role and scope of the responsibility to respect human rights, while conceding that corporations currently operate without binding or legal commitments to protect human rights, and frequently violate the norms in practice. Our perspective in this sense is an advocate's view, but one that diverse stakeholder communities have embraced and seek to implement, whether through promotion of "voluntary" corporate social responsibility programs, litigation and arbitration, national regulation, or international treaties.[17]

However, "transparency is not an ethical principle in itself," but "a pro-ethical condition that becomes a valuable tool for uncovering the ethical principles that ideally inspire organizations' decisions and those principles that are factually endorsed in their everyday activities," showing "how the ethical principles, to which they are committed, are prioritised and translated into practise and governance" (Floridi and Turilli 2009, p. 110). We argue that for the case of OSPs and telecoms, these ethical principles should be the protection of users' right to privacy, security, and expression, and a workforce able, trained, and enabled to root out bias, and transparency reporting should be a method of their demonstration.

Therefore, we present three areas where transparency currently makes an impact on ethical standards for information and communication technologies (ICTs) companies including OSPs:

1. corporate reporting on interaction with governments on privacy and free expression issues;
2. disclosures of staff demographics in ICT companies; and
3. sharing of information on digital security topics, including encryption and breach.

Gaps and challenges remain, however. To continue improving, transparency reports should be expanded to provide a fuller picture of the discretionary, internal processes that ICT companies employ to restrict or promote content and control user access to their platforms, even in the absence of government demands. According to Ranking Digital Rights, no company in their 2015 Corporate Accountability Index disclosed "any information whatsoever about the volume and type of user content that is deleted or blocked when enforcing its own terms of service" ("Findings" 2015). Without much reporting by companies on their own rule enforcement, observers are left to compile anecdotal evidence, which shows inconsistency in corporate behaviour.[18] The opaque decision-making processes appear to differ at each company, based on their different Terms of Service, community guidelines, corporate structures, and other factors ("Opening the Black Box" 2016; Majoo 2016). In some cases, companies do appear to alter their terms of service at government request, or in response to extra-legal pressures that governments apply, most notably in the recent drive to push "terrorist content" or "violent extremism" off the major internet platforms (Drozdiak 2016). With regard to such fraught topics as these, transparency reporting on terms of service enforcement, under common definitions, criteria, and methodology, could enhance user awareness of the rules of the road online, and repair relations that are currently characterized by mistrust between the companies and various user constituencies.

---

[17] Examples include UN Guiding Principles on Business & Human Rights, the UN Global Compact Ten Principles, Global Network Initiative Principles, and the efforts of various shareholder advocacy organizations as well as socially responsible investing companies (Sjöström 2010).

[18] An example is the onlinecensorhip.org project, which "seeks to encourage social media companies to operate with greater transparency and accountability toward their users as they make decisions that regulate speech" ("onlinecensorship.org" 2016).

13.5 Conclusion: Transparency Reporting Now and in the Future

Though somewhat late to the party, many ICT companies have begun releasing statistics, narrative information, policies, and position papers on a host of non-financial topics. In addition, the companies began forming multi-stakeholder initiatives and coalitions such as Global Network Initiative or Reform Government Surveillance, through which they have institutionalized their human rights commitments including those involving transparency. In this chapter, we explored the various types of information being reported and the motivations – whether ethical, legal, reputational, financial, or otherwise – behind the disclosures. By their own words, companies use the reports to spark and grow the trust of their user bases, but increasingly we also see corporations expressing their values, establishing brand identities, defying government authorities, and mollifying stakeholder concerns around new risks and compliance issues that threaten the sustainability of their investments and growth.

Further, we speculated on the utility of the reports to various stakeholder groups, noted areas where gaps remain, and posit improvements to reporting that would benefit the companies and their stakeholders.

Issuing transparency reports does not guarantee that a company operates with respect for human rights online, nor does this mechanism alone present a full understanding of digital surveillance and censorship. However, together with a guide to the legal landscape and insights into a company's internal policies enforcing terms of service, community guidelines, and requests originating from different government agencies and private parties, these reports can provide a fact-based representation of how user rights and interests are mediated online. Corollary benefits, from insights into the current state of the cross-border data sharing, to expanded reporting on corporate personnel demographics, are likewise welcome and will benefit a broad range of stakeholders, while ensuring more ethical outcomes in corporate practice.

References

Access Now Transparency Reporting Index. (2016). Retrieved from https://www.accessnow.org/tri

Baddache, F. (2012). *The five W's of France's CSR reporting law.* Retrived from https://www.bsr.org/reports/The_5_Ws_of_Frances_CSR_Reporting_Law_FINAL.pdf

Bohlin, E., Levin, S., Sung, N., & Yoon, C. (2004). Convergence phenomenon and new service development in the telecommunications industry.

Bort, J. (2011). Cisco shareholders: Stop helping government crackdowns. Retrieved from http://www.businessinsider.com/cisco-says-no-to-human-rights-sustainability-shareholder-requests-2011-12?IR=T

boyd, D., & Crawford, K. (2012). Critical questions for big data: Provocations for a cultural, technological, and scholarly phenomenon. *Information, Communication & Society, 15*(5), 662–679.

Budish, R., Woolery, L., & Bankston, K. (2016). The transparency reporting tool kit. New America Foundation. Retrieved from https://www.newamerica.org/oti/policy-papers/the-transparency-reporting-toolkit/

Carroll, A. B. (2004). Managing ethically with global stakeholders: A present and future challenge. *The Academy of Management Executive, 18*(2), 114–120.

Castro, D. (2013). How much will PRISM cost the U.S. cloud computing industry? Information Technology & Innovation Foundation. Retrieved from http://www2.itif.org/2013-cloud-computing-costs.pdf

Castro, D., & McQuinn, A. (2015). Beyond the USA freedom act: How U.S. surveillance still subverts U.S. competitiveness. Information Technology & Innovation Foundation. Retrieved from https://itif.org/publications/2015/06/09/beyond-usa-freedom-act-how-us-surveillance-still-subverts-us-competitiveness

Cheezburger Transparency Report. (2015). Retrieved from http://blog.cheezburger.com/community/cheezburger-inc-2014-transparency-report/

Cole, J. M. (2014). In Stretch C., Walker K., Rottenberg E., Smith B. & Bell R. (Eds.). Washington, DC: Office of the Deputy Attorney General. Retrieved from https://www.justice.gov/iso/opa/resources/366201412716018407143.pdf

Currents of change: The KPMG Survey of Corporate Responsibility Reporting. (2015). KPMG.

Deibert, R., Palfrey, J., Rohozinski, R., Zittrain, J., & Stein, J. G. (2008). *Access Denied: The practice and policy of global internet filtering*. Cambridge, Mass: MIT Press.

DeNardis, L., & Hackl, A. M. (2015). Internet governance by social media platforms. *Telecommunications Policy, 39*(9), 761–770.

Drozdiak, N. (2016, May 31). U.S. tech firms agree to EU code of conduct on terror and hate content. *The Wall Street Journal.* Retrieved from http://www.wsj.com/articles/u-s-tech-companies-sign-up-to-eu-code-of-conduct-on-terror-1464689959

EU directive on disclosure of non-financial and diversity information published. (2014, November 17). Retrieved from http://www.iasplus.com/en/news/2014/11/eu-esg

European privacy requests for search removals. (2016). Retrieved from https://www.google.com/transparencyreport/removals/europeprivacy/?hl=en

Extractive Industries Transparency Initiative. (2003). Paper presented at the *Statement of Principles and Agreed Actions*, London. Retrieved from http://webarchive.nationalarchives.gov.uk/+/http:/www.dfid.gov.uk/pubs/files/eitidraftreportstatement.pdf

Fact sheet: Disclosing the use of conflict minerals. (2014). Retrieved from http://www.sec.gov/News/Article/Detail/Article/1365171562058

Findings. (2015). Retrieved from https://rankingdigitalrights.org/index2015/findings/

Floridi, L., & Turilli, M. (2009). The ethics of information transparency. Ethics and Information Technology, 11(2), 105–112.

Ghonim, W. (2012). *Revolution 2.0: The power of the people is greater than the people in power: A memoir*. Boston: Houghton Mifflin Harcourt.

Global Network Initiative Principles. (2008). Retrieved from https://globalnetworkinitiative.org/principles/index.php

Global Reporting Initiative. (2016). Retrieved from https://www.globalreporting.org/standards/Pages/default.aspx

Goldberg, R. (2016, May 13). Lack of trust in internet privacy and security may deter economic and other online activities. Retrieved from https://www.ntia.doc.gov/blog/2016/lack-trust-internet-privacy-and-security-may-deter-economic-and-other-online-activities

Green, D., & Miller-Dawkins, M. (2014, March 10). How can you tell whether a multi-stakeholder initiative is a total waste of time? Retrieved from https://oxfamblogs.org/fp2p/how-can-you-tell-whether-a-multi-stakeholder-initiative-is-a-total-waste-of-time

Greene, R., Kehl, D., Morgus, R., & Bankston, K. (2014). *Surveillance costs: The NSA's impact on the economy, internet freedom & cybersecurity.* New America Foundation. Retrieved from https://www.newamerica.org/oti/policy-papers/surveillance-costs-the-nsas-impact-on-the-economy-internet-freedom-cybersecurity/

Gryta, T., Schechner, S., & Troianovski, A. (2013, October 13). NSA fallout thwarts AT&T. *Wall Street Journal.* Retrieved fromhttp://www.wsj.com/news/articles/SB10001424052702304073204579167873091999730

Guiding principles on business and human rights. (2011). United Nations. Retrieved from http://www.ohchr.org/Documents/Publications/GuidingPrinciplesBusinessHR_EN.pdf

Gullo, K. (2015, May 8). When is a justice department rule not a rule? Report from Twitter's transparency fight. Retrieved from https://www.eff.org/deeplinks/2015/05/when-justice-department-rule-not-rule-report-recent-hearing-twitters-lawsuit

Howard, P. N., & Hussain, M. M. (2013). *Democracy's Fourth Wave?: Digital Media and the Arab Spring*. Oxford: Oxford University Press on Demand.

Huang, I., Guo, R., Xie, H., & Wu, Z. (2012). The convergence of information and communication technologies gains momentum. *The Global Information Technology Report 2012* (pp. 35–45). New York: World Economic Forum. Retrieved from https://www.weforum.org/reports/global-information-technology-report-2012.

Huh, B. (2015, February 5). The transparency report is the new 404 hotness. Retrieved from https://medium.com/inside-cheezburger/the-transparency-report-is-the-new-404-hotness-c7edf40f44d5#.ggxio3u0t

Indicators - Ranking Digital Rights. (2015). Retrieved from https://rankingdigitalrights.org/project-documents/2015-indicators/

MacKinnon, R. (2006, January 14). Before congressional hearings, Yahoo! issues vague ethical "guidelines". Retrieved from http://rconversation.blogs.com/rconversation/2006/02/before_congress.html

MacKinnon, R. (2010). Networked authoritarianism in China and beyond: Implications for global internet freedom. *Liberation Technology in Authoritarian Regimes, Stanford University.*

MacKinnon, R., Hickok, E., Bar, A., & Lim, H. (2015). *Fostering freedom online: The role of internet intermediaries*. Paris: UNESCO Publishing.

Malby, S., Mace, R., Holterhof, A., Brown, C., Kascherus, S., & Ignatuschtschenko, E. (2013). *Comprehensive study on cybercrime.* Vienna: United Nations Office on Drugs and Crime. Retrieved from https://www.unodc.org/documents/organized-crime/UNODC_CCPCJ_EG.4_2013/CYBERCRIME_STUDY_210213.pdf

Manjoo, F. (2016, May 11). Facebook's bias is built-in, and bears watching. *New York Times.* Retrieved from http://www.nytimes.com/2016/05/12/technology/facebooks-bias-is-built-in--and-bears-watching.html?_r=0

Masden, C. (2016, June 1). Yahoo announces public disclosure of national security letters. Retrieved from https://yahoopolicy.tumblr.com/post/145258843473/yahoo-announces-public-disclosure-of-national

Measuring the Information Society Report. (2015). Geneva: International Telecommunication Union. Retrieved from www.itu.int/en/ITU-D/Statistics/Documents/.../MISR2015-w5.pdf

Mendel, T., Puddephatt, A., Wagner, B., Hawtin, D., & Torres, N. (2012). *Global survey on internet privacy and freedom of expression.* Paris: UNESCO. Retrieved from unesdoc.unesco.org/images/0021/002182/218273e.pdf

Miller, C. C. (2010, September 21). Google reports on government requests and censorship. *New York Times.* Retrieved from http://bits.blogs.nytimes.com/2010/09/21/google-reports-on-government-requests-and-censorship/?_r=0

Monshipouri, M., Welch Jr., C. E., & Kennedy, E. T. (2003). Multinational corporations and the ethics of global responsibility: Problems and possibilities. *Human Rights Quarterly, 25*, 965.

Net firms criticised over China. (2006). Retrieved from http://news.bbc.co.uk/1/hi/technology/4699242.stm

Oates, J. (2011). Vodafone says sorry for pro-mubarak messages. Retrieved from http://www.theregister.co.uk/2011/02/03/egypt_vodafone_propaganda/

Onlinecensorship.org. (2016). Retrieved from https://onlinecensorship.org/

Opening the black box: Understanding how companies enforce their rules. (2016). Retrieved from https://rightscon.sched.org/event/6Ih6/opening-the-black-box-understanding-how-companies-enforce-their-rules
Parsons, C. (2015, July 8). Industry Canada transparency report guidelines intensely problematic. Retrievedfromhttps://www.christopher-parsons.com/industry-canada-transparency-report-guidelines-intensely-problematic/
Personal information lifecycle. (n.d..) Retrieved from http://privacy.kakaocorp.com/en/transparence/lifeCycle
Rampell, C. (2007). Yahoo settles with Chinese families. Retrieved from http://www.washingtonpost.com/wp-dyn/content/article/2007/11/13/AR2007111300885.html
Ratute, A. (2010). *Expanding social justice knowledge with sweatshop history*. Unpublished Master's Thesis. Iowa State University, Ames. Retrieved from http://lib.dr.iastate.edu/cgi/viewcontent.cgi?article=2662&context=etd
Reform Government Surveillance. (2015). Retrieved from https://www.reformgovernmentsurveillance.com/
Roberts, J. (2014). Apple's "warrant canary" disappears, suggesting new patriot act demands. Retrievedfromhttps://gigaom.com/2014/09/18/apples-warrant-canary-disappears-suggesting-new-patriot-act-demands/
Ruggie, J. G. (2007). Business and human rights: The evolving international agenda. *The American Journal of International Law, 101*(4), 819–840.
Runhaar, H., & Lafferty, H. (2009). Governing corporate social responsibility: An assessment of the contribution of the UN Global Compact to CSR strategies in the telecommunications industry. *Journal of Business Ethics, 84*(4), 479–495.
Salgado, R., & Chavez, P. (2013, September 9). *A petition for greater transparency*. Retrieved from https://googleblog.blogspot.ie/2013/09/petitionfor-greater-transparency.html
Schonfeld, E. (2006). Analysis: Yahoo's China problem. Retrieved from http://money.cnn.com/2006/02/08/technology/yahoo_china_b20/
SEC proposes rules for resource extraction issuers under dodd-frank act. (2015). Retrieved from http://www.sec.gov/news/pressrelease/2015-277.html
Severin, T., Lopes, M. & Hudson, A. (2014). German government cancels Verizon contract in wake of U.S. spying row. Retrieved from http://www.reuters.com/article/us-germany-security-verizon-idUSKBN0F11WJ20140626
Sjöström, E. (2010). Shareholders as norm entrepreneurs for corporate social responsibility. *Journal of Business Ethics, 94*(2), 177–191.
Stempel, J. (2013). *Lawsuit accuses IBM of hiding China risks amid NSA spy scandal*. Retrieved from http://www.reuters.com/article/us-ibm-lawsuit-idUSBRE9BB1BP20131212
Taddeo, M., & Floridi, L. (2015). The debate on the moral responsibilities of online service providers. *Science and Engineering Ethics*, 1–29.
The Telecom Transparency Project. (2015). Retrieved from https://www.telecomtransparency.org/about/
Transparency reporting by private sector companies: Comparative analysis. (2015). Retrieved from https://www.priv.gc.ca/information/research-recherche/2015/transp_201506_e.asp
Transparency reporting guidelines: Innovation, science and economic development Canada. (2015). Retrieved from http://www.ic.gc.ca/eic/site/smt-gst.nsf/eng/sf11057.html
Tschopp, D., & Nastanski, M. (2014). The harmonization and convergence of corporate social responsibility reporting standards. *Journal of Business Ethics, 125*(1), 147–162.
Twitter Transparency Report. (2015). Retrieved from https://transparency.twitter.com/
U.S. National Security Orders Report. (2016). Retrieved from https://www.microsoft.com/about/csr/transparencyhub/fisa
Volz, D. (2016). Reddit deletes surveillance 'warrant canary' in transparency report. Retrieved from http://www.reuters.com/article/us-usa-cyber-reddit-idUSKCN0WX2YF

Wettstein, F. (2012). CSR and the debate on business and human rights: Bridging the great divide. *Business Ethics Quarterly, 22*(04), 739–770.
Watts, S. (2015). Corporate social responsibility reporting platforms: enabling transparency for accountability. *Information Technology and Management, 16*(1), 19–35.
Who has your back? (2015). Retrieved from https://www.eff.org/who-has-your-back-government-data-requests-2015
Williams, C. A. (1999). The securities and exchange commission and corporate social transparency. *Harvard Law Review*, 1197–1311.
Woods, A. K. (2015). Data beyond borders: Mutual legal assistance in the internet era.

Chapter 14
Should We Treat Big Data as a Public Good?

Katarzyna Śledziewska and Renata Włoch

Abstract By clicks, likes, tweets, photos, blog posts, online transactions, we (people and businesses) allow for creating Big Data. Our digital traces tell the story about our professions, preferences, choices, transactions, traveling, habits etc. They permit to get information about who we are and what we do as individuals and organisations. This process is associated with a threat but also an opportunity for the social and economic development. The key question is how to ensure that the data we create continue to improve our lives, innovation, and sustainable development. A majority of Big Data is no longer housed in statistical agencies, but housed by businesses including Online Services Providers (OSPs). We state that regulators should consider this data as a public good and build the platforms for public and private partnerships to ensure advantage of digital technology, improved connectivity, and data revolution.

14.1 Introduction: A New Bright World of Big Data

Big Data is a buzzword. And like most of buzzwords – globalization, synergy, and innovation – it remains a loosely defined notion. One of the most commonly recognized definition was offered by Gartner (2016): Big Data is "high-volume, high-velocity and/or high-variety information assets that demand cost-effective, innovative forms of information processing that enable enhanced insight, decision making, and process automation". Big Data may differ as to their structure (structured, unstructured, semi structured) or their source (internal to the organization or external). It may be collected by public institutions, but the bulk of it consists of the enterprise Big Data, collected and processed by private Online Service Providers (OSP).

The Internet offers multiple platforms for people to consume, disseminate, and produce their own content – or leave their digital traces. This results in exponential rise of data and information. The often-cited IBM source claims that 90 % of all

K. Śledziewska (✉) • R. Włoch
DELab, University of Warsaw, Warsaw, Poland
e-mail: k.sledziewska@uw.edu.pl

M. Taddeo, L. Floridi (eds.), *The Responsibilities of Online Service Providers*,
Law, Governance and Technology Series 31, DOI 10.1007/978-3-319-47852-4_14

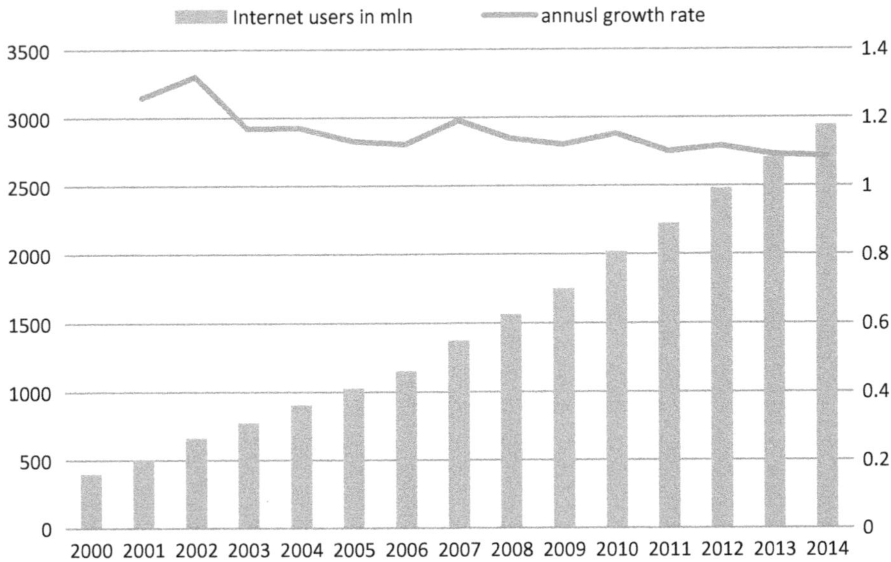

Fig. 14.1 Internet users and annual growth rate (Source: ITU 2015)

existing data has been generated in just last 2 years, daily there appears 2.5 quintillion of bytes, and up to 2020 there will be 43 trillion of gigabytes (IBM 2015). There are many different types of Big Data sources collected by OSP: social media data, personal data from tracking devices, sensor data, transactional. The total sum of data store by biggest OSPs (Google, Amazon, Microsoft and Facebook) is estimated to amount to least 1200 petabytes. Google receives over 4 million search queries per minute. Facebook users shared 2,460,000 pieces of content. This increase of Big Data is mainly related to the growth of Internet users (James 2014). In one decade, the global Internet population grew from 0.5 to 3 billion people . It means that now 3 billion users produce data as part of their everyday participation in the digital world. (Fig. 14.1).

In this chapter, we bandwagon in the trend of analysing the Big Data revolution. We try to defend our right to do so by introducing the elements of economic and sociological reflection concerning the private and public goods in order to analyse the specificity of the enterprise Big Data. We start by reviewing the basic assumptions of economics of information, and then we continue to discuss their social and economic consequences, in the light of the emergence of Big Data. We underline that there is an unlimited potential in Big Data to be realized but first there are some challenges that have to be resolved. Not only issues related to individual privacy, data protection, and data integrity should be discussed and regulated. The Big Data should be considered as a public good. On this paradigm, the cooperation between businesses, academic institutions, and statistical organizations should be built. In this process, those who generate the biggest amount of data – OSPs- play a crucial role.

14.2 Information as a Public Good

The point of departure, then, is that recently we witness the data revolution: never before humanity had so much data on their hands (Cukier and Mayer-Schoenberger 2013). But, to state the obvious, having data is not coterminous with having information, which, in turn, does not automatically lead to possessing more knowledge. Information (if it is to be reliable, needs to be extracted from data with use of rigorous, scientifically grounded methods) is always contextual and involves the process of ordering, structuring and presenting of data, while knowledge is generated in the process of culturally defined understanding of information (Elsner et al. 2014). As many social scientists proposed – e.g. American economist and Nobel Prize winner George Stigler (1961) and French thinker Michel Foucault (1980) – knowledge is power. And to take this argument further, power is equivalent with the potential to change the behaviour of other people, the functioning of the institutions, and consequently the shape of the society.

Owing to contributions from the aforementioned Stigler (1961), George Akerlof (1970) and Joseph Stiglitz (1977, 1983, 1999), the theory of economy gained valuable insights into the economics of information. The later assumes that information is a kind of a product, although by no means a typical one. It is relatively easy to generate and disseminate, and when accumulated it has more value than the sum of the individual values. What is more important, there are immense economies of scale. Information is not consumed by being used or transmitted to others: many persons may possess and use the same information, even at the same time, without diminishing its value to others.

It was Stiglitz who noticed that the economic properties of information have those characteristic of public goods. Although main epistemological approaches distinguish between knowledge and information (Floridi 2013), Stiglitz argued that for all practical reasons, information can be equated with knowledge (Stiglitz 2008, 1983). He pointed out two main critical properties that allow classification of information/knowledge as public good, precisely: nonrivarlous nature and nonexcludability. The former implies that there is zero marginal cost for an additional consumer to enjoy the benefits of the knowledge and the latter implies that the knowledge and all its benefits cannot be provided privately. While these two characteristics may seem very easy and clear to be identified, the precise understanding of these properties poses the basis for the strong economic implications.

Namely, Stiglitz (1999) emphasised that when we consider nonrivarlrous nature of the knowledge we should notice that zero marginal costs imply that the knowledge with the zero price is the knowledge that can be produced at the zero costs. And besides, we should not forget about the costs stemmed from transmission of the knowledge: private providers of knowledge can charge the transmission of information to reflect the marginal cost of transmission though, without alerting the nonrivalrous nature of the good itself. In other words, while there is a cost related to transmission, the good itself remains free of charge. Therefore, the existence of private providers of the information does not take the information out of the public good classification.

When it comes to the nonexcludability (nonexclusion), this implies that no one can be excluded from the consumption of knowledge. However, Stiglitz discusses that there are many cases when some forms of knowledge (for instance secrets of the firms to produce particular products, etc.) are excludable. In this direction, we may see patents as the tools for information owners to disclose information. More precisely, by giving rights to inventors to enjoy the fruits of their innovative activity over a limited period of time, we get the information about the details of their innovations. This process generates positive externalities since availability of new knowledge is in fact the source of further innovations. So patents basically can determine the pace of innovations. However, since nonexcludability is not fully satisfied by the phenomena of the creation and transmission of the knowledge, the latter is rather considered as an impure public good.

The proposition of Stiglitz (1999) counters the classical economic approach by Paul Samuelson that emphasizes that public goods are, by definition, generated by public sector. As Holcombe puts it, "an economist argues that a good has the characteristics of either jointness in consumption or nonexcludability, and then, because that makes the good a public good, implies that the good should be produced in the public sector" (Holcombe 1997: 1). Stiglitz's argument goes hand in hand with the approach of the researchers working in the area of development and underdevelopment who emphasize that "public goods are socially defined and constructed: the outcome of complex political processes which evolve around the definition of public need" (Wuyts 1992, p. 13–38). Stiglitz (1999) furthermore argues that besides being the public good, information could be moreover classified as a global public good. The idea is that the access to information and knowledge have to be guaranteed and safeguarded so that the asymmetry of information does not diminish the equal playing ground for individuals and firms, especially when that knowledge is crucial to internal and international development. In other words, a public good is not defined as a good that is characterised by nonexcludability and nonrivarlous nature, but as such that *should* be nonexcludable and nonrivarlous, and this includes the access to information and knowledge.

Let us assume, then, that information is a global public good. Does this classification unequivocally relate to data? Does the data, particularly Big Data, represent a public good?

Whose Big Data? The digital traces left by digital subjects in virtual activity are collected by OSPs. The Atlantic journalist Alexis Madrigal found out with the help of Mozilla's software Collusion, that in 36 hours of "standard web surfing" his data was tracked and collected by 105 companies, mostly by "smaller data and advertising businesses that form a shadow web of companies that want to help show you advertising that you're more likely to click on and products that you're more likely to purchase" (Madrigal 2012). They play the role of intermediaries for consumers and enterprises by providing them online access to the digital goods and services. At the same time, they use search engines to generate and maintain extensive databases of Internet addresses and content (web pages, images or other types of digital files) in an easily searchable format. Unsurprisingly, they use the collected data for their own benefit, e.g. in order to improve their services and monetize them by targeted advertising. The revenue from advertising makes it possible to provide valuable services that are offered to users at low (subsidized) prices or for "free". Yet there are no free lunches. The cost that the customer has to pay for those services often takes the shape of the acceptance of the fact that his or her data will be gathered and used for (mainly) marketing purposes. This way the Internet intermediaries are located at the crucial nodes of power: they have an access to an unimaginably large pool of data from which they can extract information, building themselves an informed and intimate knowledge about the preferences of their customers, trends in social, political and economic areas of life. And, only naturally, they use their privileged epistemological position to strengthen their hold on the market.

These new possibilities of data-nourished market adjustment were readily taken up generally by private entrepreneurs. The collection and use of customer data is practiced not only by OSPs, but also by traditional offline businesses. More and more firms have been taking their day-to-day operations first to computers, and then online. They collect large-scale and even real-time data on prices, employment, spending, costs, sales contracts, physical shipment and transport. They also keep digital records of collaborative work efforts, personnel evaluations, and productivity measure. Most of these data sets are easily searchable and easy to merge. What is even more important, nowadays private entrepreneurs virtually effortlessly and without burdensome costs are able to harvest data lined to advertising, social media activity, offline spending or credit history. As a result, sellers are able to tailor their offer to the individual customer needs by using systems that aggregate user feedbacks and by offering advertisements and product recommendation based on past browsing behaviour, past purchases or analysis of data on individual user preferences as evoked via social media (Hennig-Thurau et al. 2010). Generally, circa two thirds of the Big Data inflow comes from the internal sources of the companies, the rest is bought from other OSP, especially those who own social media (TCS 2013, p. 20).

Getting back to the point: ownership issues related to Big Data used by enterprises are complicated. Basically, enterprise Big Data is mostly private data produced by the individuals acting – for most of the time – privately, i.e. not in their working time and for private, often intimate, purposes. The individuals have got no effective control over their "online production". In 2012 (effectively in 2013) the European Union (EU) introduced the obligation to inform the users about online tracking. But still the only viable, though increasingly impractical, option to limit the scale of online tracking is to opt out from using the technology. OSPs are the effective possessors of these data as they collect them. We could possibly extend this argumentation to show that this relationship can be treated as a kind of Marxian clinch between the digital analogue of the exploited and the exploiters who capture the added value for their own benefit (Mosco 2015). The digital "proletariat" toils more than willingly, but nevertheless they have no tools to evaluate whether the perks they are granted, such as the use of free software, is worth it. One of the sides of the bargain acts in the context of maximum information asymmetry. OSPs control the data that make it possible to build information and knowledge about society: they know about us, and can look at us, analyse us, sell the data and information about us. This view seems to be supported by Stiglitz, who emphasised in his speech in 1999 that nowadays the ability to produce knowledge is in fact the digital analogue of means of production:

> Knowledge and information is being produced today like cars and steel were produced a hundred years ago. Those, like Bill Gates, who know how to produce knowledge and information better than others reap the rewards, just as those who knew how to produce cars and steel a hundred years ago became the magnates of that era. (Seemann 2002, p. 1.)

Effectively, then, Big Data, even though in main part produced by plethora of individuals, are treated as private good by the actors who are located in privileged positions to collect it.

14.3 Why Should We Treat Big Data as Public Good?

In the past, data was gathered, structured, interpreted and – often grudgingly or never – presented by specialized institutions, usually at the payroll of nation-state (such as National Statistics Offices) or of the market actors (e.g. R&B departments of corporations or market research agencies). The process of data collection was almost invariably tedious, time-consuming, and costly. This changed with the triumphant ascent of digital technologies that culminated in the so-called Big Data Revolution. Data available from traditional sources such as censuses and surveys are not entirely sufficient for modern research.

The consequences of Big Data Revolution were most convincingly discussed by Schoenberger and Cukier in their seminal book (2013). They emphasize that the emergence of Big Datasets opens new methodological opportunities for social sciences. The basic assumption is that we witness the increasing "datafication", that is the growing possibility to capture "many aspects of the world that have never been quantified before". This in turn leads to three immensely important changes in the attitude to data. First of all, in many areas of social life sampling, the pillar of modern statistics, is no longer that crucial. In the past social scientists had no means of analysing the bulk of the society. They had to rely on the assumption that a random sample of the society will tell them something about the whole. In other words, their probing techniques were similar to those applied by geologists who insert drills into the ground, extract core samples and then use it to describe a certain geological area. But the geological areas tend to be more structurally coherent than human societies, so the social scientists, particularly sociologists, had always taken into account the possibility of error.

Sociologists often complain that their object of research is immensely complicated, fragmented and changeable. The effects of their analyses most often described the society that was no longer there. For example, it took the National Statistics Office more than 3 years to publish the results of the Polish census of 2011. The widely approved Gartner definition of Big Data emphasizes velocity as its main feature – the data are available in almost real-time. Sociologists do not have to paint a perfected sociological portrait of the society, which takes (too) much time, they can snapchat its changing faces. In other words, they can gain new epistemological vistas of the ontologically incoherent subject of analysis, if only they will be able to make peace with two other aspects of Big Data revolution Schoenberger and Cukier emphasize. First of all, social scientists have to accept the fact that Big Data are messy: "a bit inaccuracy can be tolerated, because the benefits of using vastly more data of variable quality outweigh the costs of using smaller amounts of very exact data". Secondly, and even more importantly, they have to renounce causality and accept correlations (and with it the risk that lot of it will prove to be spurious). But this, taking their argument further, may become the realization of the big dream of social sciences from their onset: the power of predicting the future. This revolution in social science is already taking place in several departments in the world, e.g. in Oxford Internet Institute or Cambridge Big Data Strategic Research Initiative.

Big data is poised to reshape the way we live, work, and think. A worldview built on the importance of causation is being challenged by a preponderance of correlations. The possession of knowledge, which once meant an understanding of the past, is coming to mean an ability to predict the future. The challenges posed by Big Data will not be easy to resolve. Big Data can change economic and social research relied on methods such as surveys, experiments, and ethnographies. For example, real time data can be used to estimate inflation. In the MIT Billion Prices Projects called PriceStats prices are collected daily from hundreds of online retailers. Thereby changes in inflation trends can be observed sooner in PriceStats than in the monthly Consumer Price Index (CPI). Social media messages can be the source of measuring the job losses or households' sentiments on their financial situation and on the economic climate (Daas and Puts 2014). The Big Data can be the important source for research development especially in countries where accessibility to the Open Data is limited to the researchers.

The ancestors of sociology, notably August Comte, wanted social sciences to "know to predict, to predict to control". Access to Big Data may make this dream come true: it enables building multifactor knowledge about society and its problems, facilitates extrapolating trends, and makes it possible to react more precisely and quickly to those identified problems, bringing entirely new possibilities to public policies. Just like the firms can tailor their market offer to the needs of customers extrapolated from their virtual tracks, the government could tailor its undertakings to the needs of specific social groups, and react in time to prevent crises. Chris Yiu, an expert of Policy Exchange, claims (2012) that using Big Data in administration facilitates sharing (e.g. streamline transactions between different agencies and departments); learning (e.g. generating agile insights about the internal functioning of the organization in question); service personalisation (enabled by "granularity" in big data); solving (by way of examination of previously hidden patterns and correlation and enhanced prediction); and innovating for growth (by strengthening efficiency and cost-effectiveness). The traditional policy cycle of "planning – implementation – evaluation" may be, in result, replaced by "agile public policy" (Parcell and Holden 2013).

Big Data have been greeted as the "new oil" of the twenty-first century (Greenwood et al. 2014), the necessary component of future socioeconomic development. In 2011, McKinsey Global Institute experts referred to Big Data as to "The next frontier for innovation, competition and productivity" and outlined five ways in which Big Data may create additional value. First, by creating transparency: "simply making big data more easily accessible to relevant stakeholders in a timely manner"; secondly, by "enabling experimentation to discover needs, expose variability, and improve performance" of organizations; thirdly, by segmenting populations to customize actions; fourthly, by replacing or supporting human decision making with automated algorithms; fifthly, by innovating new business models, products, and services. Interestingly, the report focused not only on the productivity benefits for enterprises, but also for public sector. The research proved that "Big Data levers, such as increasing transparency and applying advanced analytics, offer the public sector a powerful arsenal of strategies and techniques for boosting productivity and achieving higher levels of efficiency and effectiveness" (McKinsey 2011, p. 54).

Detailed data on human beings can be used by policy – makers to reduce crime, improve health delivery and manage cities (Keller et al. 2012). Services can be effectively targeted to the citizens, health interventions can be better designed, and taxpayers may pay less for government services (Lohr 2012). Real time data can be used to monitor traffic or to identify infrastructural problems what permits to react efficiently and quickly. In 2011, McKinsey estimated that use of Big Data may potentially contribute 250 billion € annually to Europe's public sector administration.

14.4 Towards New Deal Between Private and Public Interest

So far we have tried to highlight the importance of accessing the Big Data. But here the main difficulty comes down to convincing the OSP that introduction of some routine mechanisms to share their commercially valuable asset should be their (corporate) social responsibility.

This new kind of negotiation to make the consensus around the private and public interests in Big Data was recently highlighted by Big Data UN Global Working Group. During conference on "Big Data for Official Statistics", in October 2015 in Abu Dhabi, the Group presented the "Draft Principles for Access to Big Data Sources" (UNSTAT 2015). The proposal tries to reconcile "the high and urgent need for access to Big Data sources for the production of official statistics" and "legitimate interest of businesses" who may become financially and organizationally burdened by requesting access to Big Data, by way of introducing eight principles:

1. **Social responsibility:** The national and international systems of official statistics need data in order to „provide relevant, high-quality information to society in an impartial way". To this end, they need free of charge inputs from individuals, including businesses which should treat this obligation as a part of their social responsibility.

2. **Fundamental principles of official statistics:** the statistical agencies will obviously guarantee the confidential use of the data.
3. **Level playing field**: No particular entrepreneurship should be particularly burdened by the obligation to provide data: "the fairness of the distribution of the burden across businesses has to be considered, in order to guarantee a level playing field".
4. **Acknowledgment of business interest**: Data have market value, so there "must be a fair balance between public and business interest when data is requested".
5. **No more than needed**: The data must be adequate and relevant in relation to the purposes for which it is requested from the business.
6. **Benefit proportionality**: The cost and effort of providing data, including possible pre-processing, must be reasonable compared to the expected public benefit of the official statistics envisaged.
7. **Non-discrimination**: Businesses should treat requests from national statistical systems in a non-discriminatory way.
8. **Transparency**: both businesses and statistical authorities should ensure full transparency about the provision of data.

To sum it up, for the practical reasons (operation costs, transaction costs) the access to Big Data cannot be unrestrictedly open. The two principal intermediaries between entrepreneurship Big Data and the public should be public institutions and, particularly, universities. The academic society should see they role in processing and explaining the data and its potential uses to the decision makers and society in general. Particularly social scientists should act as translators of abstract digits to so-called "common people". In 2010 Eric Schmidt and Jared Cohen suggested that one of the major sources of the present social, economic and political change is the diminishing importance of intermediaries in the circulation of information. They argue that we are witnessing the emergence of new types of non-hierarchical relations of power in the framework of "virtual communities" of people. Yet, the revolutionary potential that Schmidt and Jarvis saw in the sheer existence of the data and information will not be fulfilled without input of intermediaries, which will transform the data into the easily understandable format. To reap the fruits of the recent worldwide digitization, there is the need of more systematic analysis of ever-changing and tantalizingly growing bytes. "The numbers have no way of speaking for themselves. We speak for them. We imbue them with meaning. Like Ceasar, we may construe them in self-serving ways that are detached from their objective reality" (Silver 2012). And here, we consider particularly universities, above all other actors in public space to play the crucial role. Their traditional role of safeguarding the right to knowledge and protecting the autonomy of intellectual inquiry may be imbued with a new meaning. As never before, there is a room for the academia in preserving the right to knowledge and information by providing the tools and means for analysis of both public and enterprise Big Data. David Weinberger (2009) argues that Internet substitutes objectivity with transparency up to the point when "transparency becomes new objectivity". Methodological rigor and verification mechanisms upheld at and by universities legitimizes their claim to scientific objectivity and makes it possible to step up from opinion to knowledge (Lynch 2016).

Additionally, the universities have another traditional role to play. As we have already stated, the governments need more precise knowledge of their citizens. They gather it with their own hands and by looking for ways to access the enterprise Big Data. But, as the ACTA protesters emphasised, the too well informed Leviathan may easily transform into virtually omnipotent – because omniscient – Big Brother. Tailored solving of social problems may grow into social engineering that would not necessarily be mild. The academia should guard and guarantee the free flow and use of knowledge produced with the use of Big Data.

References

Akerlof, G. (1970). The market for "Lemons": Quality uncertainty and the market mechanism. *The Quarterly Journal of Economics, 84*(3), 488–500.

Cukier, K., & Mayer-Schonberger, V. (2013). *Big data: A revolution that will transform how we live, work, and think.* Boston: Houghton Mifflin Harcourt.

Daas, P., & Puts, M. (2014). Social media sentiment and consumer confidence. In *European Central Bank Workshop on Using Big Data for Forecasting and Statistics, 5, 1–26.*

Elsner, W., Heinrich, T., & Schwardt, H. (2014). *The microeconomics of complex economies: Evolutionary, institutional, neoclassical, and complexity perspectives*. Amsterdam: Academic.

Floridi, L. (2013). *Philosophy of Information*. Oxford: Oxford University Press.

Foucault, M. (1980). In C. Gordon (Ed.), *Power/knowledge: Selected interviews and other writings 1972–1977*. London: Harvester.

Gartner. (2016). *IT glossary*. Resource document. Available: http://www.gartner.com/it-glossary/big-data/. Accessed 17 May 2016.

Greenwood, D., Stopczynski, A., Sweatt, B., Hardjono, T., & Pentland, A. (2014). The new deal on data: A framework for institutional controls. In J. Lane, V. Stodden, S. Bender, & H. Nissenbaum (Eds.), *Privacy, Big Data, and the public good: Frameworks for engagement* (pp. 192–210). New York: Cambridge University Press.

Hennig-Thurau, T., Malthouse, E. C., Friege, C., Gensler, S., Lobschat, L., Rangaswamy, A., & Skiera, B. (2010). The impact of new media on customer relationships. *Journal of Service Research, 13*(3), 311–330. doi:10.1177/1094670510375460.

Holcombe, R. G. (1997). A theory of the theory of public goods. *Review of Austrian Economics, 10*(1), 1–22.

IBM. (2015). https://www.ibm.com/annualreport/2015/assets/img/2016/02/IBM-Annual-Report-2015.pdf

ITU. (2015). http://www.itu.int/en/ITU-D/Statistics/Pages/stat/default.aspx

James, J. (2014). *Data never sleeps 2.0.* Domo. Document Resource. https://www.domo.com/blog/2014/04/data-never-sleeps-2-0/ Accessed 17 May 2016.

Keller, S. A., Koonin, S. E., & Shipp, S. (2012). Big Data and city living – What can it do for us? *Significance, 9*(4), 4–7.

Lohr, S. (2012, February 11). *The age of Big Data. New York Times*.

Lynch, M. P. (2016). *The internet of us. Knowing more and understanding less in the age of Big Data*. New York: Liveright.

Madrigal, A. C. (2012). *I'm being followed: How Google—And 104 other companies—Are tracking me on the web.* Resource document. The Atlantic. http://www.theatlantic.com/technology/archive/2012/02/im-being-followed-how-google-151-and-104-other-companies-151-are-tracking-me-on-the-web/253758/. Accessed 17 May 2016.

Manyika, J., Chui, M., Brown, B., Bughin, J., Dobbs, R., Roxburgh, C., & Byers, A. H. (2011). *Big Data the next frontier for innovation, competition, productivity*. Washington, DC: *The McKinsey Global Institute*.

Mayer-Schonberger, V., & Cukier, K. (2013). *Big Data: A revolution that will transform how we live, work, and think*. Boston: Houghton Mifflin Harcourt.

Mosco, V. (2015). Marx in the Cloud. In C. Fuchs & V. Mosco (Eds.), *Marx in the age of digital capitalism* (pp. 516–535). Leiden/Boston: BRILL.

Parcell, J., & Holden, S. H. (2013). Agile policy development for digital government: An exploratory case study. *Proceedings of the 14th annual international conference on digital government research*. ACM.

Seemann, K. W. (2002). *Can our schools deliver an education in technology?*. Conference Publication.

Silver, N. (2012). *The signal and the noise: Why so many predictions fail-but some don't*. New York: The Penguin Press.

Stigler, G. J. (1961). The economics of information. *The Journal of Political Economy, 69*(3), 213–225.

Stiglitz, J. E. (1977). Theory of local public goods. In M. S. Feldstein & R. P. Inman (Eds.), *The economics of public services*. New York: Halsted Press.

Stiglitz, J. E. (1983). Public goods in open economies with heterogeneous individuals. In J.-F. Thisse & H. G. Zoller (Eds.), *Locational analysis of public facilities*. Amsterdam/New York: North- Holland.

Stiglitz, J. E. (1999). Knowledge as a global public good. In I. Kaul, I. Grunberg, & M. Stern (Eds.), *Global public goods* (pp. 308–325). New York: Oxford University Press.

Stiglitz, J. E. (2008). Economic foundations of intellectual property rights. *Duke Law Journal, 57*, 1693–1724.

TCS. (2013). *The emerging big returns on Big Data*. Resource document. http://www.tcs.com/big-data-study/Pages/default.aspx. Accessed 17 May 2016.

UNSTAT. (2015). *Draft principles for access to Big Data sources*. Resource document. http://unstats.un.org/unsd/trade/events/2015/abudhabi/Draft%20Principles%20for%20Access%20to%20Big%20Data%20Sources.pdf. Available 17 May 2016.

Weinberger, D. (2009). *Transparency: The new objectivity*. http://www.kmworld.com/Articles/Column/David-Weinberger/Transparency-the-new-objectivity-55785.aspx. Available 17 May 2016.

Wuyts, M. (1992). Deprivation and public need. In M. Wuyts, M. Mackintosh, & T. Hewitt (Eds.), *Development policy and public action*. Oxford: Oxford University Press.

Yiu, C. (2012). The big data opportunity. *Policy Exchange* http://www.policyexchange.org.uk/images/publications/the%20big%20data%20opportunity.pdf

Chapter 15
Internet Intermediaries as Responsible Actors? Why It Is Time to Rethink the E-Commerce Directive as Well

Sophie Stalla-Bourdillon

Abstract The purpose of this chapter is twofold. First, it aims to show that there is a need to review Articles 12 to 15 of the e-Commerce Directive for at least two reasons: the e-Commerce Directive does not harmonise the conditions for holding intermediaries liable (but only the conditions for exempting Internet intermediaries from liability) and recent case law both at national and European levels, e.g. Delfi v Estonia, shows that divergences still persist among Member States; in addition, Article 15 has become an empty shell despite L'Oréal v eBay, Scarlet v Sabam and Sabam v Netlog.

Second, this chapter suggests that if a review is not possible, the very rationale of Articles 12 to 15 and thereby their interpretation should be rethought. Indeed, the rationale said to underlie Articles 12 to 15– and used by the CJEU to determine the domain as well as the effects of these Articles – is ill-suited to the complexity and diversity of intermediary activities given the current trend towards more 'voluntary' (or better 'suggested') intervention. What is needed is a test that would ensure that control measures put in place by Internet intermediaries are protective of both Articles 10 and 8 of the ECHR and their equivalents within the EU Charter of Fundamental Rights.

S. Stalla-Bourdillon (✉)
School of Law, University of Southampton, Southampton, England
e-mail: S.Stalla-Bourdillon@soton.ac.uk

M. Taddeo, L. Floridi (eds.), *The Responsibilities of Online Service Providers*,
Law, Governance and Technology Series 31, DOI 10.1007/978-3-319-47852-4_15

More than 20 years after the creation of the Single Market, the European Commission is now working on the creation of a digital Single Market.[1] A digital Single Market, says the Commission, is *"one in which the free movement of goods, persons, services and capital is ensured and where individuals and businesses can seamlessly access and exercise online activities under conditions of fair competition, and a high level of consumer and personal data protection, irrespective of their nationality or place of residence"*.[2] The creation of a digital Single Market is seen as a means to make sure the European Union (EU) maintains its position as one of the world leaders of the global information society.[3] In order to reach this goal, different challenges have to be tackled, and not least among them that of "[c]*reating the right conditions for digital networks and services to flourish*" including content services.[4] Such a challenge, writes the Commission, requires appropriate "*regulatory conditions for innovation, investment, fair competition and a level playing field*".[5]

Said otherwise, one of the primary objectives of the digital Single Market is the promotion of innovation and, in particular, the burgeoning of content services. The Commission assumes that there is a direct link between the promotion of innovation and the very content of the applicable regulatory framework, the rights and duties of service providers. Is the Commission implying that a light-touch approach is therefore needed?

Reading further the Digital Single Market Strategy, it is not entirely clear whether the Commission is of the view that a lighter approach is needed, at least when it comes to the regulation of Internet intermediaries' activities. The Commission, in its May 2015 document, introduces a new term to describe the activities of a specific type of private actor: online platforms.[6] *"Certain platforms also qualify as*

---

[1] Communication from the Commission to the European Parliament, the Council, the European Economic and Social Committee and the Committee of the Regions, A Digital Single Market Strategy for Europe, Brussels, 6.5.2015, COM(2015) 192 final (Digital Single Market Strategy). See also Commission Staff Working Document, A Digital Single Market Strategy for Europe -Analysis and Evidence SWD(2015) 100 final. After the submission of this paper the Commission issued a new Communication on 'Online Platforms and the Digital Single Market -Opportunities and Challenges for Europe COM(2016) 288/2 on 25 May 2016 followed by the released of a Code of Conduct on Conduct on Countering Illegal Hate Speech Online on 31 May 2015, http://ec.europa.eu/justice/fundamental-rights/files/hate_speech_code_of_conduct_en.pdf, accessed 9 July 2016 (*Code on illegal hate speech*). For a review of the main features of the new Communication in relation to intermediary liability see Stalla-Bourdillon 2016.

[2] *Ibid*, p. 3.

[3] *Ibid*, p. 3.

[4] *Ibid*, pp. 3-4.

[5] *Ibid*, pp. 3-4.

[6] This term is defined in the consultation following the issuance of the Commission's Communication and the Staff Working Document in the following: "*"Online platform" refers to an undertaking operating in two (or multi)-sided markets, which uses the Internet to enable interactions between two or more distinct but interdependent groups of users so as to generate value for at least one of the groups"*. European Commission, Consultation on Regulatory environment for platforms, online intermediaries, data and cloud computing and the collaborative economy, 2015, https://ec.europa.eu/eusurvey/runner/Platforms/, p. 5, accessed 17.05.2016. After the submission of this paper, a slightly revised definition was proposed in the Commission's communication on online platforms of 25 May 2016.

Intermediary service providers",[7] adds the Commission, with the implication that probably not all of them do. The Commission also points to Recital 48 of the e-Commerce Directive[8] and the infamous 'duty of care' of hosting providers.[9] Besides, the Commission is in the process of looking for new measures to deal with illegal content and thereby asks the question whether Internet intermediaries should "*exercise greater responsibility and due diligence in the way they manage their networks and systems"*.[10] Should we really impose a duty of care on Internet intermediaries, or at the very least hosting providers, and, if yes, what would be the implications of such a duty of care for these private actors and their users? What is a responsible intermediary?

The European Commission does acknowledge that "[i]*t is not always easy to define the limits on what intermediaries can do with the content that they transmit, store or host before losing the possibility to benefit from the exemptions from liability set out in the e-Commerce Directive"*.[11]

While a comprehensive assessment of the role of 'Internet Intermediary Service Providers' (which seems to be the term favoured by the Commission)[12] was scheduled to take place before the end of 2015,[13] this ambition does not seem to encompass that of reviewing the e-Commerce Directive of 2000 or part thereof, to eventually amend it. In its Digital Single Market Strategy, the Commission does mention the "*need for new measures to tackle illegal content on the Internet"*[14] but whether hard law measures will follow is uncertain, if not unlikely.

The purpose of this chapter is twofold. First, it aims to show that there is a need to review Articles 12 to 15 of the e-Commerce Directive for several reasons: the e-Commerce Directive does not harmonise the conditions for holding intermediaries liable (but only the conditions for exempting Internet intermediaries from liability) and recent case law both at national and European levels, e.g. by the European

---

[7] *Ibid*, p. 5.

[8] Directive 2000/31/EC of the European Parliament and of the Council of 8 June 2000 on certain legal aspects of information society services, in particular electronic commerce, in the Internal Market ('Directive on electronic commerce') OJ L 178, 17.7.2000, pp. 1–16 (e-commerce Directive).

[9] Recital 48 reads as follows: "*This Directive does not affect the possibility for Member States of requiring service providers, who host information provided by recipients of their service, to apply duties of care, which can reasonably be expected from them and which are specified by national law, in order to detect and prevent certain types of illegal activities"*.

[10] Digital Single Market Strategy, p. 12.

[11] *Ibid*, p. 12.

[12] I used the expression 'intermediary Internet service providers' in 2010 to cover both Internet service providers and over-the-top service providers as the expression 'Internet service provider' was too much linked to that of Internet access providers. See S. Stalla-Bourdillon, Chilling ISPs... when private regulators act without adequate public framework. *Computer Law & Security Review*, 26, (3), 290–297 (2010a), p. 291. It would seem that the intention of the Commission was the same.

[13] This public consultation on the regulatory environment for platforms, online intermediaries, data and cloud computing and the collaborative economy ran from 24 September 2015 to 6 January 2016 and first brief results of the public consultation were published on 26 January 2016, https://ec.europa.eu/digital-single-market/en/news/first-brief-results-public-consultation-regulatory-environment-platforms-online-intermediaries, accessed 17 May 2016.

[14] *Ibid*, p. 12.

Court of Human Rights (ECtHR) in Delfi v Estonia[15] (*Delfi*), shows that divergences still persist among Member States. In addition, Article 15 has become an 'empty shell' despite the judgements of the Court of Justice of the European Union (CJEU) in L'Oréal v eBay[16] (*eBay*), Scarlet v Sabam[17] (*Sabam*) and Sabam v Netlog[18] (*Netlog*).

Second, this paper suggests that if a review is not possible, the very rationale of Articles 12 to 15 of the e-Commerce Directive and thereby their interpretations should be rethought. Indeed, the rationale said to underlie Articles 12 to 15 of the e-Commerce Directive – and used by the CJEU to determine the domain as well as the effects of these Articles – is ill-suited to the complexity and diversity of intermediary activities in particular given the current trend towards more 'voluntary' (or better 'suggested') intervention. Following the opinion of its Advocate General[19] in the joint cases Google v Louis Vuitton Malletier et al[20] (*Google*) the CJEU held in 2010 that the yardstick to determine whether an intermediary can be exempted from liability is whether *"the service provider is neutral, in the sense that its conduct is merely technical, automatic and passive"*[21] (i.e. whether it lacks knowledge or control over the data). Nevertheless, by stating that hosting providers within the meaning of Article 14 of the e-Commerce Directive, such as search engines or referencing services, can be characterised as data controllers, the CJEU has lately blurred its message.[22] Most of the time, intermediaries are well-placed to exercise control over the data they store or transmit, and on occasion they have actually put in place con-

---

[15] ECtHR, Grand Chamber, Delfi AS v. Estonia, 16 June 2015, Application no. 64569/09, [2015] E.M.L.R. 26. The *Delfi* case was distinguished by the Fourth Section of the ECtHR in the case of Magyar Tartalomszolgáltatók Egyesülete and Index.hu Zrt v. Hungary, 2 February 2016, Application no. 22947/13 (*MTE*). While the Fourth Section often refers to *Delfi*, it does seem that the approach taken by the judges in *MTE* is not the same as the approach taken by the Grand chamber in *Delfi*. The Fourth Section insisted in particular on the implications of holding Internet portals liable for third party content in terms of monitoring. The Fourth Section stated that this would amount *"to requiring excessing and impracticable forethought capable of undermining freedom of the right to impart information on the Internet"*. *MET*, at [82]. In this sense *MTE* could be seen as an attempt to narrow down the scope of the solution adopted in *Delfi*.

[16] CJEU, Grand Chamber, Case C-324/09 L'Oréal SA and Others v eBay International AG and Others, 12 July 2011, ECLI:EU:C:2011:474.

[17] CJEU, Third Chamber, Case C-70/10, Scarlet Extended SA v Société belge des auteurs, compositeurs et éditeurs SCRL (SABAM), 24 November 2011, ECLI:EU:C:2011:771.

[18] CJEU, Third Chamber, Case C-360/10, Belgische Vereniging van Auteurs, Componisten en Uitgevers CVBA (SABAM) v Netlog NV, 16 February 2010, ECLI:EU:C:2012:85.

[19] Opinion of Advocate General Poiares Maduro delivered on 22 September 2009 in Joined Cases, Google France SARL and Google Inc. v Louis Vuitton Malletier SA (C-236/08), Google France SARL v Viaticum SA and Luteciel SARL (C-237/08) and Google France SARL v Centre national de recherche en relations humaines (CNRRH) SARL and Others (C-238/08), ECLI:EU:C:2009:569, at [143].

[20] CJEU, Grand Chamber, Joined cases Joined cases C-236/08 to C-238/08, Google France SARL and Google Inc. v Louis Vuitton Malletier SA (C-236/08), Google France SARL v Viaticum SA and Luteciel SARL, 23 March 2010, ECLI:EU:C:2010:159.

[21] *Ibid*, at [114].

[22] CJEU, Grand Chamber, Case C-131/12, Google Spain SL and Google Inc. v Agencia Española de Protección de Datos (AEPD) and Mario Costeja González, 13 May 2014, ECLI:EU:C:2014:317, at [33] (*Costeja*).

trol measures. What is therefore needed is a mechanism that would ensure that control measures are protective of both Articles 10 and 8 of the European Convention on Human Rights (ECHR), and their equivalents within the European Charter of Fundamental Rights, i.e. Articles 11, 7, and 8.

15.1 Persisting Divergences

Recital 40 of the e-Commerce Directive aims to explain why an EU action is necessary in this field: *"Both existing and emerging disparities in Member States' legislation and case-law concerning liability of service providers acting as intermediaries prevent the smooth functioning of the internal market, in particular by impairing the development of cross-border services and producing distortions of competition."* Thus the divergence of national solutions, obstructing the proper functioning of the internal market, officially served as the justification for the adoption of Articles 12 to 15, constituting Section 4 of the e-Commerce Directive entitled 'Liability of intermediary service providers'.

In 2000, national solutions were far from being firmly established.[23] Germany was an exception on account of its law regulating teleservices, section 5 (Burmeister and Kohler 1999) of which created a horizontal exemption mechanism[24] for intermediaries. Thus, the Digital Millennium Copyright Act[25] (DMCA) was not the only source of inspiration for the European legislature, and the German law of 1997 (Jansen 1999; Burmeister and Kohler 1999; Julia-Barcelo 1998) was undoubtedly a strong argument in favor of the adoption of a horizontal – not vertical – liability regime, like that developed by the US legislature. However, it is probably true to say that the categories retained by the Directive are closer to those of the DMCA than those of the German Teleservices Act.[26] As for the UK, the country had certainly responded to the Internet revolution, but mainly only to resolve defamation issues.[27]

In reality, the law was simply in the process of being formed. The French example perfectly illustrates this process of 'creative oscillation': at the time, no (special)

[23] The fact is that at least since 1995 the impact of intermediary liability regimes was considered significant for the future development of information society services. See the Green Paper on Copyright and Related Rights in the Information Society, COM (95) 382 final, § 49. Hence, the need not only to harmonise in order to create a single system for concrete and predictable applications, but also to establish a conflict rule for the coordinated - not harmonised - domain which ensures the application of a single law.

[24] Applicable regardless of the nature of the interests violated.

[25] PL 105-304 (HR 2281) 28 October 1998 as codified in 17 U.S.C. §512.

[26] At a time when the European authorities are working on the issue of intermediary liability, the German system is based on a tripartite distinction between three types of actors with distinct activities: the content provider, the hosting provider and the access provider.

[27] See the Defamation Act of 1996. It should be noted that if this law is sectoral, in this instance the criterion for the implementation of intermediary liability was also that of awareness of the illegal activity.

legislative provision regulated the matter.[28] Thus, it was judges – not the legislature[29] – that were the first to deliver verdicts on issues of intermediary liability. The same trend is found in Italy, where the (less numerous) legal decisions were often limited to responding to questions of criminal liability (De Arcangelis 2001).[30]

The EU legislature tried to restore judicial security as quickly as possible in this area, with the aim of creating a flourishing digital market, able to compete with the US market. This said, the EU legislature clearly opted for a minimalist approach, which consisted in combining a targeted harmonisation of key provisions of substantive law and a conflict method (Stalla-Bourdillon 2009) that resulted in the insertion of an internal market clause. Therefore, Member States were still left with a certain room for maneuver.

Contrary to the wishes of the drafters of the e-Commerce Directive, the harmonised domain, limited though it is, has not always been interpreted in a consistent manner at national level.[31] To account for this, it is enough to observe the manner in which online auction platforms, and principally eBay, have been treated by national judges (Cheung and Pun 2009; Schaffner and Abello 2008). While eBay was able to benefit from Article 14 of the e-Commerce Directive in Belgium,[32] in France on occasion the online marketplace provider was condemned to pay substantial damages for neglecting its duty to prevent the commissioning of illegal acts on its system.[33]

The CJEU did step into the debate several times and in particular in 2010 and 2011 in the *Google* and *eBay* cases in an attempt to clarify the domain *rationae personae* of Article 14 of the e-Commerce Directive and the conditions for liability exemptions. However, it is far from clear whether the CJEU has managed to issue helpful guidance to the benefit of national courts. This is true for at least two reasons.

First of all, the CJEU's reasoning is inherently flawed in the sense that it uses the same criteria to determine both the domain *rationae personae* of Article 14 and the conditions for liability exemptions. To determine whether a service provider can avail itself of Article 14, says the Court, it is essential to determine whether that service provider *"has not played an active role of such a kind as to give it knowledge of, or*

[28] The e-Commerce Directive was then anticipated by a 2000 Act (Loi n° 2000-719, 1er août 2000).

[29] France was one of the countries where the case law was the most verbose on the matter.

[30] Such an extension was however heavily criticised by legal commentators (Zeno-Zencovich 1998, 1999, Sica 2001).

[31] Obviously, divergences also started to appear within the same jurisdiction. For an overview of the state of the law up until 2007 see Van Eecke and Ooms 2007.

[32] See e.g. T. com. Brussels, 7th chamber, 31 July 2008, n° A/07/06032, Lancôme Parfums & Beauté & Cie v eBay International AG, eBay Europe SARL, eBay Belgium, Comm. Com. Electr. 2009, étude 4, A. Regniault and F. Guthfreund-Roland.

[33] T. com., Paris, 1st chamber, B, 30 June 2008, Parfums Christian Dior and others v eBay Inc., eBay International AB AG, www.legalis.net ; T. com., Paris, 1st chamber, B, 30 June 2008, Christian Dior Couture v eBay Inc., eBay International AG, www.legalis.net ; T. com., Paris, 1st chamber, B, 30 June 2008, Louis Vuitton Malletier v. eBay Inc., eBay International AG, www.legalis.net. The Court of Appeal and the Supreme Court (Cour de Cassation) subsequently agreed with the Tribunal as regards the characterisation of eBay's activity. See e.g. Cour d'appel de Paris, Pole 5, ch. 2, 3 Septembre 2010, eBay v Parfums Christian Dior and others, www.legalis.net and Cour de cassation, civile, Chambre commerciale, 3 May 2012, 11-10.508, Bulletin 2012, IV, n° 89.

control over, the data stored".[34] Yet, once it is established that the activity of the service provider is covered by Article 14, it is then necessary to establish whether the service provider can be exempted from liability. And to determine whether the service provider meets the conditions for exemption, we once again have to ask the same question: whether the service provider had actual knowledge or was aware of facts or circumstances making it apparent that the third party content at stake was unlawful, and whether the service provider had exercised control over that content.[35]

Second, the reasoning of the CJEU lacks clarity as regards the requirements of 'control' and 'awareness'. The CJEU said in *eBay* that the 'awareness' requirement is met when the service provider has *"been aware of facts or circumstances on the basis of which a diligent economic operator should have identified the illegality in question and acted in accordance with Article 14(1)(b) of Directive 2000/31"*.[36] Reading this sentence in isolation, this appears to indicate that it is possible to presume 'awareness': the CJEU seems to suggest that the provision of added-value tools such as optimisation or promotion tools to one's users could allow a national judge to presume 'awareness'.[37] But, is it not the case that the CJEU is mixing up the requirement of 'awareness' with that of 'control'? Or are they the same requirement? In *Google*, the CJEU does not distinguish between these two requirements when it mentions that *"the role played by Google in the drafting of the commercial message which accompanies the advertising link or in the establishment or selection of keywords is relevant"*.[38]

There is, however, another way to read the *eBay* judgement: reading together paragraphs 120 and 122, it would seem that what the CJEU may in fact have had in mind was two, separate, situations (although the CJEU uses the expression 'in particular'): a situation in which a service provider engages into an investigation on its initiative; and, a situation in which the service provider is notified about the existence of an unlawful content on its system.

Given the evolution of the national case laws after the adoption of the e-Commerce Directive, it is not surprising that in the *Delfi* case the Estonian national courts were able to find that Delfi, as regards its comment section, could not avail itself of the national transposition of Article 14: Delfi was not for the courts a 'mere' passive intermediary.[39] The Estonian Supreme Court in 2009 seemed to characterise 'con-

[34] *Google*, at [120].

[35] Here is the exact wording of the first two paragraphs of Article 14: *"1. Where an information society service is provided that consists of the storage of information provided by a recipient of the service, Member States shall ensure that the service provider is not liable for the information stored at the request of a recipient of the service, on condition that:*

(a) the provider does not have actual knowledge of illegal activity or information and, as regards claims for damages, is not aware of facts or circumstances from which the illegal activity or information is apparent; or

(b) the provider, upon obtaining such knowledge or awareness, acts expeditiously to remove or to disable access to the information.

2. Paragraph 1 shall not apply when the recipient of the service is acting under the authority or the control of the provider".

[36] *eBay*, at [120].

[37] *"The operator plays such a*[n active] *role when it provides assistance which entails, in particular, optimising the presentation of the offers for sale in question or promoting them"*. *eBay*, at [123].

[38] *Google*, at [118].

[39] The first national decisions were issued in 2006 in this case.

trol' from the fact that *"the defendant ha*[d] *integrated the comment environment into its news portal, inviting visitors to the website to complement the news with their own judgements [hinnangud] and opinions (comments). In the comment environment the defendant actively call*[ed] *for comment on the news items appearing on the portal. The number of visits to the defendant's portal depend*[ed] *on the number of comments; the revenue earned from advertisements published on the portal, in turn, depend*[ed] *on the [number of visit]"*.[40] But that was in 2009, a year before the CJEU's *Google* judgement and 2 years before the *eBay* judgement.

Even if there is a sensible argument that a strict reading of *Google* and *eBay* should in the future guide a national court to find that it is not enough that a user-generated content section is present on a platform and that the service provider has an economic interest in the creation of user-generated content to reject the applicability of Article 14 of the e-Commerce Directive (as this does not establish assistance on the part of the service provider to the benefit of the authors of the comments), the *Delfi* case is not that straightforward. Delfi did engage into *ex ante* preventive practices. Moreover, to repeat the words of the Estonian Supreme Court, Delfi *"enact*[ed] *the rules for the comment environment and ma*[de] *changes to it (remove* [d] *a comment) if those rules* [were] *breached"*.[41] Although a generous reading of *eBay* could make it possible for a national court to disregard these facts, such an exclusion would not be necessarily obvious to all courts.[42] In other words, the CJEU's statements in *Google* and *eBay* are not sufficient to eliminate divergences, promote legal certainty, and effectively reduce the number of scenarios in which Internet intermediaries can be held liable for the activities of their users, which seemed to be the ambition of the EU legislature at the end of the nineties.

15.2 Minimum Standards

However, assuming the CJEU was given another opportunity to clarify in the future the requirements for 'control' and 'awareness', this would not necessarily make divergences disappear. This is due to two reasons. Both are consubstantial to the harmonisation process initiated by the e-Commerce Directive.

The first reason is that the EU legislature had decided to only harmonise minimum standards in the field of intermediary liability. As it will become clearer below, the conditions for liability exemption were conceived as minimum standards in the sense that Member States remained competent to determine the conditions for liability and could therefore decide to exempt intermediary providers further.

[40] *Delfi*, at [31].
[41] *Ibid*, at [31].
[42] See Van Eecke's plea for a balanced approach (Van Eecke 2011), who reviews national cases and argues that the monitoring of users' activities should not prevent internet intermediaries from benefiting from Articles 12 to 14 of the e-Commerce Directive.

The originality of the EU harmonisation method deployed through the means of the e-commerce Directive (comprising a mix of conflict rules and partial harmonising rules) made it necessary to distinguish between the harmonised and coordinated fields:[43] the former was to fall under the competence of the EU legislature, while the latter under the competence of the national legislatures. The harmonised field is comprised of Article 12 and following. Three types of activities are involved, each of which is subject to a very specific set of conditions for exemption from liability: mere conduit, caching and hosting.

As explained by one of the Commission's authorised spokespersons at the time, "[the] *Directive determines only those cases where a provider can enjoy a dispensation or a limitation of liability. This does not mean that* [the provider] *will necessarily be liable in the case that it does not meet the conditions to qualify. In this case, the national liability regimes will apply to determine the liability of providers*" (Crabit 2000, p. 812).[44] Crabit continues, stating, "[T]*his approach is explained, on the one hand, by the fact that the case law that has emerged over the past five years demonstrates that strong legal insecurity reigned concerning the possible non-liability of a purely passive intermediary. The problems are not derived from a possible non-application of national liability regimes, which could have created a sort of immunity for these new intermediary activities. On the other hand, the negative form* [of the provisions] *allows intervention at Community level to the extent that is strictly necessary for internal market purposes. Indeed,* [the Community legislature] *cannot harmonise the civil and criminal liability regimes as such, but can adjust them, only to the extent necessary to remove the legal uncertainty weighing on the intermediaries*" (Crabit 2000, p. 812).[45]

As a result, *a contrario* reasoning cannot be used to interpret Articles 12 and following.[46] Ultimately, a Member State should then be able to further exempt intermediaries from both civil and criminal liability. By way of example, this is what happened in part with the UK Defamation Act of 2013. Section 5 of this Act provides for a defence to the benefit of website operators, which can be defeated only when three conditions are met: "*(a) it was not possible for the claimant to identify the person who posted the statement, (b) the claimant gave the operator a notice of complaint in relation to the statement, and c) the operator failed to respond to the notice of complaint in accordance with any provision contained in regulations*". In other words, under this Act, the claimant cannot go after the website operator if she is able to identify the poster. In addition, the only way a website operator can gain knowledge about a defamatory act is through the means of a notification. Besides, section 5(12) makes it clear that the fact that the website operator engages in moderation practices is irrelevant.

[43] For a definition of the coordinated field, see Article 2(h) of the e-Commerce Directive.
[44] Traduction from the author.
[45] Traduction from the author.
[46] See on these questions, Stalla-Bourdillon 2010b, pp. 319 ff.

15.3 Exclusion of Data Protection Law Issues

The second reason explaining the weaknesses of the harmonisation approach embodied in the e-Commerce Directive is the way the scope of the e-Commerce Directive has been delineated. Article 1 of the e-Commerce Directive excludes from its remit *"questions relating to information society services covered by Directives 95/46/EC and 97/66/EC"*. Does it mean that when unfair processing of personal data is at stake, Internet intermediaries cannot benefit from liability exemptions anymore, even if the content ultimately complained of has been transmitted at the initiative of a third party?

The separation between the domain of the e-Commerce Directive and that of the Data Protection Directive[47] should be assessed in the light of the *Costeja* judgement, a case colloquially known as *Google Spain*.[48] In *Costeja*, the CJEU found that a search engine, i.e. the provider of a (natural) referencing service, is a data controller.[49] Such a qualification is obviously instrumental. The purpose is to impose upon Google a very specific obligation: an obligation *"to remove from the list of results displayed following a search made on the basis of a person's name links to web pages, published by third parties and containing information relating to that person"*,[50] when the processing of personal data through the means of the provision of these links is or becomes incompatible with the Data Protection Directive.[51] It is an application of Article 12(b) of the Data Protection Directive: *"Article 12(b) ... provides that Member States are to guarantee every data subject the right to obtain from the controller, as appropriate, the rectification, erasure or blocking of data the processing of which does not comply with the provisions of Directive 95/46, in particular because of the incomplete or inaccurate nature of the data"*.[52]

Generally speaking, Article 22[53] of the Data Protection Directive provides that "[w]*ithout prejudice to any administrative remedy for which provision may be made, inter alia before the supervisory authority referred to in Article 28, prior to*

[47] Directive 95/46/EC of the European Parliament and of the Council of 24 October 1995 on the protection of individuals with regard to the processing of personal data and on the free movement of such data OJ L 281 , 23/11/1995.

[48] The following of this section builds upon the author's blogpost (Stalla-Bourdillon 2015).

[49] It is worth comparing the words of the CJEU with those of the French Supreme Court in a criminal case involving the 'Google suggest' functionality. C.Cass, 1 civ., n° 12-17.591, 19 June 2013 (12-17.591), ECLI:FR:CCASS:2013:C100625.

[50] *Costeja*, at [88]. Could the CJEU have reached the same solution without characterising Google as a data controller? Looking at the way the Spanish Court had formulated its questions in its reference for a preliminary ruling and their respective order, it seems that only if Google was a data controller would it then have been possible to recognise a legal obligation upon it to delist.

[51] *Ibid*, pp. 31 – 50.

[52] *Costeja*, at [70].

[53] Article 6 of the Data Protection Directive is also problematic. Under this article, the data controller has to make sure that data quality principles are also complied with. Some then say that it does not make sense to characterise Google as a data controller because the principle of data minimisation does not work when applied to the activities of search engines. Nevertheless, it might be argued that the obligation to delist is the very means by which a search engine can comply with the principle of data minimisation.

referral to the judicial authority, Member States shall provide for the right of every person to a judicial remedy for any breach of the rights guaranteed him by the national law applicable to the processing in question" and Art. 23(1) that *"Member States shall provide that any person who has suffered damage as a result of an unlawful processing operation or of any act incompatible with the national provisions adopted pursuant to this Directive is entitled to receive compensation from the controller for the damage suffered"*. A crucial question arising in the wake of the *Costeja* case, is therefore whether a provider of a referencing service such as Google could then be held liable for a provision of links that is or has become incompatible with the Data Protection Directive. And it does not seem that Article 23(2)[54] could be of some help for a provider of a referencing service since its very activity is at stake (not that of the actual publisher of the material linked to).

This is where the English case Max Mosley v Google[55] (*Mosley*) becomes really interesting. In a nutshell, and not to relate the whole Mosley saga on the facts, the claimant (Max Mosley) had requested that Google cease the listing of a certain number of images, as well as links which directed to websites displaying his personal data from its search engine results, under section 10 of the UK's Data Protection Act 1998 (DPA). Under section 10, individuals have a (limited) right to require the data controller not to process data where such processing causes, or is likely to cause, the individual or anyone else unwarranted substantial damage or distress. Google refused to acknowledge the applicability of this right against it for a variety of reasons on the facts of the case, including arguing the fact that it was not a data controller (which after the CJEU's judgement was not really tenable) and the fact that the notices served upon it on behalf of Mosely did not *"identify the personal data in respect of which it was given or the steps required to cease processing it"*.

The claimant thus brought a suit against Google on the grounds of sections 10, 13 and 14 of the DPA. Under section 13, individuals are entitled to compensation from data controllers for damage caused by any breach of the DPA. Under section 14, an individual can obtain a court order for the rectification, blocking, erasure or destruction of data that is inaccurate. The court may also, where it considers it reasonably practicable, order the data controller to notify third parties to whom incorrect data has been passed of the rectification, blocking, erasure or destruction.

Reading Articles 12, 22 and 23 of the Data Protection Directive together, it would seem that if a search engine is a data controller it could be held financially liable for the damage suffered by the claimant as a result of a supply of links that is incompatible with the Data Protection Directive, although the listing would be comprised of links referring to material published by others. Such a solution would be odd as the e-Commerce Directive, as aforementioned, sets forth horizontal third-party liability exemptions to the benefit of intermediary service providers including caching and hosting providers.

---

[54] Article 23(2) provides that *"The controller may be exempted from this liability, in whole or in part, if he proves that he is not responsible for the event giving rise to the damage"*.

[55] Mosley v Google Inc & Anor [2015] EWHC 59 (QB).

This explains why Article 1 of the e-Commerce Directive is clearly problematic. Such an odd result might explain paragraph 12 of the Article 29 Data Protection Working Party's guidelines[56] on the implementation of the *Costeja* judgement, which states that: "[b]*y making a request to one or several search engines the individual is making an assessment of the impact of the appearance of the controverted information in one or several of the search engines and, consequently, makes a decision on the remedies that may be sufficient to diminish or eliminate that impact*". Said otherwise, it would seem (although this is *clearly* adding to the legislative text as well as to the *Costeja* judgement) that by claiming for the take-down of some links, a data subject would be waiving her right to monetary compensation. Such an effect is, however, not that straightforward.

In *Mosley*, although Justice Mitting was not asked to solve this issue, he expressed the view that *"the two Directives must be read in harmony"*.[57] The adoption of the proposed General Data Protection Regulation could have been the occasion to clarify the question. However the text of the final version of the Regulation does not seem to elucidate this point.[58] Besides, Article 82 confirms *that* "[a]*ny person who has suffered material or non-material damage as a result of an infringement of this Regulation shall have the right to receive compensation from the controller or processor for the damage suffered*".

15.4 The Distinction Between Active and Passive Intermediary Service Providers

In the end, what could be the most problematic concern with respect to the appropriateness of the e-Commerce Directive as a regulatory instrument is the way its applicability test has been formulated by the CJEU in an attempt to make sense of both its key provisions and its recitals. As stated above, a service provider is an intermediary service provider if and only if it *"has not played an active role of such a kind as to give it knowledge of, or control over, the data stored"*,[59] said the CJEU in *Google*.

---

[56] Article 29 Data Protection Party, Guidelines on the implementation of the Court of Justice of the European Union Judgement on "Google Spain and Inc. v Agencia Espanola de Proteccion de Datos (AFPD) and Mario Costeja Gonzales" C-131/12 WP 225, p. 7.

[57] *Mosley* at [45]. A Spanish Court also seems to be of the view that both Directives should be read together. JUZGADO PRIMERA INSTANCIA 8 BARCELONA, 17 July 2014, ROLLO n° 99/2012-A. For a comment see Peguera 2014.

[58] Regulation (EU) 2016/679 of the European Parliament and of the Council of 27 April 2016 on the protection of natural persons with regard to the processing of personal data and on the free movement of such data, and repealing Directive 95/46/EC (General Data Protection Regulation) OJ L 119, 4.5.2016, pp. 1–88. Recital 21 simply states that: *"This Regulation is without prejudice to the application of Directive 2000/31/EC of the European Parliament and of the Council, in particular of the liability rules of intermediary service providers in Articles 12 to 15 of that Directive. That Directive seeks to contribute to the proper functioning of the internal market by ensuring the free movement of information society services between Member States"*. See also Article 2(4) of the General Data Protection Regulation.

[59] *Google*, at [120].

First of all, such an applicability test is awkward because most (if not all intermediary provider) can potentially exercise control over the data they store or transmit (Stalla-Bourdillon 2012b). Besides, this is implicitly acknowledged by the EU legislature itself which distinguishes between two types of remedies (monetary compensation and injunctions),[60] and specifically carves out the possibility for Member States to enjoin intermediary providers to stop existing or prevent future unlawful activities on their systems or networks.

Such a confusion explains in part why Article 15 of the e-Commerce Directive has become an empty shell (Stalla-Bourdillon 2013, pp. 707 ff). Indeed, through the means of injunctive orders, national judges have imposed upon service providers (both mere conduits[61] and hosting providers)[62] an obligation to monitor the behaviour of all their subscribers. By way of example, the obligation to use the Cleanfeed technology to block access to copyright infringing websites implies that the addressee of the injunction has the obligation to process the IP addresses of all its subscribers to determine who is trying to access the infringing website.

The same confusion also explains why, at a higher level, the CJEU's judgments could seem inconsistent. Indeed, on the one hand the CJEU held in *Google* that *"it cannot be disputed that a referencing service provider transmits information from the recipient of that service, namely the advertiser, over a communications network accessible to internet users and stores, that is to say, holds in memory on its server, certain data, such as the keywords selected by the advertiser, the advertising link and the accompanying commercial message, as well as the address of the advertiser's site"*.[63] However, such data processing is not deemed sufficient by the CJEU in this case to make a provider of a referencing service lose Article 14 immunity. At this stage therefore, the provider is still passive. To determine whether it is active, one has also to consider *"the role played by* [the provider of the referencing service] *in the drafting of the commercial message which accompanies the advertising link or in the establishment or selection of keywords"*.[64]

On the other hand, in *Costeja* (which involves personal data and not 'simple' data) the CJEU seems to imply that Google, when it processes the data to create its listing, already engages in an active role irrespective of the way the messages attached to the links are drafted. *"It is the search engine operator which determines the purposes and means of that activity and thus of the processing of personal data that it itself carries out within the framework of that activity and which must, con-*

---

[60] See the last paragraphs of Article 12 to 14 of the e-Commerce Directive.

[61] See e.g. Mr. Justice Arnold's long list of blocking injunction cases and, in particular, the first one, Twentieth Century Fox Film Corp v BT [2011] EWHC 1981 (Ch) (*Newzbin 2*). See also Dramatico Entertainment Ltd and others v British Sky Broadcasting Ltd and others (No.2) [2012] EWHC 1152 (Ch); EMI Records Ltd and others v British Sky Broadcasting Ltd and others [2013] EWHC 379 (CH); Cartier International AG v British Sky Broadcasting Ltd [2014] EWHC 3354 (Ch) (*Cartier*) confirmed by the Court of Appeal in Cartier International AG & Ors v British Sky Broadcasting Ltd & Ors [2016] EWCA Civ 658.

[62] See e.g. the French case brought by Mosley against Google: TGI Paris, 17e ch., 6 November 2013, RG 11/07970, Max Mosley v Google Inc & Google France, http://droitdu.net.

[63] *Google*, at [111].

[64] *Google*, at [118].

sequently, be regarded as the 'controller' in respect of that processing pursuant to Article 2(d)", says the CJEU at paragraph 33.

Finally, the foregoing explains why the dissenting opinion of the ECtHR in *Delfi* is also subject to criticism. While the dissenting judges concluded that Delfi should not be held liable in this case, since this would amount to imposing upon the service provider an obligation to monitor its platform 24/7, they found that as regards the comment section Delfi is an active provider: *"It is certainly true that the active intermediary can exercise control over the comments that appear on its site and it is also true that by creating a space for comments, and inviting users to participate, it engages in an expressive activity that entails responsibility. But the nature of the control does not imply identification with a traditional publisher"*.[65] Such a finding is, however, difficult to accept as, strictly speaking, regarding the comment section, Delfi is simply a conduit for speech and not a speaker: its users, the authors of the comments are the speakers.[66] Besides, that finding would mean that service providers of this type would hardly have to justify take-down activities. An assessment by activity or service, (and not by provider) should be preferred, as one provider can have several types of activities or supply several types of services.

The second reason that explains why the distinction between active and passive actors should be abandoned is simply that it is wrong to consider Article 14 applicability test in isolation: it should be read and applied in the light of the regulatory goal to be achieved. But this is where things become really intricate, because three distinct regulatory goals could be said to underlie section 4 of the e-Commerce Directive, and, these goals are likely to conflict with one another: the protection of freedom of expression, the promotion of innovation and the regulation of unlawful content (Stalla-Bourdillon 2012b).

Assuming all of these three goals have to be balanced together (and not one is to significantly prevail over the other two), it would make sense to get rid of the distinction active/passive actor and adopt the following two-part test to delineate the domain of liability exemptions:

1. Is the provider (when it supplies the service at stake) a speaker or a conduit for expression? Said otherwise, is the provider exercising its own freedom of expression or is it a conduit for others to exercise their freedom of expression?
2. Are the measures put in place by the provider to mitigate the transmission of unlawful content (for the service at stake) proportionate and effective? (in particular are these measures transparent? To what extent are these measures interfering with Internet users' freedom of expression? To what extent are these measures interfering with Internet users' rights to privacy and data protection?)

---

[65] *Delfi*, at [30].

[66] It is true that such a solution might have been forced upon the judge, since Delfi was arguing that it was a speaker and that making it financially liable in such a case would amount to a breach of its freedom of expression.

Ex post moderation systems as well as *ex ante* filtering systems (such as key words filtering systems) of which the aim is to eliminate the transmission of unlawful content, should not per se make a provider a speaker.[67] However, the characterisation of 'conduit for expression' should only be retained in cases in which these systems are fully transparent (when users are truly informed about the modalities of their implementation) and appeal procedures are in place. Their implications in terms of privacy interests should also be taken into account. The provider should bear the burden of showing that the measures put in place are both transparent and balanced with Internet users' rights to freedom of expression and privacy and data protection.

If an assessment of this type is required at a pre-screening stage, *a fortiori* a similar assessment should be undertaken to determine the content of the injunctive relief to be granted. Judges would need to examine the technical measures to be put in place by service providers and their legal implications at the stage at which the order is made and carefully justify why such an order is to be granted. This is in effect the only way to assess whether the balancing between the different competing interests identified above is adequate. This is the reason why the approach adopted by the English Judge Mr. Justice Arnold in both copyright and trade mark cases should be welcome, although it is missing considerations relating to the protection of the rights to privacy and data protection[68] and his assessment of effectiveness does not seem to be consensual.[69] And this also explains why the approach adopted by the CJEU in UPC Telekabel v Constantin Film et al (*Telekabel*)[70] is not satisfactory. In that case the CJEU ruled that the court shall not engage into the review of the means used by the service provider to implement the necessary technological measures at the stage of the enforcement proceedings if such measures are not challenged.[71] Said otherwise, the CJEU held that the issuance of an injunction against an intermediary is a two-step process and it is only at the hypothetical second stage, when and if Internet users want to complain about the intrusiveness of the measures put in place to stop or prevent infringing activities, that the court will examine whether the measures excessively impede upon Internet users' fundamental rights. It is also worth mentioning that, among the list of Internet users' fundamental rights, the rights to privacy and data protection, while mentioned in the *Sabam* case, have disappeared from consideration in *Telekabel*.

[67] The French Dailymotion case is interesting in that the Court of Paris characterises Dailymotion as a hosting provider after having found that the moderation activity undertaken by the service provider is always triggered by user complaints. CA Paris, 2 December 2014, TF1 et autres c/ SA Dailymotion, www.legalis.net. Yet, it is not clear why if a content is manifestly unlawful and the service provider has actual knowledge of it, the latter should wait for a user's complaint.
[68] See *Cartier*, at [188].
[69] See *Cartier*, at [167].
[70] CJEU Case-314/12 UPC Telekabel Wien GmbH v Constantin Film Verleih GmbH and Wega Filmproduktionsgesellschaft mbH ECLI:EU:C:2014:192.
[71] *Telekabel*, at [57].

Importantly, the implementation of the applicability test proposed above should not have the consequence of making *ex ante* monitoring systems – which are costly, more or less privacy intrusive depending on data retention practices, and more or less freedom of expression chilling depending upon the availability of appeal procedures- conditions of eligibility. If one takes *Netlog* or *Sabam* seriously (although strictly speaking both *Netlog* and *Sabam* do not seem to say anything about the voluntary adoption of *ex ante* monitoring systems, which is why they are problematic), such systems are highly suspicious in particular if they target the content of communications.

As a rule of thumb "*the notice-and-take-down-system* [should] *function in many cases as an appropriate tool for balancing the rights and interests of all those involved*"[72] as stated by the Fourth Section of the ECtHR in *MTE* in an attempt to distinguish *MTE* from *Delfi*,[73] but as long as providers are not asked to substitute judges[74] and decide whether the content is lawful or not, or at the very least as long as the content is manifestly unlawful, which should mean that in case of doubts providers should not react; i.e. it is not a Good Samaritan exemption that is needed but an exemption for good faith refusal to take down. It is one thing to detect child pornography or maybe some forms of copyright infringement (i.e. exact reproductions)[75] or some forms of trade-mark infringement, it is another thing to detect defamatory or insulting content or even hate speech. The very work of Inhope hotlines[76] to which trained professionals contribute, does show that there is a lot of subjectivity involved in the determination of harmful content once one goes beyond its 'hard' core comprising child pornography content. This is not to say that automatic detection of child pornography content cannot raise serious legal issues.[77]

[72] *MTE*, at [91]. Case mentioned at fn 16.

[73] Unless the judiciary is reactive enough (as well as low-cost) so that it can handle a great number of complaints simultaneously and in a timely fashion. Privately managed notice-and-take-down-systems as opposed to court orders should therefore be considered as second-best solutions and serious thoughts should be given to the creation of accessible ad hoc judiciary institutions.

[74] Some commentators criticised Article 14 of the e-Commerce Directive for this very reason right after its adoption (see e.g. Montero 2001, p. 287).

[75] Notice-and-take-down procedures could be used for exact reproductions of copyright works for example. See Pei Zhang, Sophie Stalla-Bourdillon and Lester Gilbert, A Content-Linking-Context Model for "Notice-and-take-down" Procedures, ACM Web Science Conference, 2016. See also the second and third studies conducted by Urban, Karaganis, and Schofield who found high number of questionable notices in their second and third studies (Urban, Karaganis and Schofield 2016).

[76] "*INHOPE is an active and collaborative network of 51 hotlines in 45 countries worldwide, dealing with illegal content online and committed to stamping out child sexual abuse from the Internet*". For more information, see www.inhope.org.

[77] See e.g. the US 2004 decision in Center for Democracy and Technology v. Pappert 337 F. Supp. 2d 606 (2004). Notably, the US court considered DNS blocking, IP blocking and URL blocking and concluded that the implementation of the Pennsylvania's Internet Child Pornography Act would have led to the overblocking of innocent speech in violation of the First Amendment to the US Constitution, even if the court ascknowledged that "*URL filtering results in the least amount of overblocking*" at [143]. Today, solutions such as PhotoDNA (released by Microsoft), which consists in hashing and converting images into numerical values that are matched against databases of hashes derived from known illegal images, are now used in the US. Yet, both in *Sabam* and *Netlog* it was precisely solutions of the latter type that were at stake (i.e. CopySense by Audible Magic). See Stalla-Bourdillon 2013.

Sectoral codes of practices could be used to detail sector-specific notice-and-take-down procedures, which would be calibrated in such a way that detection of manifestly unlawful content would become easier for providers.[78] At a bare minimum, the precise location of the content should be included through the means of a list of URLs.[79] In case of copyright infringement, a link (and eventually access) to the original work would also be of great help to make comparisons possible. Notably, current automated notice-and-take-down procedures, such as Google's procedure, essentially rely on domain analysis and domain analysis only.

Besides, and this is essential, if the implementation of a notice-and-take-down procedure means that the provider has to resort to deep packet inspection technologies, only court orders should be allowed to check the validity of the practices through the means of a comprehensive assessment given the intrusiveness and chilling effect of these technologies. It is thus crucial to distinguish between access and hosting providers. Article 12 should be firmly applied, as Advocate General Szpunar did it in Tobias Mc Fabben v Sony Music Entertainment Germany GmbH (*Mc Fabben*).[80]

It should also become clear that with regard to the conditions for liability exemption for hosting providers, including the requirements of 'awareness' and 'control', while the latter loses its usefulness, the second should be interpreted restrictively, so that the answer should depend upon either the characterisation of actual knowledge or the establishment of the reception of a notification.

Furthermore, the object of the obligation to promptly remove unlawful content should depend upon an appreciation of what is reasonable to expect from the provider in the light of the circumstances of the case: e.g. the quantity of material on the platform, whether it is possible to fully or partially automate the process.[81]

Finally, judges should be careful when ordering stay-down obligations: they should take into account the nature of the content and the gravity of the violation at stake, as well as the costs, characteristics and implications of the technological measures available. Yet such considerations were completely absent from consideration in the French version of the *Mosley* case[82] for example.

---

[78] Which is not what the Code on illegal hate speech does (see fn 2).

[79] See Urban, Karaganis, and Schofield 2016, p. 38 who notes that "[f]*or several OSPs, attempting to identify allegedly infringing material based on imprecise location pointers represents the most challenging and resource-intensive aspect of takedown*". With this said, national judges have not always been of this view. See e.g. TGI Paris, 3e ch., 2e section, 10 July 2009, Bayard Presse c/ YouTube LLC, www.legalis.net; Tribunale Civile di Roma Sentenza n. 8437/2016, 27 April 2016, R.T.I c/ TMFT Enterprises, LLC – Break Media. For a comment see Rosati 2016.

[80] Advocate General Opinion of 16 March 2016, Case C-484/14 Tobias Mc Fabben v Sony Music Entertainment Germany GmbH.

[81] By way of example no contextual assessment of this type was undertaken in the French Dailymotion case. Case mentioned at fn 68.

[82] In this case, the French Tribunal ordered Google to stop referencing in its listings nine pictures of Max Mosley for a period of five years. Case mentioned at fn 63. Notably, while the French Supreme Court (Cour de Cassation) rejected the validity of stay-down obligations triggered by the reception of private notices in Cour cass, 1er civ,. case n° 827, 12 July 2012 (11-15.165 ; 11-15.188), ECLI : FR : CCASS : 2012 : C100827, court orders can contain obligations of this type.

15.5 Conclusions

To conclude, harmonising a limited number of private law ‘islands’ is always a tricky exercise, as the broader context in which the new legislative provisions are inserted must be taken into account to fully comprehend the implications of such measures. And obviously each national context is peculiar. In the United Kingdom, for example, the exemptions of financial liability introduced by the e-Commerce Directive and transposed by the Electronic Commerce (EC Directive) Regulations 2002[83] were for the most part redundant (Stalla-Bourdillon 2012a). This is less true for injunctive relief as recent case law shows it.

The interpretative role of the CJEU is thus crucial to frame subsequent national development. It is argued in this chapter, however, that the CJEU has not always given clear guidance to national courts. It is also suggested that the CJEU has not been helped by the choice made by the EU legislature to adopt a partial harmonising approach and thereby only harmonise conditions for liability exemptions and not conditions for liability.

Even if the CJEU ultimately manages to “modernise” the test for the application of intermediary liability exemptions, as this chapter suggests it should do by adopting a twofold test, a crucial question remains: whether it is more appropriate to adopt a sectoral approach tailored to the specificities of each type of unlawful content or to stick to a horizontal approach. Given the great diversity of unlawful content and the fact that the protection of freedom of expression is at least as important as the elimination of unlawful content, it is argued in this chapter that a sectoral approach is warranted to assess the mitigation measures that Internet intermediaries put in place, as long as a notice-and-take-down procedure forms the base line.

Finally, making Internet intermediary responsible actors should require, at the very least, robust oversight mechanisms, which are only possible if intermediaries are given an incentive to be truly transparent.

References

Cheung, A. S., & Pun, K. (2009). Comparative study on the liability for trade mark infringement of online auction providers. *European Intellectual Property Review, 31*(11), 559–567.

Crabit, E. (2000). La directive sur le commerce électronique. *Revue du Droit de l’Union Européenne, 4*, 749–834.

De Arcangelis, M. (2001). La responsabilité des fournisseurs de services d’hébergement sur Internet en Italie. Published on Juriscom.net, 21 July 2001.

Jansen, T. (1999). Legal aspects of doing E-commerce business in Germany. *International Company and Commercial Law Review, 10*, 39–42.

Julia-Barcelo, R. (1998). Liability for on-line intermediaries: A European perspective. *European Intellectual Property Review, 20*, 453–463.

[83] 2002 No. 2013.

Köhler, C., & Burmeister, K. (1999). Copyright liability on the internet today in Europe (Germany, France, Italy and the EU). *European Intellectual Property Review, 21*, 485–499.
Montero, E. (2001). La responsabilité des prestataires intermédiaires sur les réseaux. In E. Montero (Ed.), *Le commerce électronique sur les rails? Analyse et proposition de mise en oeuvre de la directive sur le commerce électronique* (pp. 273–295). Brussels: Bruylant.
Peguera, M. (2014). Right to be forgotten: Google sentenced to pay damages in Spain, 1 April, https://cyberlaw.stanford.edu/blog/2014/10/right-be-forgotten-google-sentenced-pay-damages-spain. Accessed 17 May 2016.
Rosati, E. (2016). Italian court says that rightholders do NOT have to indicate URLs when submitting takedown requests, 9 May 2015. http://ipkitten.blogspot.co.uk/2016/05/italian-court-says-that-rightholders-do.html. Accessed 17 May 2016.
Schaffner, M., & Abello, A. (2008). LVMH c/ eBay – Une synthèse des questions relatives à la responsabilité des plates-formes d'enchères. *Revue Lamy Droit de l'Immatériel, 41*, 75–86.
Sica, S. (2001). Il sistema delle responsabilità. In G. Commandé & S. Sica (Eds.), *Il commercio elettronico* (pp. 218–247). Torino: G. Giappichelli.
Stalla-Bourdillon, S. (2009). Re-allocating horizontal and vertical regulatory powers in the electronic marketplace: what to do with private international law. In F. Cafaggi & H. Muir-Watt (Eds.), *Regulatory strategies in European private law* (pp. 298–350). Cheltenham: Edward Elgar.
Stalla-Bourdillon, S. (2010a). Chilling ISPs... when private regulators act without adequate public framework... Computer Law & Security Review 26(3): 290–297.
Stalla-Bourdillon, S. (2010b). Responsabilité civile et stratégie de régulation: essai sur la responsabilité civile des prestataires intermédiaires de service en ligne. PhD thesis. EUI. http://cadmus.eui.eu/handle/1814/15385. Accessed 17 May 2016.
Stalla-Bourdillon, S. (2012a). Liability exemptions wanted: internet intermediaries' liability under UK law. *Journal of International Commercial Law and Technology, 7*, 289.
Stalla-Bourdillon, S. (2012b). Sometimes one is not enough! Securing freedom of expression, encouraging private regulation, or subsidizing internet intermediaries or all three at the same time: The dilemma of internet intermediaries' liability. *Journal of International Commercial Law and Technology, 7*(2), 154.
Stalla-Bourdillon, S. (2013). Online monitoring, filtering, blocking.... What is the difference? Where to draw the line? *Computer Law & Security Review, 29*(6), 702–712.
Stalla-Bourdillon, S.(2015). *Google: A data controller as well as an intermediary service provider? Does this make sense? Who cares*? Blog: https://inforrm.wordpress.com/2015/2004/2005/google-a-data-controller-as-well-as-an-intermediary-service-provider-does-this-make-sense-who-cares-sophie-stalla-bourdillon/. Accessed 17 May 2016.
Stalla-Bourdillon, S. (2016). *The EU commission and its official communication on online platforms: Is the e-commerce directive being attacked by the back door*? Blog: https://peepbeep.wordpress.com/2016/06/04/the-eu-commission-and-its-official-communication-on-online-platforms-is-the-e-commerce-directive-being-attacked-by-the-back-door/. Accessed 9 July 2016.
Urban, Jennifer M., Joe Karaganis, & Brianna L. Schofield. (2016). *Notice and takedown in everyday practice*. UC Berkeley Public Law Research Paper No. 2755628. Available at SSRN: http://ssrn.com/abstract=2755628. Accessed 17 May 2016.
Van Eecke, P., & Ooms, B. (2007). ISP liability and the e-commerce directive: A growing trend toward greater responsibility for ISPs. *Journal of Internet Law, 11*, 3–9.
Van Eecke, P. (2011). Online serive providers and liability: A plea for a balanced appraoch. *Common Market Law Review, 48*, 1455–1502.
Zencovich, Z. (1999). I rapporti fra responsabilità civile e responsabilità penale nelle comunicazioni su Internet. *Diritto dell'informazione e dell'informatica, 6*, 1049–1057.
Zeno-Zencovich, V. (1998). La pretesa estensione alla telematica del regime della stampa: note critiche. *Diritto dell'informazione e dell'informatica, 1*, 15–28.

Chapter 16
Towards Fostering Compliance by Design; Drawing Designers into the Regulatory Frame

Ewa Luger and Michael Golembewski

Abstract We are seeing the rise of a new type of socio-technical infrastructure. The systems with which we increasingly interact no longer exhibit the characteristics of the systems used even a decade ago. Today's technologies have an ambient quality, are embedded in our most intimate spheres, are data-driven, and exhibit increasing agency in our lives. The quality of our interactional experiences therefore depends upon the quality and quantity of our data and, as such, data has become a recognised global asset. With this heightened economic interest has come a concurrent review of regulation in the sphere of data protection, as the law moves towards mandating privacy by design and by default. However, for this to become a reality requires not only more inclusive regulatory practices, but also the engagement of those responsible for the formation of such systems – designers and systems architects. This raises several questions, principal of which being how one might engage these communities within the emerging discourse. This paper begins to address such issues by exploring a recognised approach to fostering creative thought within design, that of ideation, and suggests a way by which such a methodology might begin the job of drawing designers into the regulatory frame.

16.1 Introduction

Human data, drawn from both public and more intimate spheres of life, has become an asset traded in new and emerging markets. Increasingly we 'pay' for our online services through exposure and commodification of our previously private selves. Online Service Providers, having only had access to clearly parametrised user information, are now engaging us in on-going and only partially visible data-driven

---

E. Luger (✉)
University of Edinburgh, Edinburgh, UK
e-mail: eluger@exseed.ed.ac.uk

M. Golembewski
Microsoft Research, UK
e-mail: t-michgo@microsoft.com

M. Taddeo, L. Floridi (eds.), *The Responsibilities of Online Service Providers*,
Law, Governance and Technology Series 31, DOI 10.1007/978-3-319-47852-4_16

dialogues. This asymmetric and bidirectional exchange relies upon increasingly granular and aggregate user data to inform the generation of algorithmically defined versions of our 'selves'. As the popular term 'internet of things' implies, even the mundane artefacts of our day-to-day existence will have the capacity to contribute to this data-rich profiling, thus allowing for more contextual services.

Such developments have contributed to a visible shift in the regulatory climate. Currently, the control and processing of human data is broadly legitimised on the grounds of necessity, legitimate interest or subject consent. It is effectively dealt with after the fact of design itself. However the proposed EU General Data Protection Regulations potentially reset the governance of human data in a series of key ways. Principal amongst these are that data protection (DP) will be mandated 'by design and by default' through the concept of Privacy by Design (PbD). According to the UK Information Commissioners Office, taking a PbD approach brings with it a number of benefits. These include (a) early stage privacy issue identification, (b) increased organisational awareness of privacy and data protection issues, (c) the reduced likelihood that the resulting actions will be privacy intrusive and negatively impactful, and (d) the increased likelihood that organisations will meet their legal obligations (ICO). It also (e) locates the user, and their interests, as central to the design of IT systems and business practices (Cavoukian 2011).

This notion both shifts the locus of responsibility away from the user and semantically hardwires data protection to the processes underpinning systems architecture and design. In so doing, it not only explicitly invokes the role of designer within the regulatory frame, but also broadens the notion of regulation to formally include such non-state actors. In practice, however, there is little by way of collaborative interaction, particularly at the early stages of design when data protection principles might be meaningfully encoded within systems architecture.

At first blush, it might seem that the disciplines of Design and Law seem reasonably aligned as both locate the human at the centre of their practice; whilst the law enshrines and protects human rights, designers employ practice-led human-centred techniques to shape system functionality and form. Instead, designers create systems, lawyers ensure that those systems comply with current regulation, and this siloed arrangement has thus far required little in the way of common ground. Too often, as designers develop new systems and services, privacy concerns are sidelined, and left to be addressed at the end of a development cycle. The creators of systems might go through multiple iterations of conceptual development, refinement, and revision, and systems might reach a state of near completion before privacy issues are noted, and before solutions are incorporated into an overall design. This late integration of privacy matters within the system design process is, to our view, somewhat problematic. Addressing these concerns at the end of a design cycle leaves the creators of a system with little time or agency to manoeuvre, and leads to a situation where potential privacy problems are addressed – if at all – as afterthoughts, with inelegant solutions and imperfect implementations bolted on to a mostly-complete system design.

If, in contrast, designers consider legal and conceptual privacy issues at the formative stages of the design of a system, a number of advantages arise. Compliance issues related to a systems' architecture or function are much more easily avoided if addressed before development work begins, and before system functionalities are codified, refined, and 'set in stone'. Equally importantly, means of providing the user of a system with clear understandings of how these systems employ their data, and why the use of sensitive information is needed, can be included within a design from its inception. We hold that designers who begin to engage with privacy matters from the very start of a design cycle – during the ideation and concept generation activity that grounds much of the subsequent design and development of a system – are afforded the best chance to create systems featuring elegant solutions to privacy concerns, and of employing and developing graceful means of maximising system functionality whilst still preserving the privacy of users. Yet at the moment, engagement with privacy concerns at the ideation and formative stages of a designs' creation is somewhat rare. However, if data protection is to be a 'by default' consideration in the design process, some efforts are required to foster greater cross-disciplinary understanding.

This begs the question: what mechanisms or instruments might stimulate greater engagement, particularly at the early stages of the design cycle? One approach, familiar to the creative design community, is that of ideation cards. With this as a foundation, we have developed and evaluated a deck of such cards intended to raise issues of data protection and privacy at the earliest creative stages of systems design. Whilst such cards will of course not create a raft of data protection experts, they will at least sensitise designers to existing regulation and begin the journey towards more compliant online services by design and by default. This paper outlines an exploration of the key elements of this project in more depth through a description of the broader problems space, an overview of what is meant by 'privacy by design', and an overview of the relationship between regulation and design. It then outlines the notion of ideation as a bridging mechanism and describes development and use of the data protection cards in more detail as one solution to how one might motivate and operationalise privacy within the design communities.

16.2 Data and the Porosity of Our Private Sphere

The emerging class of pervasive systems not only amplifies existing concerns over the use of personal data but also raises a host of new challenges. As technologies are woven ever closer into our daily lives, our interactions with them have become increasingly personal and intimate. As such, the resulting data has come to surface aspects of our selves, which might previously have remained private or hidden. Online service providers capture and analyse our routine endeavours to better personalise the interactive experience, and mobile and ubiquitous systems use a host of sensed information to realise a model of embedded, context-driven interaction. In order to provide such services, systems must capture, analyse and exploit an extensive and ever-growing range of human data but the resulting experiences, whilst arguably richer, bring with them a district trade-off between privacy and convenience, raising concerns related to privacy and consent in the collection of personal data (Langheinrich 2002; Luger and Rodden 2013; Palen and Dourish 2003).

Whilst once, such systems were realised only within experimental research settings, we are now living with them 'in the wild'. They are interwoven and embedded within the fabric of our everyday lives, supporting our routine endeavours in ways that render their presence seemingly 'unremarkable' (Tolmie et al. 2002). At the same time, our human data, drawn from both public and more intimate spheres of life, has become an asset, traded in new and emerging markets where we are the product; effectively 'paying' service providers by exposing our private selves (Rauhofer 2013). Human Data are generated, sorted, and distributed across global boundaries, at unprecedented speed and with historically limited checks and balances, compounded by the heightened pace of technological innovation and convergence of systems with different functionalities.

This trading of human data, whilst unobserved for some time, has now attracted not only media attention but also public concern. In out post-Snowdon world, data breaches have become securitised and, as such, have risen in their media value. This has resulted in data protection cases receiving the highest levels of media coverage, as evidenced by the international interest in the recent fragmentation of Safe Harbour, a long-standing transatlantic data-transfer agreement that allowed companies to transfer citizen data from the European Union to servers located in the United States. Such transfers, however, also allowed data to become subject to access by national intelligence services and, after a legal case between privacy activist Max Schrems versus the Irish Data Protection Commissioner, they were found to have run counter to the legal requirements of the European Union (Ballard 2015).

This rising concern is not simply a metanarrative at the international level. There is a growing understanding that with the rise of data-driven systems comes the likelihood that companies can surface highly personal life events. Most notoriously, evidence has shown that companies can identify specific life events such as pregnancy (Duhigg 2012) or divorce (Burkeman 2011), and even personal characteristics such as sexual orientation, religious beliefs, political views, personality traits or use of addictive substances (Kosinski et al. 2012) can be predicted with acceptable tolerances of certainty, often without the knowledge or consent of the user. Whilst such concerns have led some service providers to seek to make their use of data more visible to users, for example Google's Transparency Report and detailed description of collected data types seen in Microsoft's Windows 10 privacy policy, these moves often raise more concerns than they allay (Merriman 2014). This is particularly true where the purposes are obscure and where there are no immediate, clear, or simple means of recourse for the data subject.

As this climate of concern continues to build, the rules as to how we might handle human data are being reset in a series of key ways, principally articulated through the upcoming EU General Data Protection Regulations, which for the first time will see a harmonisation of data protection regulations across all EU member states. Whilst the 1995 EU data protection regulations saw fairly limited subject rights, recent developments have seen this status change. The past decade, for example, has seen Europe begin to take more seriously the notion of online consent. Manifest through EU directive, this resulted in what has commonly become know as the 'EU cookie law' (2011). More recently we have seen the rise of legal narratives surrounding user control of data, most notably articulated through the 'right to be forgotten' ruling (2014), a move that now allows EU citizens to demand their personal data be removed, or made inaccessible, via a search engine under certain conditions; namely, that it is 'inaccurate, inadequate, irrelevant or excessive for the purposes of data processing' (European Commission). One further such change, core to the proposed EU General Data Protection Regulation (GDPR) is that user privacy and data protection will be required by design and by default, meaning that the locus of responsibility is moved away from the user and towards the service provider, clearly implicating the process by which systems are designed.

16.2.1 So, What Is Privacy by Design?

PbD is defined by the Information Commissioner Office as "an approach to projects that promotes privacy and data protection compliance from the start" (Information Commissioner's Office Website) including both IT systems and business practices. Whilst there has been much discussion as to its articulation, it is generally agreed to be predicated upon seven principles, as outlined by Cavoukian (2011). Specifically, these are (1) that such design is proactive rather than reactive, and preventative rather than remedial. From this perspective, PbD "begins with an explicit recognition of the value and benefits of proactively adopting strong privacy practices, early and consistently"(Cavoukian 2011: 1), (2) that privacy is embedded as the default rather than being engineered in after the fact, (3) that privacy is not a 'bolt on' solution, but embedded into the design and architecture of IT systems and business practices, (4) that the approach is positive sum, meaning that the user should be allowed full system functionality, without having to make unnecessary trade-offs, (5) that there should be end-to-end security ensuring that from the first piece of data collected, and throughout the entire lifecycle of the data, strong security measures are in place, (6) that the component parts of the business practice or technology remain visible, transparent and open to interrogation in order that one can confirm that they are operating in accordance with the stated promises and objectives, and (7) that above all, the interests of the individual should remain uppermost, including measures such as strong privacy defaults, appropriate notice, and the development of empowering user-centric options. This final tenet in particular draws attention to the notions of consent, accuracy, access, and compliance (Cavoukian 2011).

From this definition it is clear that consideration of the individual, or data subject, should be central to ensuring the development of privacy enhancing systems. In particular, principle 7 (respect for user privacy) extends beyond the reach of more traditional security practices; where security is 'done to' a user, to embrace consideration of the explicit interests and needs of users, implicitly recognising that individuals "have the greatest vested interest in management of their own personal data". This push towards user-centric approaches to privacy speaks clearly to notions of human centred design, where construction of the system is tightly coupled to user expectation, experience, and requirements (Preece et al. 2015). In turn, this begs the question as to whether the approaches commonly used within design practice might also transfer to support the PbD agenda.

Whilst this seems a reasonable conclusion, there are several tensions here. Most notable is that from a PbD perspective, privacy is viewed not as a function in competition with other legitimate interests, but as a means of system optimisation. This perspective is somewhat counter to current approaches to systems design, where privacy and regulatory considerations are often, if not always, considered after the fact of design itself (ICO). Privacy considerations, therefore, are seen as constraints to creative design in that they fundamentally limit the purposes to which data can be put, seek to make clear how, when and why data is being collected, insist upon explication to (and agreement from) the user, highlight the need for user control over the data generated through course of system use, and seek to both minimise data collection and limit data sharing and exposure as a default. In contrast, when considering the development of ambient, networked technologies, the preferred tendency amongst online service providers has, thus far, been gathering of the broadest range of data to improve products and services, ideally in ways not detectable to the user in order to create a seamless experience.

Unfortunately, developers and designers are "often unaware of statutory provisions that regulate human interactions" (Oberle et al. 2012) and currently may not see designing-in privacy as something that falls within their remit (Luger and Rodden 2014). Nevertheless the growing focus of regulators, upon information technology governance, means that this position is increasingly untenable as designers and developers are already being strongly advised by regulators to engage with the forthcoming changes surrounding issues of human data and consent and to "start thinking now as to how you gather and document this" (Smith 2013). In reality, however, encouraging all those implicated in the design process to engage with policy and legislation presents a real challenge. This is a position often reinforced by a presumption that the risk-averse nature of regulation inherently restricts designers' ability to do innovative research and development (Luger et al. 2015). The nature of policy and regulation discourse is not immediately comprehensible outside of the legal profession, and this difficulty of access has meant that the communications and transferrals of ideas between law and design have been minimal.

16.3 Regulation and Design

Despite the disciplinary disconnect, technology development does not occur in a vacuum. It is shaped and guided through complex governance and regulatory infrastructures, both formal and informal. Forms such as law, markets, norms, architecture (Lessig 2006), public perceptions and morality, and timely incentives deriving from markets and regulation (Jaffe et al. 2005) sculpt the systems we release into the world. Whilst technology remained 'in the box', and researchers in the lab, such dynamics saw relatively little change, allowing a broad reliance upon minimal (though legally admissible) adherence to regulation (Luger and Rodden 2013).

However, the systems that were once the stuff of imagination are now firmly in the marketplace. They are networked, always on, ambient, and often take the form of traditional non-networked artefacts, meaning that users are likely to apply ineffective cognitive models when using the artefact and can fail to consider the implications of data exposure in these new contexts. Issues such as the implication of third parties through data, the social stratification that can result from large-scale data analysis, and the publicity of our intimate endeavours were all, at varying points, predicted as issues that would impact future policy (Poh and McDuff 2010).

Over a decade later, it is clear that these predictions have not only come to pass, but have in some cases been underestimated, as perceived impacts are now seen to some extent within all spheres of life (Friedewald and Raabe 2011). Smart environments, for example, rely on implicit interaction and data driven inference. This has resulted in systems that are (a) modular and networked, (b) embedded in everyday objects, (c) offer always-on responsive information services that are adaptive to context, and (d) exhibit agency in the processing of repetitive tasks; characteristics that in effect distance the user from the data they produce. These developments have both enabled and are driven by the realisation that personal data is a new kind of asset (World Economic Forum 2011). Whilst use of data as a means to raise revenue is not in itself new (Lake and Crowther 2013), the extension of data collection into all spheres of life (Arvidsson 2002) has raised unprecedented opportunities for monetisation (World Economic Forum 2011). Consequently, we are seeing the emergence of companies that "collect and mine vast quantities of personal data, public and private, about us and our activities" (Haddadi et al. 2013) in order to turn that information into knowledge, firm in the belief that more data is better. Now, in a world rich with data, economic drivers are pushing beyond volume towards the need for more sophisticated algorithms and predictive models (Guazzelli et al. 2009); the locus of value being not in the scale of data but in the point at which such data is turned into information.

These new models not only enable market responsiveness (Woods 2013), but also increasingly allow inference within most areas of life, both public and private, to the extent that "algorithms are increasingly determining our collective futures" (Hickman 2013). The heightened availability of this information as a resource and its classification as an asset, essentially mean it is drawn upon by services not responsible for its initial collection, and often for purposes other than those originally proposed. In this way, and largely unintentionally, users become incognisant 'prosumers' (Hemetsberger 2004); co-producers of a product sold on their behalf and without their explicit control stimulating a call for greater regulation in this sphere. This constantly shifting data landscape is problematic not only because of the changing nature of technology, but also due to the means by which data is collected, used and shared beyond the initial context of collection.

The challenge now is how to regulate such diverse ecosystems. Philip Selznick's definition of regulation as "the sustained and focused control exercised by a public authority over activities valued by the community" (Baldwin et al. 2010: 12), whilst seminal, has been found to be somewhat lacking, particularly when one looks to the Internet. When we speak of online systems, regulation is better understood as aligned to political science notions of governance, meaning "different forms of deliberate collective action in matters of public interest, i.e. authoritative political intervention, private self-regulation, and cooperation between public and private actors" (Mayntz 2009). One such set of actors are those that constitute design communities. According to Yeung (2014), "design can be usefully understood as an instrument for implementing regulatory goals" (Yeung 2014: 1) such as data protection or privacy and, if we broaden our definition of regulation to that posited by Julia Black, that of "a process involving the sustained and focussed attempt to alter the behaviour of others according to defined standards or purposes with the intention of producing a broadly defined outcome or outcomes", then we are approaching a position that recognises the work of systems design as something akin to an instrumentation of regulation. However even this, argues Yeung, does not go far enough; we should instead take the position that regulation encompasses not only approaches to design that seek to alter behaviour, but recognise all 'design-based instruments and techniques' within the broad church of regulatory inquiry. Indeed, one might argue that design has long served as a regulatory instrument. It is possible to identify a distinct tradition of attempts to design systems which promote the goals of social policy through an alteration of the conditions influencing an individual's choice or power to act through the traditional policy instruments of command (e.g. prohibition of behaviour), competition (e.g. taxation), communication (e.g. education or awareness raising) or consensus; for example engagement of the public through benefits associated with particular behaviours, as well as more contemporary instruments such as nudging behaviour through architectural strategies such as choice architecture (Yeung 2014).

In the same way as Lessig (2006) argued that code could be law, so the form that technology takes can have an effect (intentional or otherwise) on human endeavour. Designers might, for example, 'nudge' users towards particular types of practice through design (Thaler and Sunstein 2008) and, in this way, one might consider technology as a 'regulatory tool' in itself (Brownsword and Yeung 2008). From this perspective, designers not only create systems to meet explicit regulatory purposes, such as for example criminal justice, but also implicitly regulate by affecting users through systems which "seek to, or have the effect of, shaping behavior" (Brownsword and Yeung 2008). However, if we look to current practice, there are three identifiable 'modalities of design' in support of regulation; those that (a) encourage behavioural change, (b) ameliorate the effects of harm-generating behaviour, or (c) totally prevent harmful behaviour. Whilst such approaches regulate human behaviour, they can result in a concurrent reduction of human agency, having a negative effect on accountability, transparency and participation (Brownsword and Yeung 2009).

From this perspective, designers could already be considered regulators. Even before we explicitly consider what that might mean, the mere notion of 'constructing' a user experience implicitly invokes notions of bespoke systems that both reflect and shape human behaviour, curate events in a particular way, and consider the means by which intended outcomes might be achieved. It is not therefore an unimaginable leap to consider how privacy architecture might be reframed as a component of user experience design. Despite this, preservation of user privacy is not reflected within mainstream design heuristics. Whilst designers see the value of privacy and seek to enhance interactional experiences for the user, this stops short of taking responsibility for privacy enhancing features, which are for the most part considered as ‘bolt-on’ solutions or factors thought of only after the fact of design, often described as Privacy Enhancing Technologies (PETs); an area of siloed focus. The benefits of such post hoc privacy solutions, whilst predictably limited, have yet to affect greater consideration of privacy within the design process.

16.4 Designers, Design Practice, and the Need for Flexible Instruments

When we speak of ‘designers’, we are referring to a broad range of potentially very divergent categories of practitioner. Depending on the specificities of the project at hand and the skills needed to realise that project, the designers tasked with its creation and implementation might have backgrounds in computer science, engineering, psychology, business, or else in any of a host of studio-centric design traditions – visual design, interaction design, user interface / user experience design, or product design engineering, to name a few. Most likely, in projects of any size or complexity, we will encounter groups of designers from a range of the aforementioned types of practice working in tandem. So, when we think about designers as a group, we must acknowledge that we are dealing with a collection of people with a heterogeneous set of disciplinary backgrounds.

This diversity of disciplinary background leads to a corresponding diversity of working processes and practices – of techniques for moving design projects from conception to completion. Each distinct tradition of practice has ‘its own way of doing things’, its own suggestions concerning how to formulate knowledge, discuss ideas, and implement design concepts, and its own models describing how design work moves from concept to execution. Some models of design practice are rigid, codified, and grounded in objectively-focused engineering practices or in the scientific method, such as those of Simon (Simon 1994) or Parnas and Clements (Parnas and Clements 1986); others, oftentimes those emerging from a studio tradition, are more methodologically flexible. We also encounter models of design practice that are linked to specific industries, such as architecture (Royal Institute of British Architects 1980) or application development (Wood and Silver 1995). At times, specific organisations or companies develop their own models of design, either for in-house adoption, as demonstrated by IBM (Vredenburg 2003) or as a means of quality assurance for other organisations, as in the case of the international standards organisation (ISO 9241-210, 2010).

Designers might adopt any of a huge array of models as a means of shaping their workflows and activities, or – as is more likely – adapt and combine a range of process models and techniques to suit the purposes of the tasks at hand. In addition, many of these models describe ideal practices and means of understanding design practice and activity, but are less able to describe the messy and sometimes chaotic and practically constrained activities found within real-world design and development environments. As Parnas and Clements note, in their aptly named 'A Rational Design Process: How and Why to Fake It':

> Even if we could master all of the details needed, all but the most trivial projects are subject to change for external reasons. (…) The resulting design is not one that would have been produced by a rational design process. (Parnas and Clements 1986, pp.251)

The wide range of disciplinary backgrounds found within systems design and the variations in working practices that this implies, taken in combination with broad variations of working practices as necessitated by both industry vertical and specific organizational practices, and also the inherent chaotic and context-dependent nature of real-world design activity, to some extent precludes the development of a rigidly defined methodological instrument for general use by all systems designers. A specific methodological practice that resonates effectively with working processes might be developed, perhaps, for use by a single organization, or even by designers employing a specific and highly codified workflow for systems design, but the breadth and variety of practice types encompassed by the phrase 'systems design' precludes a specific and highly targeted solution.

When we set out to develop instruments to intervene in and modify existing design practices, then, we must acknowledge that we are not attempting to create change in a single, codified workflow or method of working, but rather aim to impact an extremely broad range of potential practices and practical activities. We must take note of the fact that the target users of our instruments – the 'designers' we hope to engage with – will already have their own ways of doing things, and that the implements we create as a means of helping them to consider broad privacy concerns within their work must be open-ended, flexible, and able to be employed in an ad-hoc manner within a wide variety of professional contexts.

16.5 The Use of Ideation-Based Instruments with Designers

Ultimately, we resolved to develop a set of ideation cards catering to privacy concerns, and to frame this deck of cards as an appropriate instrument to sensitise the widest range of designers to these issues. Card-based ideation tools are a relatively common form of tool intended to support creative practice. By card based ideation tools, we refer to card-centric toolkits designed to directly influence the outcome of creative activities. It is worth noting that these are distinct from card sorting methods, which tend to be employed in order to gather user data to be interpreted (Kuniavsky 2003). Here, we briefly examine a range of card-based ideation tools intended to directly influence design practices, which we used as inspiration for the development of our own instrument.

The most commonly recognised ideation instruments are IDEO's *Method Cards.* These consist of decks of cards describing methods that might be used to engage in user-centric design practice (IDEO 2010), with each card describing a different design method that might be employed by a designer. IDEO's Method Card decks are organized in suits; in standard playing cards, the suits indicate categories that are relevant to gameplay. With *Method Cards*, suits denote broad categories relevant to different aspects of user inquiry. The specific uses for *Method Cards* were purposely left open-ended, in order to cater to a wide range of practitioners. Designers, informed by the physical affordances of playing cards (random dealing, shuffling, spreading out, etc...), were encouraged to develop and employ their own open-ended uses of the cards within their practices. *Oblique Strategies* (Taylor 2010) cards are also an open ended aid for creative practice, developed in order to promote lateral thinking during studio-centric activities. The cards in *Oblique Strategies* decks each contain obscure, cryptic, but sometimes quite effective suggestions for addressing general design problems. As various stages in a project, cards could be selected randomly and used as a means of considering various aspects the work-in-progress, as a means of both providing deeper understandings of the work in progress, and also as a means of informing subsequent design work. In their first incarnation, *Oblique Strategies* were bespoke card decks designed by Brian Eno and Peter Schmidt in order to aid their own practices; as new ideas and ways of thinking were discovered, they were used as the basis of cards, which were subsequently added to the deck. Subsequent published versions, however, fixed the content of the cards; to other users, the cards became manufactured commodities rather than bespoke, custom-tailored critical instruments.

Dan Lockton's *Design With Intent* toolkit (Lockton 2010) is a card-based tool framework that centers around a deck of cards used to pose questions about various design issues relating to behavior change; like Method Cards, they are organised into a series of suits, with each suit tailored to reflect the perspectives of a different field of practice. These cards are notable as they exist with the stated intent of aiding in the ideation process – they are instruments intended to help inform the design process at formative stages. Golembewski and Selbys' *Ideation Decks* (Golembewski and Selby 2010) describes a technique for the creation of bespoke decks of cards as a means of supporting project-specific design ideation. Using this technique, the creator of the cards first breaks down a given project space down into a range of relevant design factors deemed to be of importance to the overall design. These factors are positioned as a range of suits in a deck of cards – an ideation deck might have suits representing 'users', 'technologies', 'accessibility', and the like. Specific examples relevant to each of these broad design concerns are composed, with each example used as the basis for an individual card in the deck – a 'users' suit might contain cards for 'children', 'families', 'people of low literacy', or 'consumers', with each card representing a particular type of user. These cards are then designed and fabricated, resulting in a unique deck describing multiple factors of potential concern to a specific project space. When cards of various suits are viewed in combination, they provide the designers with a means of analyzing the design possibilities and potentials within a particular design space, and also an instrument usable for both the generation and evaluation of emergent design concepts.

16.6 Designing for Privacy: Using Ideation Cards to Articulate Privacy Concerns

It is clear therefore that use of ideation cards as a methodological design instrument is well established as a means to "help define constrained design problems within a broader overall problem space" (Golembewski and Selby 2010). However, use of cards more broadly has increasingly seen application beyond traditional designer practice to become the creative stimulus within much broader problem spaces such as (a) the sphere of human values in the design process (Friedman and Hendry 2012), (b) to support communication within and between families (Mackay 2004), (c) to encourage participants to think about security threats (Security Cards), (d) to stimulate creative methods within design practice by supporting design for diversity (IDEO), and (e) to support discussions around online privacy (Barnard-Wills 2015). Despite their different emphases, fundamental to all of these applications is the notion of cards as a mechanism to stimulate creative thought, often around subjects previously unfamiliar to the participants. Reflective of this tradition, the focus of our research was to design and test a creative method that raised awareness, amongst designers, of the intersection between their work and the proposed EU DP Legal framework. Law and policy instruments are often drafted in disciplinary-specific language, rendering them inaccessible to anyone other than lawyers and regulators. However, there is an increasing recognition within the legal community that, through their design decisions, technologists can play a key role in data protection. Our intention was to sensitise designers to this notion.

16.6.1 Developing and Testing a Data Protection Ideation Deck

In order to reflect existing practice, our data protection cards (Luger et al. 2015) included a range of features from previous card-based ideation approaches drawing most prominently from the Value Sensitive Design cards developed by Friedman and Hendry (2012) and Golembewski and Selby (2010), sharing a suit structure with both *Ideation Decks* and *Design with Intent* (Lockton 2010). The means of generating an orthogonal breakdown of design properties to consider was identical, with our cards differing in that they were created as instruments for general use by a wide pool of designers, rather than as bespoke instruments for a small group to employ. On this basis we developed a deck of four suits of cards: (a) 'system' cards that framed the activity by proposing a design brief for known systems including the function of the system, the type of data it collected, the actors who would have access to the data and the data controller, (b) 'user' cards that outlined the type of data subject, (c) 'constraints' cards, which detailed factors known to constrain the design of systems, and (d) 'regulation' cards, which presented a pared down overview of emerging data protection regulations drawn from advice from the Information Commissioners Office and validated through a survey and discussions with legal experts specialising in the sphere of EU regulation.

As with *Method Cards* (IDEO 2010) and *Oblique Strategies* (Taylor 2010) our data protection cards were designed intentionally to a more broad set of uses and contexts. The ideation exercise we developed and conducted for our preliminary experiments was relatively structured, but its constituent data protection cards could just as easily be framed as implements for casual reflection throughout the design process, or as material aids for the impromptu critical evaluation of works-in-progress. Friedman & Hendry and Mackay et al, both emphasise the importance of time in the ideation process; e.g. authors used a 3 min sand *timer "to both symbolise and facilitate the possibility for meaningful use in a brief amount of time"* [13 p.1146]. In order to focus participant attention, our activity was strictly timed to 30 min and participants were told when to turn each card. For further information on method, a fuller account of the study is presented elsewhere (Luger et al. 2015). Our initial proof-of-concept showed the approach not only successful in terms of stimulating interest in data protection issues amongst designers, but also that by employing an ideation strategy, as opposed to simply using the cards as flashcards, participants were stimulated to more creatively consider how privacy might be embedded within the fundamental design of the artefact (Luger et al. 2015). In this way, the technique moved collective thinking beyond how do 'fix' data protection 'problems' and instead invigorated discussions of how components of data protection might be considered and designed as a component of the project brief. Illustrative of the former perspective, one systems architect noted *"when you're designing a system, at least for me, you always think of regulation as an afterthought. So, if I get what I want then I see how do I protect the user afterwards"* (Luger et al. 2015).

16.6.2 Trialing the Cards with Designers

It is this perspective which in many cases prohibits designer engagement with privacy discourses. It is not only this, however. Our initial studies in this area found that knowledge of data protection was highly limited amongst participating designers. It was, for them, a practical rather than creative concern and as such their knowledge was limited in that it was drawn either from their time at university or school and therefore partial and vaguely remembered, or had come about as a direct result of their prior work. In the majority of cases, especially for the type of designer who specialized in systems architecture, compliance with data protection was very much seen as limiting system functionality and, in this way, was seen as a factor which might limit their creativity and the potential of the system. On this basis, one might be inclined to think that privacy as a concept was not valued. However, this was entirely not the case. Designers were very concerned about the user experience and some considerations of privacy, as an abstract concept, was emergent within many of the discussions. This abstraction, however, resulting in there being something of a disconnection between the notion of privacy or data protection of the user and regulation; the latter of which being cast very much in a pejorative light.

However, that was not to say that designers felt data protection oversight to be unnecessary. All participants recognised the need for data protection. Rather, their perspective stemmed from concerns over the level of legal knowledge they would be expected to bring to bear and, as such, they did not currently see themselves as instrumental within the wider regulatory system. Instead, 'regulation' was cast very much as an exogenous force against which system specifications were set. From this perspective, the current model of post hoc compliance was seen as the only way to avoid limiting creative design and, with this perspective, came the belief that regulation necessarily meant compromise. Regulation, and more broadly law, was seen as something to be respected through systems design but its practical implementation was cast as the province of legal professionals and data protection or privacy specialists. Despite this, those from backgrounds more aligned to the creative industries found the cards an effective way to stimulate creative though whereas those with backgrounds more closely aligned to.

By framing data protection concerns within the early stages of design, we were able to some extent present a challenge to this perspective. By framing such considerations within the creative process, albeit artificially, we saw clear evidence of our participants reconsidering some of the design choices they had made earlier in the activity. In this way introduction of the legal cards acted as an instrument which enabled designers to reflect upon and evaluate their proposed systems.

16.7 Conclusions

The matter of how 'privacy by design' might be achieved in practice is of pressing concern. Whilst the rhetoric surrounding the subject becomes increasingly well-developed, the matter of how one might stimulate a changed culture to support practice is less clear. The use of ideation in design is not a new idea, such practice is well-established within design culture and is used across a range of disciplinary areas, but the extent to which this technique engages designers from non-creative backgrounds is under-explored. Using such methods to stimulate evaluation of design processes is a relatively novel application and, with data protection and privacy legislation as a focus, we hope to be able to test the extent to which such a technique might engender more creative approaches to legal tenets, and afford more reflective practice amongst design teams at the very earliest stages of their process. It is our hope that this will become just one instrument in a wider toolkit available to organisations, but the broader challenge – that of ensuring privacy by design becomes commonplace – is something we as a community have only begun to explore.

References

Arvidsson, A. (2002). On the "Pre-history of the Panoptic sort": Mobility in market research. *Surveillance & Society, 1*(4), 456–474.

Baldwin, R., Cave, M., & Lodge, M. (Eds.). (2010). *The Oxford handbook of regulation*. Oxford: Oxford University Press.

Ballard, B. (2015). The safe harbour legal battle: Everything you need to know. *ITProProtal*. http://www.itproportal.com/2015/10/21/the-safe-harbour-legal-battle-everything-you-need-to-know/. Accessed 23 Oct 2015.

Barnard-Wills, D. (2015). *Privacy game*. http://surveillantidentity.blogspot.co.uk/p/privacy-card-game.html. Accessed 15 Oct 2015.

Brownsword, R., & Yeung, K. (Eds.). (2008). *Regulating technologies: Legal futures, regulatory frames and technological fixes*. Oxford: Hart Publishing.

Burkeman, O. (2011). SXSW 2011: The Internet is Over. *The Guardian*. http://www.guardian.co.uk/technology/2011/mar/15/sxsw-2011-internet-online. Accessed 15 Oct 2015.

Cavoukian, A. (2006). Information & Privacy Commissioner, Ontario, Canada. Creation of a Global Privacy Standard at www.ipc.on.ca/images/Resources/gps.pdf. Accessed 1 Oct 2015

Cavoukian, A. (2011). *Privacy by design: The 7 Foundational principles*. Implementation and Mapping of Fair Information Practices. https://www.iab.org/wp-content/IAB-uploads/2011/03/fred_carter.pdf. Accessed 1 Oct 2015

Duhigg, C. (2012). How companies learn your secrets. *New York Times*. http://www.nytimes.com/2012/02/19/magazine/shopping-habits.html?_r=0. Accessed 1 Oct 2015.

European Commission Factsheet on the 'Right to be Forgotten' ruling (c-131/12). (2015). Proposal for a Regulation of the European Parliament and of the Council on the protection of individuals with regard to the processing of personal data and on the free movement of such data (General Data Protection Regulation). At http://ec.europa.eu/justice/data-protection/files/factsheets/factsheet_data_protection_en.pdf. Accessed 1 Oct 2015.

Friedewald, M., & Raabe, O. (2011). Ubiquitous computing: An overview of technology impacts. *Telematics and Informatics, 28*(2), 55–65.

Friedman, B., & Hendry, D. (2012). The envisioning cards: A toolkit for catalyzing humanistic and technical imaginations. In Proceedings of CHI'12 ACM.

Golembewski, M., & Selby, M. (2010). Ideation decks: A card-based design ideation tool. In *Proceedings of the 8th ACM conference on designing interactive systems (DIS '10)* (pp. 89–92). New York: ACM. doi:http://dx.doi.org/10.1145/1858171.1858189

Google Transparency Report http://www.google.com/transparencyreport/

Guazzelli, A., Stathatos, K., & Zeller, M. (2009). Efficient deployment of predictive analytics through open standards and cloud computing. *ACM SIGKDD Explorations Newsletter, 11*(1), 32–38.

Gürses, S., Troncoso, C., & Diaz, C. (2014). https://www.cosic.esat.kuleuven.be/publications/article-1542.pdf. Accessed 1 Oct 2015.

Haddadi, H., Mortier, R., MacAuley, D., & Crowcroft, J. (2013). *Human-data interaction* (Cambridge University Working Paper).

Hemetsberger, A. (2004). *When consumers produce on the internet: The relationship between cognitive-affective, socially-based, and behavioral involvement of prosumers* (Working paper). http://flosshub.org/system/files/hemetsberger1.pdf

Hemetsberger, A., & Pieters, R. (2001). *When consumers produce on the internet: An inquiry into motivational sources of contribution to joint-innovation.* In: Ch. Derbaix, et al. (Eds.), Proceedings of the Fourth International Research Seminar on Marketing Communications and Consumer Behavior, La Londe, pp. 274–291.

Hickman, L. (2013). *How algorithms rule the world.* www.theguardian.com, http://www.guardian.co.uk/science/2013/jul/01/how-algorithms-rule-world-nsa. Accessed 2 Oct 2015.

IDEO. (2010). *Method cards – Case studies – IDEO. IDEO | A Design and Innovation Consulting Firm.* IDEO. http://www.ideo.com/work/item/method-cards/. Accessed 1 June 2010.

Information Commissioner's Office. (2015). *Privacy by design.* https://ico.org.uk/for-organisations/guide-to-data-protection/privacy-by-design/. Accessed Oct 2015.

ISO. (2010). 9241-210: Ergonomics of human-system interaction – part 210: human centred design for interactive systems.

Jaffe, A. B., Newell, R. G., & Stavins, R. N. (2005). A tale of two market failures: Technology and environmental policy. *Ecological Economics, 54*(2-3), 164–174.

Kosinski, M., Stillwell, D., & Graepel, T. (2012). Private traits and attributes are predictable from digital records of human behaviour. *PNAS, 110*(15), 5802–5805.

Kuniavsky, M. (2003). *Observing the user experience: A practitioner's guide to user research.* San Francisco: Morgan Kaufmann.

Lake, P., & Crowther, P. (2013). *Concise guide to databases.* London: Springer.

Langheinrich, M. (2002). *A privacy awareness system for Ubiquitous computing environments.* In Lecture Notes in Computer Science 2498 (pp. 237–245). Springer.

Lessig, L. (2006). *Code V2.0. Basic Books.*

Lockton, D. (2010). *Design with Intent | Design with Intent Toolkit 1.0 Now Online. Design with Intent.* At http://architectures.danlockton.co.uk/2010/04/10/design-with-intent-toolkit-1-0-now-online/. Acessed 1 June 2010.

Luger, E., & Rodden, T. (2013). *An informed view on consent for UbiComp.* Proceedings of the UbiComp '13. ACM, pp. 529–538.

Luger. E., & Rodden, T. (2014). *The value of consent: Discussions with designers of ubiquitous computing systems.* PERCOM (Workshops IEEE international conference on pervasive computing and communications. IEEE, pp. 388–393.

Luger, E., Urquhart, L., Rodden, T., & Golembewski, M. (2015). *Playing the legal card: Using ideation cards to raise data protection issues within the design process.* Proceedings CHI '15. ACM, New York, NY, pp. 457–466.

Mackay, W. (2004). *The interactive thread: Exploring methods for multi-disciplinary design.* In Proceedings DIS'04. ACM.

Mayntz, R. (2009). The changing governance of large technical infrastructure systems. In: R. Mayntz (Ed.), *Über Governance.* In-stitutionen und Prozesse politischer Regelung, Schriften aus dem Max-Planck-Institut für Gesellschaftsfor-schung, Campus (pp. 121–150).

Merriman, C. (2014). Microsoft's Windows 10 preview has permission to watch your every move. *The Inquirer.* http://www.theinquirer.net/inquirer/news/2373838/microsofts-windows-10-preview-has-permission-to-watch-your-every-move. Accessed 20 Sept 2015.

Oberle, D., Drefs, F., Wacker, R., Baumann, C., & Raabe, O. (2012). Engineering compliant software: Advising developers by automating legal reasoning. *SCRIPTEd, 9*(3), 280–313.

Palen, L., & Dourish, P. (2003). *Unpacking privacy for a networked world.* In Proceedings CHI'03 (pp. 129–136)

Parnas, D. L., & Clements, P. C. (1986). A rational design process: How and why to fake it. *IEEE Transactions on Software Engineering, 12*(2), 251–257.

Poh, M. Z., McDuff, D. J., & Picard, R. W. (2010). *Non-contact, automated cardiac pulse measurements using video imaging and blind source separation.* Optics Express.

Preece, J., Rogers, Y., & Sharp, H. (2015). *Interaction design: Beyond human- computer interaction.* New York: Wiley.

Privacy Ideation Cards. (2015). At www.designingforprivacy.co.uk. Accessed 7 Oct 2015.

Rauhofer, J. (2013). One step forward, two steps back? Critical observations on the proposed reform of the EU data protection framework. *Journal of Law & Economic Regulation., 6*, 1.

Royal Institute of British Architects. (1980). *RIBA handbook of architectural practice and management* (4th ed. Revised). London: RIBA Publications.
Silver, D., & Wood, J. (1995). *Joint application development* (2nd ed.). New York: Wiley.
Simon, H. (1994). *The sciences of the artificial*. Cambridge, MA: The MIT Press.
Smith, D. (2013). *One small step for EU Parliament could prove one giant leap for data protection*. ICO At. https://iconewsblog.wordpress.com/2013/10/31/one-small-step-for-eu--parliament-could-prove-one-giant-leap-for-data-protection/. Accessed 13 Mar 2014.
Taylor, G. (2010). The Oblique Strategies Web Site. *Oblique Strategies*. http://www.rtqe.net/ObliqueStrategies/. Accessed 1 June 2010.
Thaler, R. H., & Sunstein, C. R. (2008). *Nudge*. New Haven: Yale University Press.
The Security Cards. (2015). At http://securitycards.cs.washington.edu/index.html. Accessed 7 Oct 2015.
Tolmie, P., Pycock, J., Diggins, T., MacLean, A., & Karsenty, A. (2002). Unremarkable computing. In Proceedings CHI' 02 (pp. 399–406). ACM Press.
Vredenburg, K. (2003). Building ease of use into the IBM user experience. *IBM Systems Journal, 42*(4), 517–531.
Woods, D. (2013). Why building a distributed data supply chain is more important than big data. *Forbes*. http://www.forbes.com/sites/danwoods/2013/06/27/why-building-a-distributed-data--supply-chain-is-more-important-than-big-data/. Accessed 5 May 2015.
World Economic Forum. (2011). *Personal data: The emergence of a new asset class*. World Economic Forum.
Yeung, K. (2014). *Designing for regulation*. Kings College Dickinson School of Law Legal Studies Research Paper Series, paper no.2014-2. http://papers.ssrn.com/sol3/papers.cfm?abstract_id=2377254. Accessed 2 Oct 2015.

Part IV
Commentaries

Chapter 17
Does Great Power Come with Great Responsibility? The Need to Talk About Corporate Political Responsibility

Dennis Broeders and Linnet Taylor

Abstract The global information society is shaped by the platforms and innovation of large internet companies such as Google, Facebook, Dropbox, Twitter and Instagram. These Online Service Providers (OSPs) shape our view of the world through the organisation and management of our information. OSPs are becoming part of processes that were previously regarded as the preserve of national governments – such as administration and 'national statistics' – and sometimes acquire a state-like position with regard to information and power. Yet two of the main academic disciplines that analyse the international arena, International Relations theory and the study of international law, both have a blind spot for corporate power, let alone the large Internet companies that shape the global information society. We situate OSPs in the international political domain, arguing that it is necessary to define what Corporate Political Responsibility should consist of in this context, and how it might be exercised.

17.1 Introduction: Does the Power of OSPs Go Unrecognized?

The global information society by and large runs on private, corporate hardware and software and is shaped by the platforms and innovation of large internet companies such as Google, Facebook, Dropbox, Twitter and Instagram. These Online Service Providers (OSPs) shape our view of the world through the organisation and management of our information. They do so on a global scale. If information is power – as the cliché goes – then OSPs must be powerful actors in the international arena.

D. Broeders
Wetenschappelijke Raad voor het Regeringsbeleid, The Hague, Netherlands

Erasmus University Rotterdam, Rotterdam, Netherlands

L. Taylor (✉)
Tilburg Institute for Technology, Law and Society, Tilburg University, Tilburg, Netherlands
e-mail: l.e.m.taylor@uvt.nl

M. Taddeo, L. Floridi (eds.), *The Responsibilities of Online Service Providers*,
Law, Governance and Technology Series 31, DOI 10.1007/978-3-319-47852-4_17

Interestingly enough, two of the main academic disciplines that analyse the international arena, International Relations theory and the study of international law, both have a blind spot for corporate power. Both disciplines are firmly rooted in a Westphalian view of the world and mainly concern themselves with the behaviour of sovereign nation states. Obviously there are exceptions, such as the work of Susan Strange (1988) and Joseph Nye's (2011) thinking on soft power in which corporations have a role to play, but overall transnational corporations do not figure prominently in the study of international politics and law, let alone the large Internet companies that shape the global information society.

Yet, it seems obvious that corporate power is growing, not in the least in the online world, which may not be exempt from the Westphalian order but is at times and places severely at odds with it. The global user base of many OSPs dwarfs those of smaller and larger countries and their influence on their users is significant. Internet security expert, Bruce Schneier (2013) characterises the present day as an age of digital feudalism in which users – in exchange for applications and services – are dependent on OSPs for the de facto (lack of) protection of their privacy and the security and ownership of their data. OSPs (in addition to governments) have become the feudal lords of the information society. Scott's (1998) notion of 'seeing like state' – the bureaucratic view on, and the administration of the nation – is now increasingly paralleled by a corporate big data view of the world that almost amounts to a 'God's eye view' of humanity (Pentland 2011). In both the developed and especially the developing world, Internet and information companies – often in public-private consortia – are laying out the digital infrastructure for communication, identification and parts of administration and population management (Taylor and Broeders 2015). OSPs are becoming part of processes that were previously regarded as the preserve of national governments – such as administration and 'national statistics' – and sometimes acquire a state-like position with regard to information and power. They also transform and dominate sectors that play a political role, but that were traditionally more nationally oriented, such as the news function that in many countries has always been private and has been under pressure of globalisation for a longer time.

Because of their position, OSPs also have a presence in the domain of international politics – whether they actively seek it out or not. The power to organise information and 'speech' – free or otherwise – has public and political elements. Depending on the government and the issue in question that power may align with government interest or may go against it. Both may get corporate actors in political trouble and both require political decisions in which corporate interests, the interests of their user base, political interests – at home and/or abroad – and ethical considerations play a part. In other words, the corporate power of OSPs has political and foreign policy dimensions, as can be shown with three brief examples.

17.2 OSP Power in the Global Information Society

Google entered China in order to capture part of one of the world's biggest and fastest growing markets for Internet search. The price to pay for competing in that market – considered to be too big to miss – was complying with the censorship of the Chinese government. Obviously, for a company like Google that has been flying the flag of ethical business strategy – 'Don't be evil'[1] – dedicated to organising the world's information and making it accessible to all, this was a problematic and publicly criticised step. The tension between ethics and business is obvious. During the 2006 US Congressional hearings Microsoft's associate general counsel Jack Krumholz stated a simple truism in defence of Microsoft's behaviour in China: 'companies who would try to take the moral high ground would be at a competitive disadvantage' (Dann and Haddow 2008: 228). When Google decided to pull out of mainland China in 2010 it was for both corporate and political reasons. At the level of business there were performance setbacks and much more severe competition from domestic competitors that expected (Tan and Tan 2012: 472). Business in China was nowhere near as good as expected. At the political level – and publicly quoted as the reason for the exit – there was a cyber-attack dubbed Operation Aurora against Google and other companies that targeted both their source code and the Gmail accounts of human rights activists. The ensuing public battle between the Chinese government and Google played out in the political arena of free speech versus censorship. Moreover, Google received backing from the US State Department, where Secretary of State Hillary Clinton was actively pursuing a foreign policy agenda on the issue of internet freedom, lifting the fight to the level of a diplomatic conflict between the US and China. Here, the foreign policy interests of the US and Google's agenda temporarily aligned. The US wanted to send a message to both China and to American transnational companies, the message to the latter being that ethics should be part of the national brand (Clinton 2010). This stands as a good example of exercising 'soft power' (Nye 2011) through ethical branding. For Google, it was a way to negotiate the tension between their corporate ethics and the search for profit that had placed them in, and in complicity with, China and return to the moral high ground of ethical business.

That the interests of Silicon Valley and the US government do not always align became clear in the aftermath of the Snowden revelations (see Greenwald 2014). Although it is unclear whether and to what extent large American OSPs such as Microsoft, Yahoo, Google and Facebook were knowingly contributing to the NSA's global mass surveillance efforts, there is no doubt that the international public outcry was bad for their reputations and bad for business. The difference between domestic politics and global business further aggravated the situation when the White House's first line of defence was the argument that 'there was no spying on

[1] In October 2015 Google became the largest subsidiary of a new umbrella company, Alphabet, founded to accommodate the various new businesses that were previously under the umbrella of Google. The new company did not take up 'don't be evil' as a corporate motto, replacing it with 'do the right thing'.

Americans. We don't have a domestic spying program'.[2] To companies that have a global user base and need to compete on international markets for cloud computing, for example, the American voter is not their only concern. Facebook CEO Mark Zuckerberg articulated the sentiment as follows: 'Oh, we only spy on non-Americans. Gee, thanks! We're trying to provide an international service and not get crushed in those places either.'[3] One of the things that happened in the post-Snowden information society is that large OSPs began to pick battles with the American government and with the Western intelligence community (Broeders 2015). OSPs started to encrypt their users' data traffic to prevent easy access for surveillance. Intelligence and security agencies, such as the FBI and GCHQ, protested and demanded privileged access and/or back doors into the data systems, recalling the crypto wars of the 1990s (Van Hoboken and Rubinstein 2014). They also increased their transparency reporting, showing – within the limits of the (national) law and limited to legal requests – data requests from law enforcement worldwide. Microsoft has even become an unlikely champion of end-user privacy as it challenges the US government over the question of whether the data held on Microsoft servers in Ireland fall under Irish law, or whether the US has the right to subpoena data held by a US company regardless of where they keep it. The main point here is not the moral agenda of these companies – the blend of corporate interest and ethics will be weighted heavily toward the former – but the fact that these companies can exercise at least some counter-power in the information society when faced with pressure from government, even when it comes from the intelligence and security agencies.

Sometimes the search for and capture of new markets by OSPs takes on the cloak of philanthropy or the language of international development (Taylor and Broeders 2015). Both Google and Facebook want to 'connect the world to the internet'. Facebook's internet.org aims to do so by extending Internet connectivity, primarily via mobile phones, 'to the two thirds of the world's population that doesn't have it' (internet.org 2014). The project will use 'drones, satellites and lasers to deliver the internet to everyone'. Moreover, Facebook promises mobile subscribers in low- and middle-income countries free data service on their phones when they access a text-only version of Facebook and selected 'partner sites'. Facebook markets this 'Facebook Zero' service (now renamed Free Basics) as a way to enable low-income people in areas with low levels of technological infrastructure to go online in a simpler Internet environment. Even though the used language is that of development and the public good the programme is being rolled out first in populous emerging economies, where the famous 'next billion internet users' are about to come online. Critics in India, the chief battleground for the service where regulators have threatened to ban it on the basis of net neutrality, say that Facebook's move is not

[2] President Obama said this on the Tonight Show with Jay Leno on August 6, 2013.

[3] http://www.thedailybeast.com/articles/2013/09/18/mark-zuckerberg-blasts-obama-on-spying.html

generosity but a land-grab.[4] Other critics argue that it reduces the Internet to a 'walled garden' of tethered partner services, which restricts the possibility for creative use of the web, and for making it impossible for users to use secure connections. In the words of the Electronic Frontier Foundation (2015) 'Internet.org is not neutral, not secure, and not the internet'.

Facebook has positioned itself as a development actor, invoking the discourse of international development in the marketing of its Free Basics initiative by claiming it will address 'social inclusion' and 'the digital divide'.[5] This claim to status as an international benefactor has been used to put pressure on the Indian government: when its regulators threatened to ban the initiative's rollout. Facebook launched an appeal, which, though targeted at Indian users, was not limited to them,[6] to email the Indian government in support of Free Basics. In this case, Facebook acted more like a national state than a firm engaged in capturing a new market, drawing on the language and imagery of bilateral political negotiation (Mark Zuckerberg had been repeatedly photographed in the preceding months debating Indian Premier Modi on development issues, and spoke at the United Nations in 2015 on the right to connectivity[7]), and also on the notion of its 'population' of users whose views on development it was voicing. Zuckerberg's international activities promoting Facebook's campaign for universal connectivity border on diplomacy: to add its next billion members, the company must gain access to airspace for drones and other mechanisms of provision, negotiate (effectively) monopolies on data provision to zero-rate its services for low-income mobile users, and connect to the development and economic agendas of both national governments and multilateral organisations to do so.

These examples all highlight the fact that corporations wield considerable power in both the global and national information societies. Moreover, the dynamic of power and counter-power with government and the fact that the interests of OSPs and governments sometimes align and sometimes oppose gives that power a political character. Shaping the information environment of citizens, and sometimes even shaping and designing those of governments, also sets OSPs politically aside from other transnational corporations such as oil companies or producers of soft drinks.

---

[4] See: http://www.economist.com/news/business-and-finance/21685292-critics-argue-mark-zuckerbergs-generosity-cover-landgrab-facebooks-free-internet

[5] Blog post by Mark Zuckerberg, https://www.facebook.com/zuck/posts/10102033678947881, accessed 10.1.2016

[6] https://www.facebook.com/savefreebasics, accessed 10.1.2016, asks Facebook users to 'Send a message to the Telecom Regulatory Authority of India (TRAI) and tell them that you support Free Basics in India', providing a template email.

[7] http://www.wired.com/2015/09/zuckerberg-to-un-internet-belongs-to-everyone/

17.3 Global Power, No Global Regulation and Limited Accountability

The growing influence and power of OSPs in the international political domain raises the question of what structures are in place to debate and deal with corporate power. Currently, three brands of accountability mechanisms, including some limited regulation, are in place but they all reflect the fact that the international Westphalian order does not recognize corporations as political actors.

The first one packs the biggest punch, but is usually not structured in a formal way. Public opinion mobilised through users, citizens, NGOs and the press can seriously damage the reputation of a company and make it reverse decisions and policies. Especially for those OSPs that thrive on having large numbers of users to sustain their business model – analysing and marketing the data of their users – are vulnerable to reputation damage. Facebook has had to backtrack on changes in its privacy policy more than once. Branding yourself as an ethical company – as Google does – increases the vulnerability. Also the mobilisation of public opinion through the activities of NGOs, and sometimes even challenges in court, can put pressure on companies and may alarm shareholders. Even though public opinion can be very influential it is also fickle: it is not easy to create a perfect storm of public resistance, let alone sustain it. Moreover, public opinion does not deal well with 'salami tactics', in which changes are implemented in small incremental steps that are hard to rally around individually.

Government regulation, either domestic or international, is the second mechanism but is at best underdeveloped at both levels. A government could use domestic law to hold transnational companies that are headquartered in their country to account. For example, in the wake of the 2006 congressional hearings on the activities of a number of US Internet companies in China legislation (the Global Online Freedom Act) was proposed to do just that, but was never passed (Miller 2009). Sometimes a combination of individual activism and the activism of a court may change the political landscape for OSPs: the Snowden revelations starting in 2013, the privacy case against Facebook won by Max Schrems in 2015, and the ensuing decision of the European Court of Justice to declare the EU-US Safe Harbour Framework invalid caused upheavals that highlighted both OSPs' position as political actors and the tension between governmental regulation and their international responsibilities.

At the international level, 'regulation' is still very much in its infancy. The main instrument is the so-called Ruggie framework, developed by and named after Harvard professor John Gerard Ruggie during his tenure as the UN Secretary-General's Special Representative for Business and Human Rights. This framework is a rare international instrument – though not legally binding – that connects transnational business with human rights obligations. However, even this framework does not escape the logic of the Westphalian system as it 'defines human right obligations of corporations exclusively in negative terms as duties to *respect* human rights while assigning all duties in the positive realm – i.e. to *protect* human rights –

to states alone' (Wettstein 2012: 745). The active role in the international system falls to the state, the passive role to the company.

The third and last means of holding transnational corporations to account are the corporations themselves. There is a large and growing literature on Corporate Social Responsibility (CSR) that analyses whether, how, and why corporations take on social responsibility whilst doing international business. CSR is popular in business circles but is generally considered a sideshow to transnational corporations' primary aims of profit and shareholder value maximization. It is at the core a voluntary activity – grounded in the idea of 'doing good' and/or being seen to do good – and usually complements core business rather than creating actual, enforceable restrictions for it. It does not create rights for those outside, but is considered a virtue or even a charity by those on the inside (Wettstein 2012: 749). Also, the emphasis on the social and societal character of CSR steers the concept away from any political role and responsibility for corporations (Scherer and Palazzo 2011; Wettstein 2012), even though it is clear that multinational OSPs are using the cover of CSR to engage in market expansion in ways that have significant political implications (Taylor and Broeders 2015).

17.4 Why We Need to Talk About Corporate Political Responsibility

The idea of CSR – in a broad sense – is (a) in need of more serious mechanisms for accountability and (b) in need of politicisation, in the sense of a recognition of the political role of corporations. As argued above, the political role of OSPs is clear, even though it is doubtful whether the CEOs of these companies would ever (voluntarily) phrase it in those terms. That political role should be recognized. Scherer and Palazzo (2011: 922) argue for a *political* CSR in which companies on the global playing field are addressed as both economic and political actors. In the context of CSR and human rights Wettstein (2012: 774) stresses that it is the actual public or political power that corporations have – their capabilities – that de facto create genuine political responsibility.

When talking about CSR we need to rethink what the 'R' actually stands for. Responsibility can only live up to its potential if it is not merely something corporations assign themselves on a voluntary basis, in the shadow of the far larger and more pressing imperative of profit maximization. Responsibility should be embedded in a system of externally organised *responsiveness*. There is a lot of work to be done to create genuine forms of popular, political and legal accountability in order to hold corporate political power to account. A start would be to talk about Corporate Political Responsibility (CPR) instead of CSR. This would recognize the political role and power of OSPs and call direct attention to the responsibilities that they entail. Whether or not to censor speech or enable Egyptians to express themselves by means of 'speak to tweet technology' during the protest and internet shutdown in

January 2011 – as Google and Twitter did – are political as much as business decisions. OSPs exercise power over their users and are a counter power to state power in all corners of the world. This exercise of power and the political responsibility that results from it should, again, be embedded in a broader framework of responsiveness.

One way to do this is for national governments and international organisations to accept the political nature of OSPs and approach them though the channels and language of diplomacy and international relations. Approaching OSPs though these channels would create and strengthen forms of political accountability that are currently underused. These companies are more than potential investors that must be recruited, more than violators of privacy that must be tackled: they are also political actors who merit serious diplomatic attention owing to their vital role in digital life, with all the contradictions inherent in diplomacy (Broeders 2015). That means that positions may align on some issues, and contradict on others, just as they do with other friendly and not-so-friendly nations. It does require serious thinking from a national and international political perspective about a diplomatic framework from which to interact with OSPs: diplomacy does not equal being on the receiving end of a lobbying campaign.[8] Such a diplomatic framework may necessitate new institutional configurations and new spaces for discussion. However, adding this dimension is both necessary and overdue, and may help to resolve some of the grey areas currently apparent in state-OSP relations by acknowledging that both parties have their own political rationales and agendas, and that they must be made explicit.

References

Broeders, D. (2015). *The public core of the Internet. An international agenda for Internet governance*. Amsterdam: Amsterdam University Press.

Clinton, H. (2010). *Remarks on Internet freedom*. Speech at the Newseum, Washington D.C., 21 January 2010. http://www.state.gov/secretary/20092013clinton/rm/2010/01/135519.htm. Accessed 17 May 2016.

Dann, G., & Haddow, N. (2008). Just doing business or doing just business: Google, Microsoft, Yahoo! And the business of censoring China's internet. *Journal of Business Ethics, 79*, 219–234.

EFF. (2015). Internet.org is not neutral, not secure, and not the internet. Electronic Frontier Foundation, May 18, 2015. https://www.eff.org/deeplinks/2015/05/internetorg-not-neutral-not-secure-and-notinternet. Accessed 15 June 2015.

Greenwald, G. (2014). *No place to hide. Edward Snowden, the NSA and the US Surveillance State*. New York: Metropolitan Books.

Internet.org (2014). *Announcing the Connectivity Lab at Facebook*. http://www.internet.org/press/announcing-the-connectivity-lab-atfacebook. Accessed 7 Aug 2014.

[8] Especially not when large OSPs have substantial and targeted lobbying power such Google that has both frequent access to the White House and deep pockets when it comes to their lobbying activities at the US Federal level. See for example: http://www.wsj.com/articles/google-makes-most-of-close-ties-to-white-house-1427242076. Another example is Facebook's lobby campaign on the Indian government described above.

Miller, J. (2009). Soft power and state-firm diplomacy: Congress and IT corporate activity in China. *International Studies Perspectives, 10*, 285–302.

Nye, J. (2011). *The future of power*. New York: Public Affairs.

Pentland, A. (2011). Society's nervous system: Building effective government, energy, and public health systems. *Pervasive and Mobile Computing, 7*(6), 643–659.

Scherer, A., & Palazzo, G. (2011). The new political role of business in a globalized world: Review of a new perspective on CSR and its implications for the firm, governance, and democracy. *Journal of Management Studies, 48*(4), 899–931.

Schneier, B. (2013). Power in age of the feudal Internet. In U. Gasser, R. Faris, & R. Heacock (Eds.), *Internet monitor 2013: Reflections on the digital world* (pp. 10–14). Cambridge, MA: The Berkman Center for Internet and Society.

Scott, J. (1998). *Seeing like a state: How certain schemes to improve the human condition have failed*. New Haven: Yale University Press.

Strange, S. (1988). *States and markets: An introduction to international political economy*. London: Pinter.

Tan, J., & Tan, A. (2012). Business under threat, technology under attack, ethics under fire: The experience of Google in China. *Journal of Business Ethics, 110*, 469–479.

Taylor, L., & Broeders, D. (2015). In the name of development: Power, profit and the datafication of the global South. *Geoforum, 64*(4), 229–237.

Van Hoboken, J., & Rubinstein, I. (2014). Privacy and security in the cloud: Some realism about technical solutions to transnational surveillance in the Post-Snowden Era. *Maine Law Review, 66*(2), 487–534.

Wettstein, F. (2012). CSR and the debate on business and human rights: Bridging the great divide. *Business Ethics Quarterly, 22*(4), 739–770.

Chapter 18
The Economic Impact of Online Intermediaries

Hosuk Lee-Makiyama and Rositsa Georgieva

Abstract Given the severe lack of sufficiently detailed data, there is very little measurement or quantification of the contributions by online intermediaries in the academic literature. Intermediaries and platforms are loosely defined by lawmakers and national statistics for any meaningful analysis; available data suggests online intermediaries do not yet interact with many of the sectors often mentioned as adversely affected by digitalisation, while available data on productivity and trade leads to counterintuitive conclusions. We examine to what extent the economic impact of intermediaries may differ in the US and the EU – two markets with comparable set up of market institutions and actors – and conclude that the impact is similar in terms of input, outputs, productivity and trade. Yet the online intermediaries have created different incentives in the US and the EU. Conclusion is the political economy of advertising: Online advertising continues to grow at the expense of TV, magazines and newspapers. Europe is the only geographic area in the world that has seen the advertising expenditures fall (–0.8 % compared to +0.6 % in North America). Furthermore, traditional media is the most important supplier of inputs to online intermediaries in the US, which turned a zero-sum game between into a condition of co-dependence.

This redistributive mechanism – where online intermediaries channel profits back into traditional media through advertising – could not be confirmed to exist in the EU. This lack of co-dependency is one possible determinant to why stakeholder interests are differently aligned in the EU compared to the US.

18.1 The Political Economy of Platforms

We can assume that the invention of online intermediaries has led to disruption of business models trade and changed firm-level competition. The role of Internet intermediaries to drive economic, social and political development has been

H. Lee-Makiyama • R. Georgieva (✉)
ECIPE (European Centre for International Political Economy), Brussels, Belgium
e-mail: rositsa.georgieva@ecipe.org

M. Taddeo, L. Floridi (eds.), *The Responsibilities of Online Service Providers*,
Law, Governance and Technology Series 31, DOI 10.1007/978-3-319-47852-4_18

acknowledged by policy makers and led to various policy responses, notably in the European Union (EU) with the Digital Single Market (DSM).[1]

There seems to be a number of economic factors, often associated with the discipline of political economy, that are possible determinants to how a society responds to these market changes. Such factors include redistribution of economic gains between the existing actors and newly entrants, which in turn is shaped by the behaviour and interests of existing market actors and their influence on market rules.

However, this analysis severely impaired by the lack of sufficiently detailed data. As a result, there is very little measurement or quantification of the contributions by online intermediaries in the academic literature. With the important exception of Copenhagen Economics (2013, 2012),[2] the traditional disciplines of economics or law and economics are largely absent from the topic; existing literature tends to follow methods more commonly associated with international trade, using computable general equilibrium (CGE) models on macroeconomic data and interactions on the country-level, rather than intra-economy analysis between various sectors.[3]

Trade economics confirms however, that national economic interests are at play. Recent EU policy responses shows a deep concern about its relative competitiveness vis-à-vis the US economy, especially in the digital economy: Reports of European efforts to strengthen its position have produced growing concerns of protectionism,[4] that are sufficient to cause a transatlantic drift.[5] Strong statements at highest possible political levels have been registered, with the US President Obama calling the EU digital agenda a new 'digital protectionism', and the scrutiny of US intermediaries being driven by 'commercial interests'.[6]

This seem quite remarkable given that non-indigenous technologies and industries have entered the European market throughout history without causing geo-economic implications. The competitiveness of others (in this case the US) is insufficient, if not dissatisfactory, to explain the shifts in the political economy caused by the market entry of online intermediaries. In this note, we examine to what extent the economic impact of intermediaries may differ in the US and the EU, despite being two markets with comparable set up of market institutions and actors. The comparative analysis concludes that the impact from online intermediaries are similar in terms of input, outputs, productivity and trade in the EU and the

[1] European Commission, A Digital Single Market Strategy for Europe, 2015.

[2] Copenhagen Economics, The impact of Online Intermediaries on the EU Economy, 2012; revised 2013.

[3] See Bauer, Lee-Makiyama, van der Marel, Verschelde, 'A Methodology to Estimate the Costs of Data Regulation', International Economics, 2015; also Bauer, Lee-Makiyama et al, The Cost of Data Localisation, ECIPE, 2014; The Economic Importance of Getting Data Protection Right, US Chamber of Commerce, 2013.

[4] Robinson, Duncan, Concerns grow over EU digital rules targeting American companies, Financial Times, 26 July 2015.

[5] Nocera, Joe, GAFA: Europe's Term for America's 'Evil' Internet Empire, New York Times, 28 April 2015.

[6] Swisher, Kara, White House. Red Chair. Obama Meets Swisher, <Re/Code>, February 2015 accessed at: http://recode.net/2015/02/15/white-house-red-chair-obama-meets-swisher/

US. However, the two-sided market actors may create divergences through different political economy incentives, although the limited availability of economic data on online intermediaries (especially in the EU) restrain the author from a conclusive answer.

18.2 Defining Online Intermediaries and Platforms

Although the term online intermediaries – or its derivative term, platforms – are often heard in the contexts of policy regulation, it is not entirely given what the legal and economic characteristics are that define them. OECD (2010) describes online intermediaries as entities that "bring together or facilitate transactions between third parties on the Internet. They give access to, host, transmit and index content, products and services originated by third parties on the Internet or provide Internet-based services to third parties".

The term 'platforms' is even more loosely defined, or not defined at all. In the Digital Single Market Strategy,[7] the European Commission describes some common characteristics of digital platforms, and names a number of examples, while refraining from defining it in binding terms; an independent study commissioned by the European Parliament defines digital platforms as 'a (technological) basis for delivering or aggregating services and content from service and content providers to end users'.[8]

Existing EU law makes no definition of online intermediaries, but gives broader and sweeping definitions of intermediaries with all types of online services in the category 'information society services'–[9] recognising different forms of intermediaries, including agent, distributor, commissionaire, broker, 'regie publicitaire', payment service provider, and web platforms. The only explicit differentiation of online intermediaries from other actors appear in the E-Commerce Directive, under its provisions on limitations of intermediary liability for mere conduit, caching, hosting. The liability exception is deemed to apply to 'the transmission in a communication network of information provided by a recipient of the service, or the provision of access to a communication network' –[10] a definition that, by and large, predates the popular dissemination of intermediaries (given the law was drafted before year 2000), and primarily aimed at capturing infrastructural services.

The lack of clear definition is hardly surprising. Defining online intermediaries is complicated by the variety sectors and business model they operate. The definition of 'platforms' could encompass online retailers and redistributors of third party commodities and services (Amazon, Apples iTunes, Expedia, eBay), internet por-

[7] See note 1.

[8] van Gorp, Nicholai, Challenges of Competition Policy in a Digitalised Economy, European Parliament, 2014.

[9] Directive 2000/31/EC, On Certain Legal Aspects of Information Society Services, in Particular Electronic Commerce (E-Commerce Directive).

[10] ibid., art.14.

tals (Yahoo, AOL), search engines (Google, Bing), content aggregation (Alltop.com), social media (Facebook) and user generated audio-visual services (YouTube). Some of these online intermediaries compete or replace directly with existing businesses, or mediate between end-users and only indirectly impacting pre-existing economic actors. Other online intermediaries have created entirely new markets that did not exist before. Moreover, by changing the pre-existing business or revenue models, the new market entrants are assumed change the firm-level competition: Some platforms that use direct payment for services they offer to their consumers or producers, while some use advertising as their primary source of revenues, able to supply their services without charge, and there are also several which use a mix or alteration of these revenue models.[11]

The lack of a coherent and consistent definitions have led policy makers to use approximate – sometimes innovative, often derogatory – terminology in their attempts to define a group of firms with little commonalities between them, *prima facie*. In critical policy areas (such as taxation, competition and privacy policy) a new acronym with a negative connotation related to the Internet intermediaries have recently appeared –[12] namely GAFA – which is made out of the initial letter of four US tech companies of a search engine, consumer retailer, social media and hardware manufacturer with little commonality between them. From the perspective of an economist, they have only two things in common: Namely their foreign origin, and their success in scaling their operations globally.

How online intermediaries are defined by market regulators is pertinent to a political economy analysis. Firstly, it impacts on the market regulators' ability to differentiate the treatment of market actors. For instance, the liberalisation of the Single Market and consequent deregulation makes most EU legislation 'erga omnes' per default: In plain words, per default, the Single Market is unable to differentiate the treatment of firms that originates from the EU vis-à-vis a non-EU entity who are legally established in its territory on the basis of their origins.[13] Therefore, if accusations of 'digital protectionism' are true, the legislators would need to narrowly define online intermediaries in such manner that would distinguish activities of foreign online intermediaries from local domestic counterparts.

Secondly, definitions are not insignificant for how economic data is presented, as definitions are pivotal to how the economic activity is registered in national statistics. A 'two-sided' online intermediary (say, a search engine) could be placed under advertising, information or technology related categories, affecting how the economic flows are routed. If the search engine provides its service for free, there may be no record of the transaction at all in the online processing or technology categories. Similarly, an online intermediary may act as an agent for third party content or

[11] Van Eijk, Niko, Digital platforms: An Analytical Framework for Identifying and Evaluating Policy Options, TNO report R11271, 9 November 2015, pp. 12–13.

[12] Chibber, Kabir, American cultural imperialism has a new name: GAFA, Quartz, December 01, 2014.

[13] Krol, Lee-Makiyama, Macyra, International Services Agreement: From an European Vantage Point, ECIPE, 2012.

goods in a revenue-sharing arrangement. In both cases, the real direction of inputs and outputs (i.e. payment flows) may be difficult to distinguish, or to estimate fully.

Further to this point, the definitions used in national statistics differ. The sectoral definitions used in the North American Industry Classification (NAICS) are not directly comparable to Statistical Classification of Economic Activities in the European Community (NACE) without imprecise concordances. While the US NAICS classification system is sufficiently detailed, the EU NACE proves to be insufficiently disaggregated for meaningful analysis on online intermediaries, while any commercial transactions taking place between two economies – i.e. trade – are reported using an international standard for balance of payments (extended balance of payments, or EBOP). More recently, services trade data has become available through new set of economic data through trade in value-added,[14] that are still not sufficiently detailed to isolate the economic activities of online intermediaries.

18.3 Online Intermediaries in Economic Inputs and Outputs

Even Europe's Commissioner on digital economy – Gerhard Oettinger – have admitted that policymakers need 'deeper understanding of the role that they play and of the way they interact with other businesses and with consumers'.[15] To date, the only estimate of the economic impact of online intermediaries was published by Copenhagen Economics (2013, 2012), that estimated the direct contribution of online intermediaries to €220 bn. This estimate incorporates e-commerce, and in addition the economic value of free services that cannot be captured in traditional GDP estimates, as latter is made at market cost. According to the study, the economic-wide contribution is equivalent to 1.6 % of EU GDP. Although such numbers on economy-wide impact may establish the contributions of the online intermediaries on a macroeconomic and societal level, such numbers must be disaggregated to its components in order to comprehensively understand the full dynamic of economic interests within an economy and between various sector for a political economy analysis.

As mentioned above, the difficulty in this regard arise from lack of sufficiently disaggregated data. The US data records 'internet publishing' and 'online intermediaries' grouped together (under NAICS category 51913). The category is defined under 'information services', rather than under 'data processing', 'infrastructural' or 'broadcasting' activities, thereby solely capturing the industrial activities of 'on the top' players. In contrast, the EU definitions (NACE) groups web portals with 'data processing, hosting and related activities', whether they are online or offline.[16]

[14] World Input-Output Database, 2013; Trade in Value-Added (TiVA), OECD-WTO, 2015.

[15] European Commission, Have Your Say on Geo-Blocking and the Role of Platforms in the Online Economy, Press Release, 24 September 2015.

[16] European Commission, List of NACE codes, March 2010, accessed at: http://ec.europa.eu/competition/mergers/cases/index/nace_all.html

The lack of adequate data in Europe constrains severely on meaningful analysis. At best, the level of disaggregation used in the EU allows for a general overview of the 'information economy' at large: In 2013, the category generated EUR 487.9 billion of economic value-added,[17] a considerable number. This is more than twice the number provided by Copenhagen Economics (2013, 2012), although online intermediaries ought to contribute with only a fraction of the number.

The US data is able to provide further level of detail. In the statistic from the US Bureau of Economic Analysis (US BEA) over the past 15 years, the online intermediaries account for 0.6 % of the economic inputs to all industries,[18] a relatively modest input in the economy. Some specific sectors rely more heavily on the usage of online intermediaries, such as computer systems design, software and professional services, and their reliance on online intermediaries have increased severalfolds over the past 15 years.

Inputs from online intermediaries

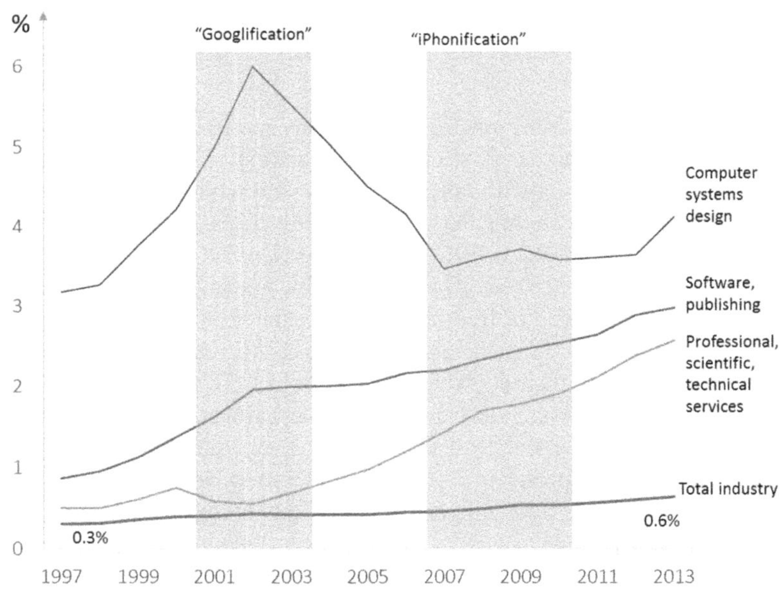

Own analysis, based on data provided by US BEA (2015)

The first conclusion seems to be that the online intermediaries do not yet interact with many of the sectors often mentioned as adversely affected by digitalisation. It including all of manufacturing (e.g. motor vehicles) – although electronics are playing a major role in manufacturing, notably in the automotive industry. Approximately half of the value of an assembled passenger car comes from various ICT components and electronics, a share that is even higher for electric or hybrid electric vehicles, at

[17] ibid., Information and communication service statistics, NACE Rev. 2, April 2013.

[18] Based henceforth on activities relating to what NAICS defines as internet publishing, broadcasting and search portals.

approximately 75 %.[19] While online intermediaries does not provide any inputs into manufacturing, it is not unforeseeable given that the main conduits for online intermediaries – software and connectivity – are being incorporated into passenger cars and other manufactured goods.

It is interesting to note that the firms more dependent on online intermediaries than any other are in computer system design. Their dependency on online intermediaries reached its peak at 6 % input contribution in 2002, after which it rapidly declined until 2007. This period which coincide with the introduction of a number of online services, including mapping services, email and other business apps, that made some of their commercially available services obsolete by two-sided market actors delivering them for free.

18.4 Productivity Gains from Online Intermediaries

As noted, quantifying the usage of online intermediaries is more ambiguous in economic terms than traditional inputs, when the industry pay for raw materials, air fares, consultancy services, or telecommunication services. Moreover, the online intermediaries function as a general purpose technology that allows other sectors to work more efficiently, similar to the traditional role of economic intermediaries (e.g. financial intermediaries, travel agents and real estate brokers). As conduits of distribution or overbridging information asymmetries, they either match supply and demand or create efficiency gains to parties and by digital means, 'enable, empowers, and equalizes trade, as well as makes companies and products more efficient'.[20] Copenhagen Economics (2013, 2012) has quantified the economy-wide impact of such indirect productivity gains to €210 bn – in addition to the aforementioned €220 bn in direct GDP effect generated.

In macroeconomic terms, such efficiency of production processes is measured by summing up the value-added created in a single process, considering multiple inputs (such as labour, capital, land). Total Factor Productivity (TFP) represents the residual growth that cannot be explained by the rate of change in the efficiency of labour, capital and intermediate inputs – and is often interpreted as the contribution to economic growth created by technical and organisational innovation.[21]

In other words, this could entail that if a particular sector of the economy utilises online intermediaries more than the others, it could be more efficient – proven by a higher rate of TFP. However, a sectoral comparison of the use of online intermediaries and TFP is quite striking: Where there should be a positive correlation between online intermediary inputs and productivity gains, the results are actually negative – in other words the more a sector uses intermediaries, the less productive it becomes – as illustrated by the red regression line.

---

[19] Scuro, Giorgio, Automotive industry: Innovation driven by electronics, Embedded-Computing, 2012.

[20] Borga, Maria, Trends in Digitally-Enabled Trade in Services, US Bureau of Economic Analysis (US BEA).

[21] Description taken from OECD, Compendium of Productivity Indicators, April 2008.

TFP (vertical); Inputs from platforms (horizontal)

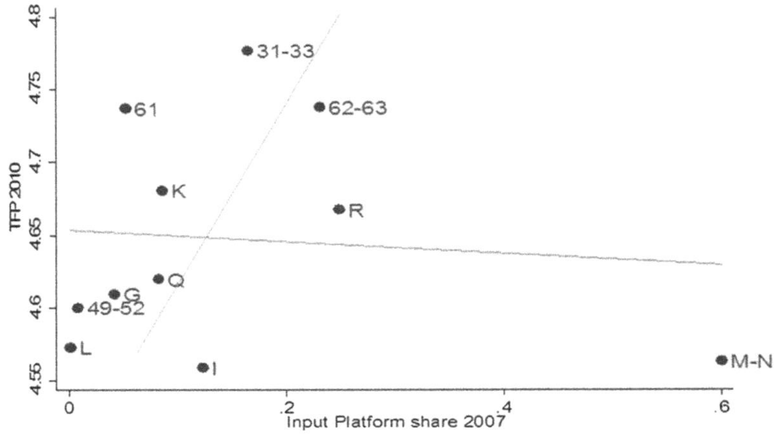

Own analysis, based on US BEA (2007), EU KLEMS (2010)

It should not be precluded that it could be a question of "Facebook at work" effect, where online intermediaries are actually negatively affecting productivity.[22] This is highly unlikely, given factor productivity of labour are increasing amongst the economies examined.[23] Instead, other explanations must be at play. There are several explanations to this counterintuitive result.

Firstly, the observation period is relatively short in terms of productivity effects. Technological shifts impact the economy gradually, through different channels, over a prolonged period of time, and indirectly through dynamic effects that takes more than a decade to fully disseminate throughout the economy. Moreover, the last available data on inputs (2007) is collected before the full transition into mobile and cloud based applications.

Secondly, the changes in TFP include other effects besides technological change, including organisational change. The period on productivity data coincides with the deepest economic crisis affecting the organisational productivity in the EU through layoffs and other transitional costs that have a diminutive effect on productivity.

Thirdly, the sector that is most heavily impacted by online intermediaries is the business services sector (denoted M-N in graph) is an extreme outlier. It is one of the smallest and least productive sectors in the service industry, making the observation overrepresented compared to major tertiary sectors in the analysis. If business services were eliminated from the sample, the correlation between inputs from online intermediaries and total factor productivity is instead strongly positive, as illustrated by the blue regression line, rather than the original line that is slightly negative.

---

[22] Mitchell, Alastair, Facebook at Work'? There's Certainly Some Work to Be Done, WIRED, November 2014.

[23] US BEA, 2007; EU KLEMS Growth and Productivity Accounts, 2010.

Fourthly, seminal research by productivity economists (Van Ark et al) have concluded that one of the biggest contributors and opportunities for productivity growth in Europe is through the ongoing digitalization of its economy –[24] ICT services, information services, telecommunication services, and producers of electrical and optical equipment accounted for more than 40 % of aggregate TFP growth in Europe, despite representing only 8 % of total GDP. Although online intermediaries are accountable for just a sub-set of the aggregated productivity gains generated by digitalisation, the entire ICT value-chain consisting of intermediaries, telecom providers, device manufacturers, content providers form an integral system necessary for the digitalisation to take place.

18.5 Online Intermediaries and Cross-Border Trade

The productivity gains from online intermediaries is not limited to productivity gains from technology – as a market maker, intermediaries enable primarily service providers to distribute their goods and services more widely, beyond the reach of traditional marketing, sales and distribution. Recent trade policy literature has highlighted the trade over internet-based platforms (such as eBay, Amazon or Alibaba) have created 'micro exporters' through the provision of global market access of the platforms that are able to export to 20–40 different countries, whereas the offline average is just three countries.[25] Although micro-exporters and SMEs are unlikely to shift the trade balance between the EU and the US (or the rest of the world), they could potentially change the competitive field for individual retailers in Europe who may be collecting disproportionate economic rents (i.e. profit margins), or choose not to supply certain type of goods.

Trade effects occur also outside of e-commerce (that generally concerns goods), but also in trade in services and intangibles. Research conducted by UN Conference on Trade and Development (UNCTAD) and service economists show that half of global trade in services (i.e. a trade worth $2.3 trillion) is enabled by information technology –[26] a share which in the EU/US bilateral trade is closer to 60 %.[27] A ratio

[24] Van Ark, Bart, Total Factor Productivity: Lessons From the Past and Directions for the Future, Working Paper No 271, National Bank of Belgium, October 2014.

[25] Rentzhog, Magnus, The New Services Era – Is GATS up to the Task?, E15 Expert Group on Services Overview Paper, November 2014, p. 6.

[26] UNCTAD, Information Economy Report: Trends and Outlook in Turbulent Times, United Nations, 2009, p. 77.

[27] Hosuk Lee-Makiyama, Digital Trade in the U.S. and Global Economies, US International Trade Commission, 2014.

that is most likely underestimated given that intermediation is possible in increasing number of services sectors, even in local transports (as evident with car sharing apps) and online intermediaries also enable knowledge process outsourcing (KPO) amongst both skilled and unskilled workers.[28]

However, a number of territorial limitations limits possibilities for intermediaries to actually trade and introduce new competition: Audio-visual services are restricted by national copyright regimes that legally forces online intermediaries to localise their services to the local jurisdiction. Similarly, online intermediation across borders of financial services is restricted by market directives on financial instruments and institutional or consumer supervision. EU data privacy rules have similar territorialisation effects, effectively limiting or disabling cross-border supply of services from a non-EU jurisdiction, further limited by recent court cases.[29] Indeed, prominent economists (Messerlin 2015) have concluded that either the EU or the US are expected to oppose further competition and liberalisation in all sectors of services.[30]

18.6 The Political Economy of Online Advertising

As described in the previous sections, some online intermediaries act as two-sided market actors, relying on advertising as its true means of income. For example, search engines are built on an advertising-based business models, and thereby compete with other media and advertising services, rather than against e.g. library services or data processing services. This two-sidedness are nearly impossible to capture accurately or consistently in national accounting, while advertising data is often business proprietary data that is rarely disclosed to academic research. However, some general conclusions can still be drawn using publicly available data on the two-sided actors and the advertising market from non-academic sources.

To begin, internet advertising seems to be displacing traditional media as online advertising continues to grow at the expense of TV, magazines and newspapers. Internet advertising seems to have surpassed all other media channels to become the largest channel for advertisement, growing from zero to more than a quarter (25.2 %) of the total advertising market during its relatively short existence.

---

[28] ibid.

[29] C-362/14, Schrems v Data Protection Commissioner, European Court of Justice.

[30] Messerlin, Patrick, The Transatlantic Trade and Investment Partnership: The Services Dimension, CEPS, 2015.

Advertising expenditure by marketing channel, Europe, 2014

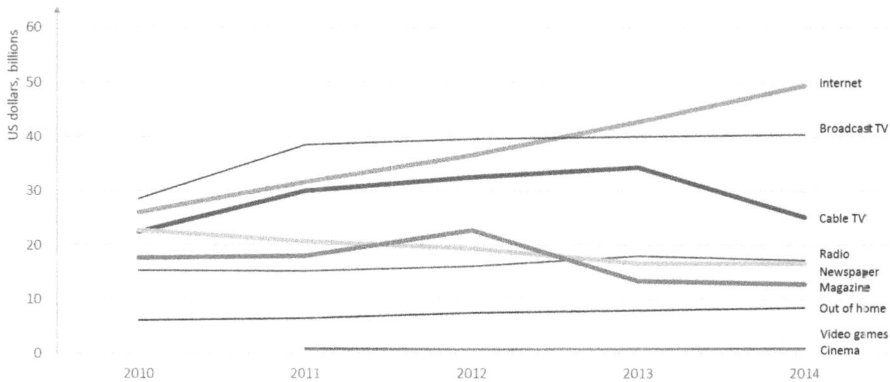

IAB/PwC, AdSpend Report, 2014

This growth of online media occurs as the advertising market in the EU is stagnant: Europe is the only geographic area in the world that has seen the advertising expenditures fall in the 5 years between 2008 and 2013, at −0.8 % (cf. +0.6 % growth in North America),[31] effectively making the competition on the advertising market a zero-sum game between traditional and online advertising channels for shrinking marketing budgets. Future market projects seem to support the conclusion that these developments will continue, given that spending on digital platforms in Western Europe were projected increase 9.1 %,[32] while the comparable year-on-year growth for TV advertising was less than 1.5 %.[33]

Non-academic literature therefore supports the hypothesis that online advertising is displacing broadcasting and print media. Nonetheless, given the technical advantages of online advertising (behavioural targeting, accountability with performance-based payments), traditional media have also transitioned to online advertising to complement their traditional product – most newspapers and broadcasters maintain a considerable online presence. At least some of the internet advertising turnover could be attributable to such examples of traditional media crossing over to new technologies.

Although it is not possible to determine the market shares of online intermediaries against the online presence of traditional media in the EU through publicly available and verifiable data, non-academic sources in the US shows that the largest online supplier of advertising services in the US was Google, which was the recipient of 10.0 % of total advertising expenditure in 2013. In falling order, Google

[31] McKinsey & Company, Wilkofsky Gruen Associates, Global Media Report, 2014.

[32] eMarketer, quoted through Chief Marketing Officer Council World Wide, Facts & Stats on Marketing Spending, November 2015.

[33] Statista, Television advertising expenditure in Europe from 2007 to 2015, by region (in million U.S. dollars), accessed at: http://www.statista.com/statistics/428893/tv-ad-spend-in-europe-by-region

followed by Facebook, Yahoo, Microsoft, AOL, Amazon and Twitter, where the smallest amongst them accounting for merely 0.3 % of the total advertising market in the US.[34] Thus, online advertising provided by traditional media must account for less than 0.3 % of the total advertising expenditure, or possibly integrated as subcontractor to the aforementioned online intermediaries. Assuming this dominance of online intermediaries over internet also holds true for the EU advertising markets, it follows that the indigenous European print and broadcasting media could not compensate for their revenue losses by launching their own internet presence.

However, the relation between online intermediaries and traditional media are not exclusively competitive. In a similar fashion that various sectors depend on online intermediaries, online intermediaries depend on purchases from other sectors, including their competitors in traditional broadcasting and print. Whereas we used US BEA data to examine the economic dependency on online intermediaries, we examine the reverse relation – the inputs used by online intermediaries.[35] Online intermediaries rely more than any other sector on economic inputs from advertising services, representing more than a quarter of total inputs. In other words, the primary source of 'fuel' for US online intermediaries are their competitors in traditional media – with only radio and television broadcasting (i.e. advertisement by media in other media) and travel services ever coming close to the advertising dependency of online intermediaries.

Inputs used by selected US sectors

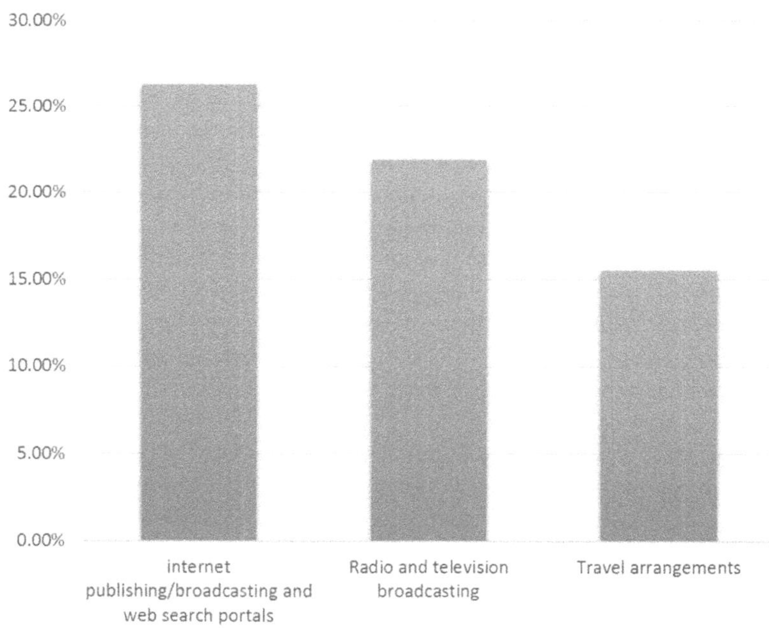

Source: US BEA 2012

[34] Net US Digital Ad Revenue Share (% of total media ad spending), based on eMarketer, 2014; IAB/PwC, 2014.

[35] Consistently with previous sections, online intermediaries are approximated to 'internet publishing, broadcasting and search portals'.

The US BEA data suggests that the online intermediaries are in a state of co-dependency with traditional media where publishers depend on the intermediaries for visitor traffic as well as advertising revenues – a reciprocal relationship that is more pronounced than with any other sector. This is also confirmed by secondary firm-level observations: the largest online advertising provider, Google, spent approximately one-third of its own advertising expenditure on television ads in 2011, amounting to approximately $70 million.[36]

Given the lack of sufficiently disaggregated sectoral information in the EU, it is not possible to establish whether this co-dependency between online intermediaries and traditional media applies also in Europe. Market data from at least two EU Member States seems to suggest the contrary: Out of the hundred largest advertisers in the UK in 2013, only three (Google, Amazon and eBay) were online intermediaries.[37] Given the maturity of the UK markets relative to the other EU Member States, the online intermediaries can be assumed to be even less prevalent amongst other national markets in Europe than in the UK. In fact, in the same market data available for the Netherlands show that no foreign online intermediaries were amongst the hundred largest advertisers same year.[38]

In conclusion, the comparison of the EU and the US advertising markets points to the following chain of hypotheses. To begin, we may almost certainly assert that the traditional media was displaced by internet advertising, a category that is likely to be dominated by online intermediaries. Moreover, this took displacement took place on an already declining market, further aggravating the impact.

Second hypothesis is even more certain as it is based on verified data using input-output tables. It shows that traditional media are the largest supplier of inputs to online intermediaries in the US, which allows online intermediaries to offset some of its negative impact on traditional media – or at least turn a zero-sum game into a condition of co-dependence. In a scenario where market regulators become too prohibitive against online intermediaries and restrict their presence, traditional media would incur further losses than limiting them.

Third hypothesis is based on a counterfactual conditional to the second: Given the lack of data, we cannot assume that same redistributive mechanism to exist between online intermediaries and traditional media in the EU. Selected data points suggest rather that traditional media is not a major supplier to foreign online intermediaries in Europe, which in turn leads to the conclusion that the European media industry – unlike their US counterparts – would have less to lose from a protectionist commercial policy that would limit the presence of foreign online intermediaries.

---

[36] McGee, Matt, Report: Google Spent $213M Advertising Itself In 2011, Including $70M On TV Ads, Marketing Land, March 2012; Efrati, Amir, Once Shunning Ad Promos, Google Now Flaunts Itself, The Wall Street Journal, March 2012.

[37] Nielsen Media Research, Marketing's Top 100 Advertisers (UK), 2013; Durrani, Arif, Google becomes major UK advertiser after boosting spend by 50% in 2013, Campaign, April 2014.

[38] Nielsen Media Research, Marketing's Top 100 Advertisers (NL), 2013.

18.7 Conclusions

If the internet and digitalisation leads to the development of a 'platform economy' where online intermediaries become network orchestrators of consumer demand, it follows that the key determinant of market power is shifting towards the demand side. In other words, the new economy would reward those who are able to achieve economy of scale on the demand side by aggregating users. This is contrary to the the traditional post-war model of industrial policy that emphasise the economies of scale in production – providing the rationale for a number of supply-side measures, such as creating national champions, imposing various form of regulation-based 'soft protectionism' against foreign firms and handing out various form of sectoral state aid to domestic industries. Such policy measures ought to be effectively defunct in a platform economy, severely limiting the governments' ability to govern through 'picking the winners'.

Commercial policies of both Europe and the US tend to be aligned to their core industrial interests. As any mercantilist entities, they use both offensive and defensive strategies, sometimes simultaneously, to resist any changes to their existing industrial structures. Online intermediaries introduced a number of technologically superior market entrants on a constant (or declining) market in Europe, providing a classic political economy set-up for a zero-sum game where industrial actors seek to influence market regulations in pursuant to their self-interest. As such regulations could limit or even bar market access for newly entrants, the reward for such strategy is potentially high.

This analysis has shown that online intermediaries have not yet seriously disrupted any supply-chains outside of software, publishing and professional services. Trade and productivity gains were also inconclusive. Nonetheless, with the exception of an outlier, online intermediaries seem to have provided tangible productivity gains in those sectors who are heavy users of them. Although the data lacks sufficient detail (on the EU side), there are no observations amongst the market data that suggest these conclusions would be disparate between the EU and the US.

The point of departure appears in the comparative analysis of inputs into online intermediaries. Public data (supported by commercial market data) reveal that the US economy shows a high co-dependency on traditional advertising services, effectively introducing an opportunity cost against market actions on intermediaries initiated by traditional media. However, similar co-dependency could not be proven on the EU market. Data points reported in non-academic publications seem to support the contrary – that the online intermediaries are not purchasing traditional advertising services from publishers in Europe.

This lack of co-dependency between the stakeholders is one possible determinant to why market powers of the stakeholders are differently distributed, and their interests are differently aligned in the EU compared to the US: while there are a number of horizontal market measures in the EU aimed at online intermediaries (notably *ex post* enforcement on privacy or antitrust), to date, new sectoral measures *ex ante* concerns publishing in various forms. Such policies include ancillary copyright laws, enacted in Spain and Germany in favour of local media,[39] or the ruling on the right to be forgotten.[40]

References

Bauer, M., Lee-Makiyama, H., & Krol, M. (2013). *The economic importance of getting data protection right*. US Chamber of Commerce.

Bauer, M., Lee-Makiyama, H., & van der Marel, E. (2014). *The cost of data localisation*. ECIPE.

Bauer M., Lee-Makiyama H., van der Marel E., & Verschelde, B. (2015). *A methodology to estimate the costs of data regulation'*. International Economics.

Borga, M. (2012). *Trends in digitally-enabled trade in services*. US Bureau of Economic Analysis (US BEA)

CCIA. (2015, February). Understanding "Ancillary Copyright" in the global intellectual property environment.

Chibber, K. (2014, December 1). American cultural imperialism has a new name: GAFA. Quartz.

Chief Marketing Officer Council World Wide. (2015, November). Facts & stats on marketing spending.

Copenhagen Economics. (2013). The impact of Online Intermediaries on the EU Economy, 2012; revised.

Durrani, A. (2014, April). Google becomes major UK advertiser after boosting spend by 50% in 2013, Campaign.

Efrati, A. (2012, March). Once shunning Ad promos, Google now flaunts itself. The Wall Street Journal.

EU KLEMS. (2010). Growth and productivity accounts.

European Commission. (2015a). A digital single market strategy for Europe

European Commission. (2015b). Have your say on Geo-blocking and the role of platforms in the Online Economy, Press release, 24 September

European Court of Justice. (n.d.). C-362/14, Schrems v Data Protection Commissioner

Floridi. (n.d.). Should you have the right to be forgotten on Google? Nationally? No.

IAB/PwC. (2014). Online AdSpend Study

Krol, Lee-Makiyama, H., & Macyra, N. (2012). International services agreement: From an European vantage point, ECIPE.

Lee-Makiyama, H. (2014). Digital trade in the U.S. and global economies, US International Trade Commission.

McGee, M. (2012, March). Report: Google Spent $213M Advertising Itself In 2011, Including $70M On TV Ads, Marketing Land.

McKinsey & Company. (2014). Wilkofsky Gruen Associates, Global Media Report

Messerlin, P. (2015). The transatlantic trade and investment partnership: The services dimension, CEPS

Mitchell, A. (2014, November). Facebook at Work'? There's certainly some work to be done, WIRED.

Nielsen Media Research. (2013a). Marketing's Top 100 Advertisers (UK)

[39] See CCIA, Understanding "Ancillary Copyright" in the Global Intellectual Property Environment, February 2015.

[40] See Floridi, Should You Have the Right to Be Forgotten on Google? Nationally, Yes. Globally, No.

Nielsen Media Research. (2013b). Marketing's Top 100 Advertisers (NL).
Nocera, J. (2015, 28 April). GAFA: Europe's term for America's 'Evil' internet empire, *New York Times*.
OECD. (2008, April). Compendium of productivity indicators.
OECD-WTO. (2015). Trade in Value-Added (TiVA).
Rentzhog, M. (2014, November). The new services era – Is GATS up to the task?, E15 Expert Group on Services Overview Paper, p. 6.
Robinson, D. (2015, July 26) Concerns grow over EU digital rules targeting American companies. *Financial Times*.
Scuro, G. (2012). *Automotive industry: Innovation driven by electronics*. Embedded Computing Design.
Statista. (n.d.). Television advertising expenditure in Europe from 2007 to 2015, by region (in million U.S. dollars), accessed at: http://www.statista.com/statistics/428893/tv-ad-spend-in-europe-by-region
Swisher, K. (2015). White House. Red Chair. Obama Meets Swisher, <Re/Code>, February accessed at: http://recode.net/2015/02/15/white-house-red-chair-obama-meets-swisher/
UNCTAD. (2009). *Information economy report: Trends and outlook in turbulent times*. United Nations.
U.S. Bureau of Economic Analysis. (2007). *Benchmark input-output accounts*.
US Bureau of Economic Analysis. (2015). *National economic accounts*.
Van Ark, B. (2014, October). *Total factor productivity: Lessons from the past and directions for the future* (Working Paper No 271). National Bank of Belgium.
Van Eijk, N. (2015, November 9). Digital platforms: An analytical framework for identifying and evaluating policy options (TNO report R11271).
Van Gorp, N. (2014). *Challenges of competition policy in a digitalised economy*. European Parliament.
World Input-output Database. (2013). *Input-output tables release 2013*.

Chapter 19
Online Data Privacy and the Justification of the Market

Jennifer Baker

Abstract In this commentary I argue that the ethical issues surrounding online data privacy are not unique and instead involve timeless features of transactions in the market. I explain how we can turn to what justifies the market most generally in analysing such issues. The use of any truncated account of the justification of the market, I argue, will mislead us into coming to either the wrong conclusions or to no conclusions about the ethics of online data privacy.

19.1 The Federal Communication Commission's New Proposed Rules

The Federal Communication Commission (FCC) has recently announced that it will issue new rules for broadband internet service providers (ISP).[1] These rules aim to "provide consumers with more control over the use of their personal information" (FCC 2016a). These rules would require certain security measures to be taken and that notifications will be sent upon any breach of security. Users will be asked to affirmatively consent to any use of their information beyond what is necessary to provide the broadband service. The FCC's justification of these rules is that they will promote customer "choice," "transparency," and "privacy." The effort is described as "pro-consumer" (FCC 2016b).

In response, Larry Downes has argued that the FCC is "proceeding under a false premise". He writes, "histrionics aside, the reality is that ISPs don't control important or extensive conduits of consumer information.[2] Thanks to the encryption

[1] See https://apps.fcc.gov/edocs_public/attachmatch/FCC-16-39A1.pdf. Also https://apps.fcc.gov/edocs_public/attachmatch/DOC-338679A1.pdf

[2] To be more specific, in most cases, the ISPs cannot see a user's PII (personally identifiable information), a user's session identity and the data they are requesting or receiving from websites. This is because virtually all sites use SSL (Secure Socket Layer) encryption. What ISPs can see are the

J. Baker (✉)
Department of Philosophy, College of Charleston, Charleston, South Carolina, United States
e-mail: bakerja@cofc.edu

M. Taddeo, L. Floridi (eds.), *The Responsibilities of Online Service Providers*, Law, Governance and Technology Series 31, DOI 10.1007/978-3-319-47852-4_19

campaign, they can't even see most of it. While leading internet companies including Google, Facebook, and Netflix have continued to collect and harvest more personal information for ever-more innovative commercial uses, broadband providers have done little customization, largely because they don't have access to transactional data, personal or otherwise" (Downes 2016).

Downes insists that not only is the FCC targeting broadband providers unduly, the cost of the new rules will interfere with the innovation that benefits customers, casting doubt on the FCC's "pro-consumer" positioning. He writes, "requiring consumers to affirmatively agree to every potential use of information — for example, to personalize ads or consolidate information for "big data" analytics — dramatically increases transaction costs for everyone, with little or no benefit. That's a source of inefficiency very few consumers will find at all helpful; more likely, quite the opposite" (Downes 2016).

How can two such opposed views be reconciled? Or how are we to decide between them? Methodologically, one option is arguing for the right balance between choice, transparency, and privacy against efficiency. Another option would be to settle the criteria both approaches invoke: what consumers want. In the rest of this commentary I will propose a different approach altogether, which is that the most general justification of the market be articulated in terms of a norm that practitioners (producers, sellers, and buyers) are expected to apply and test to their own behaviours. Three benefits will be that this justification can encompass all of the specified good identified in accounts that are less general; it moves our attention away from idealized transactions, which are, I will argue, a poor measure of what the market does well; and it can generate definitive conclusions.

19.2 An Ethical Justification of Market Practices

Several frameworks already exist in the academic literature, which focus on idealized transactions (Elia 2009; Uslaner 2004; Pollach 2011). These take what I would call a "metaphysical" approach to services funded through online data transactions. For example, theorists have argued the merits of recognizing the culling of user data as a form of transaction, where norms of transactions (such as ownership and fairness) come into play (Elia 2009; Uslaner 2004; Pollach 2011). Other theorists have suggested that classifying such data-gathering as a transaction is inappropriate both as a description and as a guide to the larger social obligations required of ISPs (Martin 2015). These social obligations include "protecting the integrity of the boundaries of the contracting community and their moral free space" (Martin 2015). This social contract approach recognizes that the criterion of "mutually beneficial and sustainable local norms" that fit with a local community's "goals or purpose" so

sites users visit. This is because the SSL certificate includes the site's SSL registration (like Amazon, Facebook, Gmail etc.). It is in this way that they can collect data about what sites you visit at specific times of the day and sell that data to third parties.

that users can meet their interests in "discriminately" sharing information" (Martin 2015).

We would need to turn to a less-nuanced approach to find the reasoning capable of outright condemning as ethical online transactions that are financed through the culling of user data. There is an earlier approach in the business ethics literature that could support the FCC's proposed rules. By turning to it I hope to establish that such condemnations are going to invoke interpersonal ethical norms rather than mere business norms.

David Holley's framework was designed to apply to "any type of sales activity." It can indeed be readily applied to condemn the practice of Google, Facebook, and Netflix retaining personally identifiable information (a user's session identity and the data they are requesting or receiving) for commercial use (1998). Its use supports the conclusion that retaining such information, whether in accordance with a stated privacy policy or not, is ethically unjustified. I will use Holley's account to point to the need for business, rather than interpersonal ethical norms, in the ethical evaluation of commercial activity. But first, let me give a brief indication of his reasoning.

According to Holley (1998), if the user's interaction with an ISP is an ethical and "acceptable exchange" it must be, in all significant dimensions, *voluntary*, which takes to mean the following criteria must be met:

- Both buyer and seller understand what they are giving up and what they are receiving in return.
- Neither buyer nor seller is compelled to enter into the exchange as a result of coercion, severely restricted alternatives, or other constraints on the ability to choose.
- Both buyer and seller are able at the time of the exchange to make rational judgments about its costs and benefits.

Even if we are mindful of that ISPs are kept, due to SSL (Secure Socket Layer) encryption, from learning a user's session identity or gathering the data they are requesting or receiving from websites, even their ability to track the sites that their users visits runs afoul of Holley's framework. It is still generally the case that users do not understand what they are giving up or what they are receiving in return when an ISP retains even depersonalized information about online habits. Holley (1998), like Martin (2015), would suggest that these users are not in a position to make a sound judgment about their use of Internet service. There are other ethical issues: ISPs have been caught blocking or degrading service from content providers. Holley's description of sale techniques that give buyers the impression of restricted alternatives as coercive backs up Net neutrality rules that forbid creating 'fast lanes' or 'slow lanes' for different traffic. Paid prioritization, throttling, and blocking will be classified as unethical.

An approach like Holley's encourages us to look beyond the customer's choice. What consumers choose is not represented of the ethical standards required of sales. The ethics is determined by the genuine satisfaction of a consumers' true needs and desires. Even if (some) consumers indicate that they have a preference for seamless internet services that are paid for in indirect ways, this is no basis on which to determine that privacy is not a more significant need or an important limitation to place on services. Though Holley's particular reasoning could recommend something even more strict, the FCC's policy in regard to ISP could be justified by any approach similar enough to Holley's.

The problem is that direct ethical justifications do not account for even the simplest of sales. What explains the limitations of his early approach is that Holley (1998) offered only a truncated account of what justifies market practices. It was this: a market "provides an efficient procedure for meeting people's needs and desires for goods and services" and that "people will efficiently serve each other's needs if they are allowed to engage in voluntary exchanges" (Holley 1998). The standard of voluntary that Holley uses depends on the notion that a buyer's needs and desires are met directly, and through the product itself (and not related interchanges). I want to suggest that voluntariness cannot be the raison d'être for markets. Neither can the values the FCC cites: choice, transparency, and privacy. The case I am making is not new, however. It is ancient.

19.3 The Indirect Justification of the Market

The Ancients puzzled over what to disclose in market transaction, just as business ethicists do today. Turning to Cicero's report of those debates is illuminating, because it advertises the role the justification of the market plays. Cicero's own analysis of ethical disclosure in sales is how I will make a case for "business norms" existing as distinct from "ethical norms." Business norms can be justified through ethical considerations but the benefits to others are less direct than those intended by us when we follow, instead, the norms of ethics itself. With this distinction in place, I aim to separate the ethical norms from the business norms that have been marshalled in analyses of the ethical obligations of ISP.

One case Cicero describes is the "Famine at Rhodes." He reports that two Stoics had maintained different views on what it would be best to do if trying to sell grain to a famine-torn island. One side of the debate is represented by Antipater, and the other by his teacher Diogenes. Antipater argues that, given the famine, any information the grain seller has about ships behind him must be disclosed to the residents of the island. According to Diogenes, a good person selling grain need only follow the "common law of the land," which includes the expectation that a seller not actively deceive, but if "has goods to sell, he may try to sell them to the best possible advantage, provided he is guilty of no misrepresentation."

Antipater responds:

> it is your duty to consider the interests of your fellow-men and to serve society; you were brought into the world under these conditions and have these inborn principles which you are in duty bound to obey and follow, that your interest shall be the interest of the community and conversely that the interest of the community shall be your interest as well. (Cicero 1971)

Diogenes then explains that the seller can follow standard business norms without ever forgetting "the bonds of fellowship forged by Nature and existing between man and man."[3] (Cicero 1971) Cicero notes how extraordinary it is that the Stoics may disagree but need never refer to what is expedient as being worth pursuing if it is morally wrong. To Diogenes and Antipater, the value of the grain remains relevant—but it cannot solve the ethical dilemma.

To agree is to recognize that we cannot directly apply ethical considerations to market transactions. Interpersonal ethics are too high a bar to apply to market transactions: the market as we know it would not exist if interpersonal ethics were controlling. To recognize this does not dismantle our interpersonal ethical obligations. First, it tells us nothing about what ought to be our personal standards for behaviour. Second, the business norms that ethical business practice ought to be concordant with are themselves justified in terms of the greater good.[4]

Particular, descriptive business norms are themselves justified by their ability to be associated with the justification for market-based behaviour: it increases general affluence over time through short-term signalling. This is a far lower bar than meeting each customer's needs, but it is still a bar many business practices do not have in sight or meet. Business norms that contribute to the affluence can be justified in terms of concern for the human community, just as Diogenes expected.

I hope it is becoming clear that we cannot, without unduly "truncating" our explanation, justify business behaviours (or their condemnation) on the basis of the standards of interpersonal transactions. To do this is to fail to recognize three features of market transactions. The first of these is epistemic and we might term it "outcome humility." Businesses cannot predict the use to which we put their products. Our experiences can turn bad even if the inputs have been carefully vetted for us. This is not to suggest that providers might not be held accountable for negative outcomes, it is instead to recognize that providers cannot justify their behaviours in terms of some prediction about customers' true needs or welfare. We can only reliably predict that in an indirect way, society is generally benefitted by market processes themselves. The point is epistemically conservative.

Second, the market is very unlike the practice of medicine, a field that has identified its controlling values, determining that patient autonomy ought to be honoured foremost. This type of absolute recommendation is violated in any one case of a physician ignoring patient autonomy. The market does not represent its values in this way. The experimentation with various levels of privacy protection that has already gone on is not condemned as unethical, it is considered part of the natural process by which we come to recognize an ethical concern for market behaviours.

---

[3] Cicero, *Selected Works*, ed. Michael Grant (London: Penguin 1971), 177–180.

[4] Joseph Heath has recently identified some of the norms that underlie appropriate market behavior. He describes four: (a) do not exploit market failure, (b) do not cheat, (c) do not game the rules, and (d) refuse to lobby against regulations designed to correct market imperfections (Heath 2014, 113). This is one step closer, but the generality of the justification I am seeking also offers an explanation of *why* following these norms brings value.

Third, we want to share the justification of the market so that it can be tested and improvements can be identified. There are already candidate norms that are used in this type of fashion. The common ones do not seem adequate. One is, of course, "increase general affluence". Given the limitations on our ability to see in the future, to be guided by this norm would let in even deceptive business behaviour. Norms like "share" or "cooperate" do not capture the operations of business in a way they would recognize as descriptive of all of their behaviour. Candidate norms that we hear rehearsed, but that rightly repel us are not even empirically adequate. The market is not a matter of "survival of the fittest" nor is profit some type of competition. The economist's "maximize surplus" is not likely to be easy to integrate into our personal ethical commitments, as the goal of this is left obscure. None of these include reference to the general good of the market. So I would suggest this norm does the work of connecting an assessment of business behaviour to the justification of the market: *we are behaving justifiably in the market when we are aware that it is the right time and place to treat resources as scarce for the sake of contributing to a market system that generates information that can be used by others immediately, and long term as a means to general affluence.*

The benefits of this norm comprise a long list. It is not at odds with our actual motivations in business. It is readily applicable to behaviour we can recognize; behaviour we engage in regularly. It is appropriately modest about what we have provided to others in the market due to a transaction (a priced opportunity rather than a positive outcome). It does not put the burden of recognizing or caring for customers' needs on sellers. It does, however, recognize some predictable features of business practice. It will treat commodities as limited resources. The meticulousness with which OSP and websites collect the information we provide is one way of conserving resources that contribute to general affluence. In doing this, sellers play small, replaceable roles in an overall system that generates affluence, affluence shared with participants in various but necessary ways. No particular transaction is regarded, on the basis of this norm, as essential to this process.

Reliance on a norm like I have proposed can illuminate features of our transactions online that have been overlooked. For example, why are our personal data not priced for us? This is a market transaction and if our shared data is not publicly priced, we are failing to signal the opportunity afforded to competitors of the services we currently use. I would suggest that ISPs need to be more transparent about the value of users' personal information, and exchange with us in terms of prices.

On the other hand, the norm I recommend puts the information that ISPs can exploit into the same category as prices. Information can help us make efficient decisions of the sort that lead to general affluence. The information that ISPs can glean from our online activity (and again, this will not be personally identifiable information) is our contribution to others and ourselves, as we take back from the market some of what we put in. This way of thinking about market transactions moves us from thinking exclusively about only our own, or only our own potential, interests to take into consideration also the requirements to keep a phenomenon like the market afloat. Like Downes, I would argue that ISPs are being unduly targeted and that the increased transaction costs due to regulations cannot be counted as "pro-consumer."

Idealized individual transactions are impossible to standardize, and also, I want to suggest, mislead us about the nature of markets. Encryption already risks aging some of the careful metaphysical work done on online data sharing. Google is ready to block ads across devices; Apple's is close to depersonalizing the user information it accesses through IPhones. We do not need to assume that the market itself cannot accommodate our concerns. The FCC's interventions are unable to change the nature of market transactions, and their argument maintains a truncated view of what we participate in markets for. We can commit ourselves to the idea that transactions in the market ought to contribute in the short and long term to general affluence, as we best understand its generation.

References

Material they disseminate? *Richmond Journal of Law & Technology*, *5*, 7, Winter, at http://jolt.richmond.edu/v5i2/boehm.html

Cicero, M. 1971. *Selected Works*, (M. Grant, ed.). London: Penguin.

Downes, L. (2016). The Downside of the FCC's New Privacy Rules. *Harvard Business Review.* Retrieved November 4, 2016: https://hbr.org/2016/05/the-downside-of-the-fccs-new-internet-privacy-rules

Elia, J. (2009). Transparency rights, technology and trust. *Ethics and Information Technology, 11*(2), 145–153.

Federal Trade Commission. (2009). FTC shuts down notorious rogue internet service provider: 3FN service specializes in hosting spam-spewing botnets, phishing web sites, child pornography, and other illegal, malicious web content. Federal Trade Commission, http://www.ftc.gov/opa/2009/06/3fn.shtm

Federal Communications Commissions (2016a). Protecting the Privacy of Customers of Broadband and Other Telecommunications Services. Retrieved November 4, 2016: https://apps.fcc.gov/edocs_public/attachmatch/FCC-16-39A1.pdf

Federal Communications Commissions (2016b). Press Release, FCC Proposes to Give Broadband Consumers Increased Choice, Transparency and Security for their Personal Data. Retrieved November 4, 2016: https://apps.fcc.gov/edocs_public/attachmatch/DOC-338679A1.pdf.

Heath, J. (2014). *Morality, competition, and the firm: The market failures approach to business ethics* (p. 113). Oxford: Oxford University Press.

Holley, D. M. (1998). Information disclosure in sales. *Journal of Business Ethics, 17*(6), 631–641.

Martin, K. (2015). Understanding privacy online: Development of a social contract approach to privacy. *Journal of Business Ethics.*

Pollach, I. (2011). Online privacy as a corporate social responsibility: An empirical study. *Business Ethics, 20*(1), 88–102.

Uslaner, E. M. (2004). Trust, civic engagement, and the internet. *Political Communication, 21*(2), 223–242.

GPSR Compliance
The European Union's (EU) General Product Safety Regulation (GPSR) is a set of rules that requires consumer products to be safe and our obligations to ensure this.

If you have any concerns about our products, you can contact us on

ProductSafety@springernature.com

In case Publisher is established outside the EU, the EU authorized representative is:

Springer Nature Customer Service Center GmbH
Europaplatz 3
69115 Heidelberg, Germany

www.ingramcontent.com/pod-product-compliance
Ingram Content Group UK Ltd.
Pitfield, Milton Keynes, MK11 3LW, UK
UKHW020236290926
15470UKWH00002B/571
* 9 7 8 3 3 1 9 4 7 8 5 1 7 *